Biological Sciences

Innovations & Dynamics

Biological Sciences

Innovations & Dynamics

Rajeshwar P. Sinha
Richa
Rajesh P. Rastogi

NEW INDIA PUBLISHING AGENCY
New Delhi – 110 034

NEW INDIA PUBLISHING AGENCY
101, Vikas Surya Plaza, CU Block, LSC Market
Pitam Pura, New Delhi 110 034, India
Phone: + 91 (11)27 34 17 17 Fax: + 91(11) 27 34 16 16
Email: info@nipabooks.com
Web: www.nipabooks.com

Feedback at feedbacks@nipabooks.com

ISBN: 978-93-83305-87-2

Composed, Designed and Printed in India

Preface

Biological science is concerned with the study of life and living organisms, including their structure, function, growth, evolution, distribution, taxonomy and metabolism. All living organisms undergo metabolism, maintain homeostasis, have capacity to grow, respond to stimuli, communicate through various means, reproduce and adapt to their environment through natural selection. The organisms that photosynthesize are the primary producers and represent the life support system from aesthetics to food to medicine. As a result, they are in the center of the attraction of man from the very beginning. The book encompasses topics on molecular biology, biotechnology, microbiology, stress biology, genomics and proteomics of lower to higher organisms. Continual advancements in technology, instrumentation and knowledge base has led to the emergence of new disciplines with narrow specialization. The requirement of food, medicines, natural products, quest to resolve and develop better understanding of life and other biological processes, need to conserve genes, plants and ecosystem, and the apprehension of land, water and environment have led to the strengthening of traditional disciplines as well as emergence of diverse disciplines such as stress biology, molecular biology, biotechnology and bioinformatics.

Contributions in this book impart the state-of-the-art knowledge and the changing trends in the approach of biological sciences. In addition, this book provides newer techniques and use of these tools in achieving the potential of biotechnology to understand some of the basic problems in biological sciences.

We have great pleasure in bringing forward this book, and express our sincere gratitude to all the contributors. Finally, it is a profound pleasure to thank the staff of New India Publishing Agency (NIPA), New Delhi, India, for taking up the publication of this book.

Rajeshwar P. Sinha
Richa
December, 2014 **Rajesh P. Rastogi**

Contents

List of Contributors

Dr. Vishal Gupta, Division of Microbiology, Indian Agricultural Research Institute, New Delhi-110012, India

Dr. Saswati Nayak, Division of Microbiology, Indian Agricultural Research Institute, New Delhi-110012, India

Dr. Radha Prasanna, Division of Microbiology, Indian Agricultural Research Institute, New Delhi-110012, India

Dr. Jaoudat Faddoul, Damascus University, Faculty of Agriculture, Plant protection Department, Damascus, Syria

Dr. Shailendra P. Singh, Department of Biology, Cell Biology Division, Friedrich-Alexander University, Staudtstr. 5, D-91058 Erlangen, Germany

Dr. Azizullah Azizullah, Department of Biology, Cell Biology Division, Friedrich-Alexander University, Staudtstr. 5, D-91058 Erlangen, Germany

Dr. Adeel Nasir, Department of Biology, Cell Biology Division, Friedrich-Alexander University, Staudtstr. 5, D-91058 Erlangen, Germany

Dr. Roland Richter, Department of Biology, Cell Biology Division, Friedrich-Alexander University, Staudtstr. 5, D-91058 Erlangen, Germany

Dr. Peter R. Richter, Department für Biologie, Friedrich-Alexander-Universität, Staudtstr. 5, D-91058 Erlangen, Germany

Prof. Dr. Donat-P. Häder, Neue Str. 9, 91096 Möhrendorf, Germany

Dr. Michael Lebert, Department of Biology, Cell Biology Division, Friedrich-Alexander University, Staudtstr. 5, D-91058 Erlangen, Germany

Dr. O. Montero, Centre for Biotechnology Development (CDB). Spanish Council for Scientific Research (CSIC). Av. Francisco Vallés, 8. 47151- Boecillo, Valladolid, Spain. *

Dr. M. Velasco, Centre for Biotechnology Development (CDB). Spanish Council for Scientific Research (CSIC). Av. Francisco Vallés, 8. 47151- Boecillo, Valladolid, Spain.

Dr. R. P. Rastogi, Centre of Advanced study in Botany, Banaras Hindu University, Varanasi-221005, India

Prof. Dr. R. P. Sinha, Centre of Advanced study in Botany, Banaras Hindu University, Varanasi-221005, India

Prof. A. Incharoensakdi, Department of Biochemistry, Chulalongkorn University, Bangkok-10330, Thailand

Dr. Richa, Centre of Advanced study in Botany, Banaras Hindu University, Varanasi-221005, India

Dr. Vinod K. Kannaujiya, Centre of Advanced study in Botany, Banaras Hindu University, Varanasi-221005, India

Dr. Shweta Yadav, Department of Botany, Mahila Mahavidyalay, Banaras Hindu University, Varanasi-221005, India

Dr. Gaurav Raj Dwivedi, Molecular Bioprospection Department, CSIR-Central Institute of Medicinal and Aromatic Plants (CIMAP) Lucknow, 226015, India

Dr. Vigyasa Singh, Molecular Bioprospection Department, CSIR-Central Institute of Medicinal and Aromatic Plants (CIMAP) Lucknow, 226015, India

Dr. M. P. Darokar, Molecular Bioprospection Department, CSIR-Central Institute of Medicinal and Aromatic Plants (CIMAP) Lucknow, 226015, India

Dr. Rakesh Pandey, Microbial Technology & Nematology, CSIR-Central Institute of Medicinal and Aromatic Plants (CIMAP) Lucknow, 226015, India

Dr. Rajesh Kumar, Centre of Advanced study in Botany, Banaras Hindu University, Varanasi-221005, India

Dr. Kamlesh K Mishra, Department of Biochemistry, University of Allahabad, Allahabad 211002, UP, India

Dr. Hoda Ahmed, Department für Biologie, Friedrich-Alexander-Universität, Staudtstr. 5, D-91058 Erlangen, Germany

Mr. Jainendra Pathak, Centre of Advanced study in Botany, Banaras Hindu University, Varanasi-221005, India

Mr. Rajneesh, Centre of Advanced study in Botany, Banaras Hindu University, Varanasi-221005, India

Mr. Arun S. Sonker, Centre of Advanced study in Botany, Banaras Hindu University, Varanasi-221005, India

Dr. Minu Kesheri, Centre of Advanced study in Botany, Banaras Hindu University, Varanasi-221005, India

Mr. Swarna Kanchan, Biological Science Department, Birla Institute of Technology and Science, Pilani, Rajasthan-333031, India

1

Exploring the Morpho-taxonomic, Biochemical and Molecular Diversity in the Genus *Anabaena*

Vishal Gupta, Saswati Nayak and Radha Prasanna

Abstract

Cyanobacteria have gained a lot of attention in recent years because of their novel applications in biotechnology, besides their well recognized role as inoculants for rice crop. They represent favourite model organisms for a deeper understanding of several plant and microbial metabolic processes and efficient systems for the production of recombinant compounds of pharmaceutical and commercial value. They have been identified as a rich source of biologically active compounds with antiviral, antibacterial, antifungal and anticancer activities, which illustrate their potential as biocontrol agents. The genus *Anabaena* assumes significance not only as a diazotroph, but also as a rich source of pharmacologically active and structurally interesting secondary metabolites with diverse applications in agriculture and industry. The present compilation aims to present an overview of the literature available globally, and studies undertaken in India illustrating the diversity and potential of *Anabaena* strains, besides their significance in biology and evolution.

Keywords: Biocontrol; Cyanobacteria; Diversity; Enzymes; Toxins

Introduction

Cyanobacteria comprise a morphologically and physiologically diverse group within the domain Prokarya, which form a monophyletic group among eubacteria (Wilmotte 1994). Cyanobacteria are often referred to as "miniature factories" of the biological world, as they represent a promising source of a variety of bioactive compounds, lipids/ fatty acids, proteins, enzymes, pigments and compounds of pharmaceutical and nutraceutical value (Rezanka and Dembitsky 2006; Prasanna et al. 2008, 2010a). Cyanobacteria proliferate in a wide range of habitats including terrestrial, freshwater, marine and saline environments, exhibiting several diverse metabolic activities. Under conditions of combined nitrogen deprivation, a small fraction of the vegetative cells, in a defined spatial pattern along the filaments, differentiate into heterocysts, which are specialized for nitrogen fixation in oxic environment (Wolk et al.1993). Such heterocystous cyanobacteria, including the genus *Anabaena,* are found in a great diversity of habitats as a result of their ability to fix atmospheric nitrogen under aerobic environments. Like several other genera, *Anabaena* also causes toxic blooms in fresh water, and intense coloration of water bodies (Wiegand and Pflugmacher 2005), which are often considered as nuisance, however can be tapped as rich sources of secondary metabolites.

Taxonomic treatises represent useful compilations of biodiversity, which improve our understanding of the Earth's biota based on phylogenetic background, ecological data, and morphological variation. Modern approaches, therefore, need to apply all available and important methodological procedures and criteria in taxonomic analyses, and the classification must include information about the organisms and their evolutionary relationships (Flechtner et al. 2002; Komarek 2005a). However, reliable classification is impossible without the careful combination of genetic data with morphological diversity and variation, ecological and eco-physiological characteristics, ultrastructural studies, and appropriate applications of convenient formal prescriptions for designation of taxa and strains (nomenclatoric rules). Hence, a "polyphasic approach" has been considered as essential for reliable cyanobacterial taxonomy (Komarek 2005b).

Molecular tools hold promise in resolving much of the controversies in cyanobacterial taxonomy, identification and the species concept. Approaches such as cataloguing based on sequences of 16S rRNA (Marquardt and Palinska 2007) or internal transcribed spacer regions between phycocyanin subunit of *cpcB* and *cpcA*, conserved structural and regulatory domains within divergent 16S – 23S rRNA spacer sequences (Chen et al. 2006) have been employed in molecular taxonomic studies of cyanobacteria. The genome size of these organisms ranges from 3.17 to 8.58 G da and G+C% is observed in the range of 38.3 to 46.7 %. This phylum comprises more than 1000 species whose members are classified by using both botanical and bacteriological taxonomic codes (Castenholz 2001). The heterocystous cyanobacteria are taxonomically placed in sub section IV (formerly Nostocales) comprising 2 families (Castenholtz 2001). The first family comprises most of the Nostocacean members including the form genus *Anabaena.* Derived from a Greek word *anabaino*, "to rise", the name *Anabaena* consists of 2 words - '*ana*' and '*baino*' which mean 'up again' and 'to walk or to step' i.e. to step up again or repeating line of units/cells.

Taxonomic status of the genus *Anabaena*

The attributes utilized for the morphological identification of *Anabaena* genus, as described by different experts is given in Table 1.

Kingdom: Bacteria

Division: Cyanophyta

Class: Cyanophyceae

Order: Nostocales

Family: Nostocaceae

Genus: *Anabaena*

The genus *Anabaena* was established by Bory in 1922. In the first revision of this genus, Bornet and Flahault (1888) described 11 species. Geitler (1932) described 57 European species of *Anabaena*- considered as the most satisfactory work on blue green algal taxonomy, though at present this taxonomy is chaotic at the genus and species level (Rippka et al. 2001). Rippka et al. (2001) described the form-genus *Anabaena* Bory de St. Vincent 1822 sensu Rippka, Deruelles, Waterbury, Herdman et al (1979a, b) in the Bergey's Manual of Systematic Bacteriology (2001) with three different clusters, each with a reference strain. Fritsch (1949) carried out taxonomic studies on this genus using species isolated from India and areas adjacent to Asiatic mainland. Kantz and Bold (1969) developed a taxonomy based on new criteria such as plant mass morphology and motility. Rippka et al. (1979) and Herdmann et al. (1979a, b) proposed a taxonomy mainly based on biochemical characters such as DNA base composition. A morphological approach adopted by Drouet (1968) led to a dramatic reduction in the number of *Anabaena* species. Drouet (1968) recognized one taxonomic criterion and two species and observed that most morphological characters in *Anabaena* are subject to a wide phenotypic plasticity, which leads to erroneous identification and nomenclature. The present confusion is mainly due to scepticism about the taxonomic value of criteria used by Bornet and Flahault (1988) and adopted by Geitler (1932).

Table 1: Morphological identification of *Anabaena* based on different investigations

Desikachary (1959)	The genus *Anabaena* has been identified based on the "Presence of uniform trichomes, absence of sheath or presence of more or less diffluent sheath forming free or floccose or soft mucilaginous thallus. Heterocysts generally intercalary and presence of a single or series of spores near the heterocyst or between the heterocysts".
Bornet and Flahault (1888)	Trichomata aequalia vel ad apices sensim attenuata, evaginata vel vaginis saepius evanescentibus cincta, libera vel in stratum mucosum congesta. Cellulae apicales difformes. Heterocystae intercalares plures. Sporae situs varii, modo heterocystis utrinque contiguae, modo inter cellulas vegetativas ortae, solitariae vel catenatae. (Conf. Wittrock in Wittrock et Nordstedt, Algae exsicc., n° 496.)
Komarek (1992)	Described that Filamentous; filaments solitary or in free clusters (subg. *Dolichospermum*) or in macroscopic mats on the substrate (subg. *Anabaena*), usually with coiled and tangled, sometimes up to screw-like coiled, rarely with more or less parallel oriented trichomes, along the length of the same width or slightly narrowed to the ends, with more or less deep constrictions at cross walls, always without firm sheaths, but sometimes with mucilaginous, hyaline, colourless, often diffluent envelopes; trichomes uniserial, often moniliform, isopolar, always metameric (heterocytes develop solitary and intercalary in more or less certain distances one from another; in one developed trichome always 3-9, rarely more, heterocytes). Cells cylindrical, barrel-shaped or spherical, shorter up to longer than wide, pale or bright blue-green or olive-green, containing aerotopes (subg. *Dolichospermum*) or without aerotopes, but sometimes with granular content (subg. *Anabaena*); terminal cells sometimes slightly elongated, conical, conical rounded or spherical, not vacuolized. Heterocytes spherical, widely oval or cylindrical, sometimes elongated, usually slightly greater than vegetative cells. Akinetes spherical, oval or cylindrical, solitary or several in a row, intercalary, developing always paraheterocytic, close to heterocytes from their both sides, or slightly distant from them.
Rippka et al. (2001)	Described as: "Trichomes straight or slightly sinuous, vegetative cells spherical, cylindrical, or barrel shaped, and separated by conspicuous constrictions at the cross walls. Heterocysts predominantly intercalary, though terminal heterocysts may also occur. Akinetes located, singly or in groups (2-3), adjacent to heterocysts or occurring in chains, initiated distant from heterocysts. A firm sheath is absent, but mucilage production may be abundant. False branching generally not observed. Reproduction by random trichome fragmentation. Hormogonia, distinct in morphology from the parental trichomes, not produced".

Most of the taxonomic studies have been based on protologues (field materials and preserved specimens). A number of workers have collected and investigated samples from different ecological habitats and given taxonomical assignments (Desikachary 1959; Starmach 1966). Desikachary and Anand (1974) identified a large number of isolates with different geographical distribution under identical cultural conditions and emphasized the need for more number of similar cultural studies.

Morphological attributes

The genus *Anabaena* exhibits tremendous diversity and has been the model system for cyanoprokaryotes. Numerous morphospecies were described, which differ in a few (sometimes transient) characters and further studies are needed to establish their identity. Stulp and Stam (1984) designated eight species among 16 cultures investigated, based on traditional cyto-morphological criteria and suggested the close genetic relatedness of these species and the uniformity of the genus. Identification of this genus is based on morphological features such as cultural behavior including growth pattern, color of the thallus and microscopic features such as shape and size of vegetative cell, relative position of heterocyst and akinetes and presence/ absence of gas vacuoles, sheath (Desikachary 1959; Komarek 2005b). Desikachary (1970) stressed that cultural studies are essential to understand the behavioral pattern of blue green algae, exhibiting a wide range of variability. In our previous study (Nayak et al. 2009), 70 *Anabaena* strains were isolated from different agro-ecologies of India and among them, most of the *Anabaena* strains exhibited planktonic-floccose thalli and the colour of the thallus ranged from dark green to brown and bluish green in colour, while filaments were mostly straight, and rarely exhibited coiling (Fig. 1). The colour of the plant mass has not been observed to have much taxonomic value as there are too many resemblances among the strains (Stulp and Stam 1982).

Incubation under different light and temperature conditions hardly affects the taxonomically important characters such as shape of vegetative cell, heterocyst and akinetes, position of akinetes (Stulp 1982). Rajaniemi et al. (2005) reported that coiling of the trichome was not useful for classification of *Anabaena* morphophyte. Earlier workers (Stulp and Stam 1982) suggested much value to the shape of the terminal cell and its utility for taxonomy. Nayak et al. (2009) observed characteristic pointed/tapered end cells in *A. vaginicola, A. ballyganglii, A. iyengarii, A. spiroides,* which was helpful in the species identification. However, this feature can be mis-identified as the result of breakage of trichomes and intermediary shape. Among all the numerical criteria utilized in taxonomic investigations (Stulp and Stam 1982), vegetative cell width seems to be the most stable and useful one, and found useful for the identification of *Anabaena* stains. Hiroki et al. (1998) also supported these observations and concluded that cell width seems to be the most useful parameter. Interestingly, the cell width of botanical 'species' of *Anabaena* ranges from 2-15 µm.

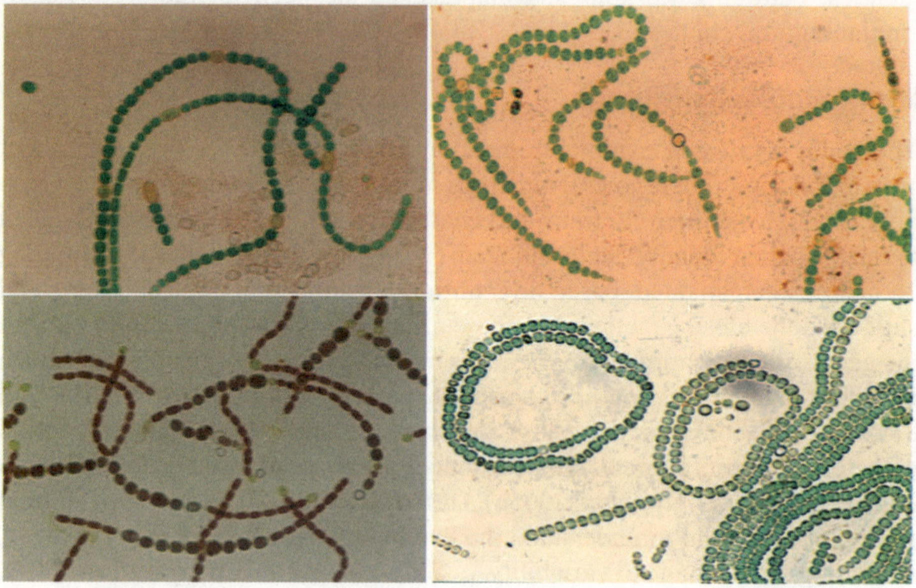

Fig. 1: Morphological diversity within the genus *Anabaena.*

Anabaena strains have been classified in many ways – based on habitat, shape, size of trichome and cells. Desikachary (1959) has divided the genus *Anabaena* into 25 species and in Bergey's Manual of Systematic Bacteriology (Rippka et al. 2001), three clusters with Type strains have been designated.

Stulp and Stam (1982) evaluated 16 strains of *Anabaena* under uniform laboratory conditions to test the value of taxonomic criteria such as cell dimensions, position of akinetes, shape of terminal cell, akinete germination and plant mass morphology and found certain characteristics taxonomically useful. A database system has been designed for identification of planktonic *Anabaena* species based on morphological characteristics with a reference database containing 19 planktonic *Anabaena* species along with their morphological characteristics for 26 attributes (Hiroki et al. 1998). Komarek (2005a) described nineteen traditional *Anabaena* morphospecies from freshwater habitats of Cuba. However, the identification of this genus is complex, as some morphological features change and may not be expressed in the culture (Li 1998). Rippka et al. (2001) in their bacteriological classification system included *Anabaena* PCC 7120 and other strains showing a high degree of DNA-DNA similarity with *Nostoc,* within the genus *Nostoc*. Recently, *Trichormus* was separated from the traditional genus *Anabaena,* on the basis of akinete development and transferred into the subfamily Nostocoidae (Rajaniemi et al. 2005). Rippka et al. (1979), on the basis of culture strains alone, differentiate *Anabaena* from *Nostoc* primarily on the basis of motility. In *Anabaena*, all vegetative trichomes are normally motile. Colony formation should continue to be a usable criterion to distinguish *Nostoc* from *Anabaena* as it has been in the past (Bourrelly 1985). Nayak et al. (2009) evaluated 70 *Anabaena* strains and found that in all the strains, mainly intercalary heterocysts were present. According to Komarek and Anagnostidis (1989), heterocyst differentiation in *Anabaena* always occurs in an intercalary position, but in *Anabaena cylindrica,* conical end cells sometimes give rise to similarly shaped heterocysts which may become terminal after trichome breakage.

The position of akinetes in the trichomes has been universally considered an important criterion in traditional taxonomy of *Anabaena* due to its stable and reliable nature (Stulp and Stam 1982; Hiroki et al. 1998). However the problems in handling this criterion include the need for akinetes, which is observed only in late stages of growth and bonding between adjacent cells and akinetes being generally weak. In our investigations, problems were faced in studying the relative position of akinetes with respect to heterocyst as frequent breakage next to akinete was commonly observed. But, these morphological features were retained in different light and temperature conditions and even brackish water conditions. An important feature which is utilized for distinguishing *Anabaena* from *Nostoc* is the mode of development of akinetes with respect to the heterocyst. *Anabaena* strains are differentiated from *Nostoc* on the basis of their development of akinetes away (distant) from heterocyst. *Anabaena variabilis* and other species exhibiting a different mode of akinete development were placed in *Trichormus* a new genus developed by Komarek and Anagnostidis (1989). Stulp and Stam (1982) observed that the position of akinetes i.e. whether it is adjacent to or away from heterocyst, as a very stable and important criterion for species identification. Akinete morphology has been observed to be a steady characteristic in light microscopic examination of cultures for their identification (Hiroki et al. 1998). The position of spore, whether adjacent to or away from heterocyst was found to be a stable character in *Anabaena* species (Stulp and Stam 1982; Nayak et al. 2007, 2009).

Akinetes of *Anabaena* species have not just been of taxonomic interest, but have also been used to analyse primary processes of germination i.e. of the physiological reactivation of these resting cells. Although structural aspects of akinete formation/development have been investigated by a number of workers (Adams 1992), akinete germination has not been very critically analyzed. Stulp and Stam (1982) also suggested that the morphology of akinete germination can be of significant value in *Anabaena* taxonomy. They recognized six morphologically well defined groups among *Anabaena* strains.

The assignment of some species either to *Anabaena* or to *Aphanizomenon* has also been discussed (Komarek and Anagnostidis 1989). Also, previous phylogenetic studies of Lyra et al. (2001), Gugger et al. (2002b) and Iteman et al. (2002) have shown that the genera *Anabaena* and *Aphanizomenon* are not monophyletic. Henson et al. (2002) in their investigations also discussed the separation of genera *Nostoc* and *Anabaena*. Gugger et al. (2002a, b) and Rajaniemi et al. (2005) studied planktic strains (morphotypes) of *Anabaena*, which differ in genotype from benthic *Anabaena, Trichormus, Cylindrospermum* and *Nostoc*-morphotypes. The planktic *Anabaena* strains are very closely related to *Aphanizomenon*-species. However, in nature, there is a rather wide morphological variation within or between *Anabaena* and *Aphanizomenon* (Komarek and Kovacik, 1989). A number of research workers have attempted to utilize alternative methods such as numericulture (numerical code based on morphological and physiological characters) or DNA analysis combined with morphological traits (Hiroki et al. 1998) for identification and characterization of *Anabaena*. A numerical taxonomy study of several species of *Anabaena* and *Nostoc* was undertaken by McGuire (1984) using 28-30 morphological characteristics. Currently, several researchers are utilizing and relying on a polyphasic approach that considers different phenotypic and genotypic features (Wilmotte and Herdmann 2001; Prasanna et al. 2006; Nayak et al. 2009) for reliable classification of cyanobacteria. Komarek (2005a) studied nineteen traditional *Anabaena* morphospecies from freshwater habitats in Cuba and concluded that the traditional system used in this study needs to be confirmed by modern molecular methods.

Distributional based on soil types

Extensive reviews exist on the geographic and regional distribution and role of cyanobacteria in irrigated rice fields (Nayak et al. 2004; Nayak and Prasanna 2007; Prasanna and Nayak 2007). Among the several genera of cyanobacteria found in rice fields, *Nostoc* and *Anabaena* have been recorded predominantly in all most rice field soils from many localities (Nayak and Prasanna 2007; Prasanna and Nayak 2007). The distributional pattern of *Anabaena* strains in various soil types has provided very interesting observations. A significant correlation was observed between the species identified and the different soil types from which the strains isolated and one can summarize that a particular soil type may select and allow the growth of a particular species of *Anabaena*. Nayak et al. (2009) isolated *Anabaena* strains from different rice agro ecologies, several soil types were also surveyed mainly alluvium, red and lateritic, sandy loam, desert-saline, red-loamy, red-black and brown forest-podozoic. It was observed that in alluvium type, representatives of all species types were found, except *A. oscillarioides*. Some species like *A. ballyganglii*, *A. flos-aquae, A. circinalis* were restricted to only alluvium type. On the other hand, red and black soil type supported the growth of only one species- *A. sphaerica*. Interestingly *A. iyengarii* was the only species present in desert soils.

The rhizosphere is well recognized as the hot spot for microbial abundance, diversity and metabolic activities, intimately related to the successful production of crops and sustenance of soil fertility. In this context, the potential of associative microorganisms, in contrast to the symbiotic systems, in promoting the growth and yields of cereals is a promising area of research, and plant growth promoting rhizobacteria or PGPR have become important partners in integrated nutrient management strategies in various crops. However, the rhizosphere of crop plants is a relatively unexplored frontier in terms of cyanobacterial abundance and diversity, which can be due to the overriding emphasis on their role as photosynthetic diazotrophs; perhaps diverting the attention of researchers from exploring the rhizosphere, which has been an exclusive domain of heterotrophic microflora. In the last decade, reports on their growth and diversity in the rice rhizosphere has been extensively analyzed, which revealed that 80-90% were belonging to the genus *Anabaena/Nostoc* (Prasanna et al., 2009a, b). *Anabaena* (40%) and *Nostoc* (38%) followed by *Hapalosiphon, Calothrix* and *Westiellopsis* represented the dominant genera. Rhizosphere samples from rice cultivars grown at Aduthurai (India) exhibited low generic richness i.e. only two genera *Anabaena* and *Nostoc* were observed.

Biochemical attributes

Biochemical studies in cyanobacteria demonstrated a significant correlation between the morphological complexity of species and their fatty acid composition. Characterization of 70 *Anabaena* strains for their growth attributes provided interesting information regarding the wide variation in terms of chlorophyll and protein accumulation (Nayak et al. 2007; 2009). In terms of chlorophyll accumulation- tremendous diversity was revealed and values ranged from 0.107- 4.89 x 10^2 mg/ml. The five top ranked strains for chlorophyll accumulation based on mean performance using DMRT analysis were *A. spiroides*, *A. variabilis*, *A. oryzae*, *A. spiroides* and *A. vaginicola.* Such wide variations within a genus and also within species have been observed in earlier investigations (Mekonnen et al. 2002). The rhizosphere isolates from diverse agroecologies of India (Prasanna et al. 2009b) were also efficient in enhancing the germination and growth of wheat and rice seeds and exhibited significantly high protein accumulation and IAA production, after incubation in light and dark (with 0.5% glucose). A novel concept of *Anabaena* sp. based biofilms through associations with agriculturally important bacteria have been standardized which have shown promise in terms of their metabolic activity, for use as diazotrophs, plant growth promoting agents and biocontrol agents (Prasanna et al. 2011a).

A set of *Anabaena* and *Nostoc* strains were tested for their response to forty antibiotics, which aided in developing selectable antibiotic markers for each of the cyanobacterial strains. Biochemical profiles generated using standardized chromogenic identification system, which included carbohydrate utilization tests revealed that 53% of the carbohydrates tested were not utilized by *Anabaena s*trains, while some strains exhibited unique abilities for utilization of saccharides such as melibiose, cellobiose, maltose and glucosamine. PCR based amplification profiles developed using a number of primers based on repeat sequences revealed the utility of three primer sets in providing unique fingerprints for the *Anabaena* and *Nostoc* strains (Prasanna et al. 2010c).

Evaluation of *Anabaena* strains as inoculants in the rice-wheat cropping sequence

Biological nitrogen fixation through cyanobacteria contributes 25-30 kg N ha^{-1} per cropping season and reports indicate increase of 10-15 per cent in rice yields (Venkataraman 1972). Nitrogen fixation is carried out in specialized cells known as heterocysts, which have thick walls and hence, physically prevent the entry of oxygen and provide necessary anaerobic conditions for the activity of nitrogenase enzyme. The evidence for the nitrogen fixation has come from long-term fertility trials, N balance studies and experiments involving acetylene reduction assay (Yoshida and Ancajas 1971). ARA (Acetylene Reducing Assay) is a sensitive tool to detect nitrogenase activity but its accuracy for quantification studies has been much debated. A simple and modified technique was developed based on collection of soil cores which can provide rapid and reliable measurements of nitrogen fixation under field conditions and experiments carried out over a period of three consecutive years revealed that the estimates of nitrogen fixation were comparable with the results of other workers (Prasanna et al. 2003a; Nayak et al. 2004).

Nitrogen fixation by cyanobacteria is an economically viable input in rice cultivation and selection and manipulation of high nitrogen fixing strains is an ongoing process in most rice growing countries. Nitrogen fixation by blue green algae varied from a few to 80 kg N ha^{-1} $crop^{-1}$ mean of 27 kg (Roger and Kulasooriya 1980). The germplasm of *Anabaena* isolates from various agro-ecologies of India (Nayak et al. 2007), evaluated exhibited a wide diversity in terms of ARA (as an index of nitrogen fixation). An investigation was undertaken to developing methods for quantifying their establishment; using physiological activity (chlorophyll as a growth index and nitrogen-fixing potential as a measure of their biofertilizing capacity), along with evaluation based on DNA fingerprints generated using repeat sequences/palindromes. Time course studies were undertaken in liquid and soil microcosm experiments inoculated with a set of four rhizosphere cyanobacterial strains (BF1 *Anabaena* sp., BF2 *Nostoc* sp., BF3 *Nostoc* sp., BF4 *Anabaena* sp.), followed by pot experiments with rice crop. The DNA fingerprints of pure cultures and soil samples revealed distinct bands with the three selected primers STRR 1A, $STRR_{mod}$ and $HIP_{AT.}$ Such specific bands were complementary with other physiological attributes used for evaluating establishment of cyanobacterial strains in soil (Prasanna et al. 2011b). Further evaluation at pot and field level under 60%WHC through intermittent irrigation, using a combination of microbiological and DNA based approaches revealed their superiority over other treatments in terms of plant growth promotion and soil microbiological parameters. Microscopic examination of soil enrichment cultures raised from the treatments revealed their comparative abundance over native flora. A significant enhancement in terms of soil microbial biomass carbon and humus content was also recorded in these treatments. The soil enzymatic profile of the inoculated treatments involving FDA, dehydrogenase, acid phosphates, alkaline phosphatase, aryl esterase and β-glucosidase also revealed the promise of the inoculated strains. 16SrDNA sequencing, followed by BLAST analyses provided valuable information regarding the establishment of *Anabaena* strains. RFLP analyses of the 16SrDNA sequences of soil DNA and phylogenetic analyses vis a vis sequences of the inoculated cyanobacterial strains revealed the promise of selected combinations of *Anabaena* and *Nostoc* strains. To the best of our knowledge, this represents a first report on evaluating the establishment of cyanobacterial inoculants in rice crop grown under non-flooded conditions using a combination of agronomic, microbiological and soil DNA related attributes (Prasanna et al. 2013a). The rhizosphere dynamics and plant growth promoting ability of a set of inoculated strains in rice crop revealed the persistence of strains on roots upto harvest stage, entry into roots and enhancement in plant growth and yield attributes, besides improved soil fertility/ microbiological parameters. Our studies carried out up to harvest stage of crop clearly demonstrated that cyanobacteria enhanced plant growth parameters (plant height, dry weight, grain yields) besides bringing about significant changes in soil microbial biomass carbon (Prasanna et al. 2009a).

Karthikeyan et al. (2007, 2009) demonstrated that cyanobacteria excrete IAA, amino acids and other growth promoting compounds into their immediate environment, which in turn can stimulate the growth of microbial populations in soil. For the first time, the agronomic potential and plant growth promoting effect of these strains, providing the support as ideal candidates for development of inoculants for wheat crop. Root sections of wheat seedlings co-cultured with *Calothrix ghosei* and cells of *Nostoc sp.,* revealed the presence of short filaments inside the root hairs and cortical region (Prasanna et al. 2012b, c). Such strains can be promising candidates for developing plant growth promoting associations for wheat crop, besides serving as model systems for understanding the metabolic interactions of cyanobacteria with host plant, such as wheat.

In another pot experiment with wheat variety HD 2687, comprising a total of 51 treatments, various combinations of a set of three bacterial (P1, P2 and P3) and three cyanobacterial isolates (C1, C2 and C3 from wheat rhizosphere) were used along with 1/3 N and full dose of P and K fertilizers (Nain et al. 2010). Significant enhancement in the soil microbiological (Dehydrogenase activity, FDA, Alkaline phosphatase and microbial biomass) and plant growth/yield parameters were observed. Observations revealed a two-fold increase in panicle weight and alkaline phosphatase activity in selected combinations (P1+P3 +C3; P1 + C1+C2/C1+C3; C2+C3), as compared to control treatment involving full dose of chemical fertilizers. Electron microscopic observations of root sections of wheat plants revealed the mode of entry and intracellular presence of short filaments and single cells of *Nostoc* sp.; emphasizing the colonization potential of the cyanobacterial strains. Synergistic interactions among the PGPR strains (bacteria-cyanobacteria) were also observed in terms of significant enhancement in the soil microbiological and plant growth/yield parameters in both rice and wheat pot experiments (Prasanna et al. 2012a). Cyanobacterial consortia, using different rhizospheric bacteria have also shown promise as inoculants for rice-wheat cropping systems, with significant enhancement in soil fertility and crop yields, besides proving efficient in C-N sequestration in soil and micronutrient enrichment (Nain et al. 2010; Manjunath et al. 2011; Prasanna et al. 2012a, b).

Anabaena germplasm as sources of pigments

Cyanobacterial pigments comprise the most colourful and attractive components, which are already extensively utilized in bioindustry. Screening programs all over the world have further confirmed the diversity and rich repertoire of pigments, which can revolutionize the industrial uses of "colours" with their nutraceutical and pharmaceutical value. The most attractive feature of cyanobacteria is the presence of vividly shaded accessory pigments, the phycobiliproteins, which account for 40% of the total proteins in *Anabaena cylindrica* and 1-10 per cent of the cell dry weight (Chapman 1973). Out of the total cell proteins, phycobiliproteins may constitute about 50 %. An *Anabaena* strain having high proportion of phycoerythrin (8.3% of dry weight) was isolated from coastal lagoon of Spain (Rodriguez et al. 1989). A large collection of cyanobacteria, including unicellular, colonial, non-heterocystous filamentous, heterocystous filamentous and heterotrichous strains have been screened for phycobiliprotein content and the heterocystous strains were found to be the superior producers (Kaushik 2000; Prasanna et al. 2003b).

Among these phycobiliproteins (comprising phycocyanin, phycoerythrin and allophycocyanin), which account for about 20% of total dry weight, are the most important (Borowitzka 1994). Cyanobacteria generally have all three types of constituent pigments: allophycocyanin and phycocyanin are always present, while phycoerythrin is found in some organisms and not in others (Samsonoff and MacColl 2001. Kaushik (2000) and Prasanna et al. (2003b) reported that among various cyanobacterial strains, heterocystous forms - *Anabaena variabilis, Aulosira fertilissima, Hapalosiphon* sp. and *Tolypothris tenuis* can be exploited for phycobilins as their phycobiliprotein content varies between 14.72 to 17.52%. The germplasm of *Anabaena* evaluated exhibited a wide diversity type and amount of pigments (chlorophyll, phycobilins and carotenoids). *Anabaena* strains: *A. fertilissima, A. iyengarii, A. sphaerica, A. spiroides, A. oryzae, A. variabilis*, were identified as potential sources of pigments and can be exploited as inoculants (Nayak et al. 2007).

Carotenoids act as antioxidants, hence, protect cells from damage by unstable oxygen molecules. These pigments can boost the immune system and possibly lower the risk of heart disease, prevent onset of cancers and protect against age related diseases such as cataracts and macular degradation, multiple sclerosis (Larsson et al. 2007). Nayak et al. (2007) characterized the *Anabaena* strains and observed significant variability among the *Anabaena* strains. More than 60 % of the strains accumulated 1.01-1.96 μg ml^{-1} of carotenoids. *A. sphaerica* recorded the highest value of 5.033 x 10^2 mg/ml. The wide range in chlorophyll: carotenoids values observed in that study further emphasized the extent of diversity, especially in relation to the photosynthetic efficiency, among these isolates. Carotenoids protect the cyanobacterial cell from photooxidative damage and are found associated with proteins. Phycobiliproteins, comprising phycocyanin (PC), phycoerythrin (PE) and allophycocyanin (APC) was measured and total phycobiliproteins (TBPS) estimated individually in 15d old culture growing in nitrogen free BG-11 medium. *A. sphaerica, A. variabilis, A. fertilissima, A. spiroides* and *A. sphaerica* were the top ranked strains for PC content while for APC content, *A. sphaerica* and *A. spiroides* ranked highest

(Nayak et al. 2007). Rippka and coworkers (1979) observed that such biochemical diversity allowed the distinction of subgroups within several genera including *Nostoc* and *Anabaena.* All naturally occurring cyanobacteria have been found to produce PC and in *Anabaena* PCC 7120 only a single set of PC genes have been identified, which are constitutive in nature (Sidler 1994). An *Anabaena* strain having high proportion of phycoerythrin (8.3% of dry weight) was isolated from coastal lagoon of Spain (Rodriguez et al. 1989). Tremendous diversity has been observed, not only at the intergeneric level (Kaushik 2000) but also at the intra generic and inter/intra specific level (Prasanna et al. 2003b) in the phycobilin composition and relative proportion of PC, PE and APC. This has helped to not only identify potential strains for pigment production, but also understand the role of different environmental and nutritional conditions in enhancing pigment accumulation (Prasanna et al. 2004; Prasanna and Nayak 2007).

Cyanobacteria are known to inhabit environments subject to intense solar radiation. They are often the primary colonizers of bare rock surfaces and soils. Under such conditions, they must possess efficient mechanisms to prevent or to counteract the harmful effects of strong solar radiation, including those caused by the particularly deleterious UV region of the spectrum. Mycosporine-like amino acids (MAAs) are thought to protect from absorbing harmful UV radiation in cyanobacteria (Bandaranayake 1998). Sinha et al. (2001) investigated the presence of ultraviolet-absorbing mycosporine-like amino acids (MAAs) and their induction by solar ultraviolet-B (UV-B) radiation in three filamentous and heterocystous N_2-fixing cyanobacteria, *Anabaena* sp., *Nostoc commune* and *Scytonema* sp. Rastogi and Sinha (2009) reviewed various aspects of cyanobacterial scytonemin and MAA production. Nayak et al. (2007) screened 70 *Anabaena* strains for the presence of UV absorbing pigments by scanning in the range of 200-700nm. Among them, 13 strains of *Anabaena* showed peaks between 300-380 nm wavelengths, which can be identified as rich source of UV-absorbing pigments. The presence of substantial amounts of UV-absorbing compounds in cyanobacteria has been documented in several instances (Castenholz 2001), and it has frequently been hypothesized that these substances may be playing a UV-photoprotective role in the organisms growing under stress conditions.

Other biochemical markers for *Anabaena*

Whole cell proteins pattern analysis has been found to be a good phenotypic method for grouping large numbers of closely related strains of bacteria. The *Anabaena* strains examined in our previous investigation exhibited distinct polymorphism with respect to their protein profiles (Nayak et al. 2007).

However, previous workers had shown that whole cell protein patterns can be useful in grouping large number of closely related organisms, including cyanobacteria (Lyra et al. 1997) or *Azolla* accessions (Sood et al. 2007). Lyra et al. (1997) characterized filamentous cyanobacterial strains from different genera using SDS-PAGE of whole cell proteins and PCR-RFLP of the 16S rRNA gene. Palinska et al. (1996) emphasized the taxonomic value of cell protein profiles at the generic and even species level.

Molecular attributes and polyphasic approaches

The evolution of the genus *Anabaena* is known to be polyphyletic (Wilmotte 1994) and researchers have utilized a wide number of molecular tools to understand its diversity. These include the 16S rRNA gene (Halinen et al. 2007, 2008; Ezhilarasi and Anand 2009), 16S-23S internal transcribed spacer region (ITS) and the intergenic spacer region (IGS) of the phycocyanin (PC) locus (Ballot et al. 2004; Premanandh et al. 2006), M13 fingerprinting (Muralitharan and Thajuddin 2010), *nifD* (Henson et al. 2004), *rpoC1* (Halinen et al. 2008), *rbcLX* and *rpoB* (Rajaniemi et al. 2005), repeat sequences (Nayak et al. 2009) genes, as well as sodium dodecyl sulphate polyacrylamide gel electrophoresis of whole cell proteins (Lyra et al. 1997) and fatty acid content (Gugger et al. 2002b) have also been used to resolve phylogenetic relationships within the genus *Anabaena*. Extensive diversity was recorded within the genus *Anabaena,* using both morphological and molecular markers (Rajaniemi et al. 2005).

Stulp and Stam (1982) evaluated genotypic relationships between a number of *Anabaena* strains, belonging to different species based on comparison of their DNA base composition, DNA-DNA reassociation rates, thermal stability values of DNA-DNA hybrids and electrophoretic behaviour of five different enzymes. A good correlation with the observed morphological similarities/differences among *Anabaena* strains was seen in relation to the DNA based attributes. Li and Watanabe (2002) examined fifty axenic strains of planktonic *Anabaena*, for their DNA base composition (GC content). The taxonomic value of their GC content at species level was evaluated by comparing their morphological, physiological and biochemical properties. However, besides the DNA rearrangements occurring during heterocyst differentiation, the discovery of some mobile DNA elements and short direct repeats, provided insights into the dynamic organization of the cyanobacterial genome (Mulligan and Haselkorn 1989). The discovery of short repetitive sequence structures in cyanobacterial DNA during the last decade has led to the development of PCR-based methods for typing, i.e., distinguishing and identifying bacterial species and strains. The number of copies of such repeated DNA sequences generally varies from 2 to 40 per genome, and range in size from 500 to 10,000 base pairs (Pfeifer et al. 1983).

Filamentous heterocystous cyanobacteria are known to have several distinct families of highly conserved repetitive sequences, such as STRR (Short tandemly repetitive repeat) sequences, LTRR (Long tandemly repetitive repeat) sequences (Mazel et al. 1990) and HIP1 (Highly Iterated Palindromic sequences), to distinguish at intra generic and inter/intra species level (Smith et al. 1998). Earlier studies (Prasanna et al. 2006) had shown that molecular polymorphisms can aid in discerning differences among *Anabaena* strains, however genetic relationships were based only on molecular datasets of a small set of 13 *Anabaena* strains. PCR amplification of 70 *Anabaena* strains with STRR1A generated 23 bands, amplification with $STRR_{mod}$ resulted in 21 amplicons and HipTG generated a total of 21 amplicons which were all polymorphic (Nayak et al. 2009). The 70 *Anabaena* strains analysed using the molecular profiles of these three primers led to a total of 65 polymorphic bands and when further analyzed using Jaccard's Similarity coefficient to generate a dendrogram that depict the relationships among them a high genetic divergence observed in the dendrogram. These three primers were highly effective to differentiating among the *Anabaena* strains but in terms of high level of discrimination, STRR1A primer was observed to be more informative. Analyses of these molecular marker profiles helped in effectively discriminating all the 70 *Anabaena* strains in our previous investigation (Nayak et al. 2009). All profiles obtained from the *Anabaena* using the STRR1A, $STRR_{mod}$ and HipTG showed distinct banding patterns, which could serve as fingerprints for the strains and all the profiles were reproducible. It is well noticed that a combination of just two primers (STRR1A and $STRR_{mod}$) could unambiguously differentiate all the 70 *Anabaena* strains taken in our previous study (Nayak et al. 2009).

Diversity analyses based on *16S rDNA*, Internal transcribed spacer sequences (ITS) and microcystin sequences

The use of *16S rRNA* in genotypic studies is widespread, but the resolution is sometimes too low to distinguish closely related species. It has been shown that bacteria with identical 16S rRNA sequences can still show considerable variations in the rest of the genome, implying that they do not belong to the same species (Fox et al. 1992; Stackebrandt and Gebel 1994). Restriction Fragment Length Polymorphism (RFLP) provides a powerful tool for taxonomy as determined by genetic relatedness, bit it is tedious and expensive, especially when one needs to classify a large number of strains. Neurotoxic *Anabaena* and non toxic *Aphanizomenon* strains have recently been found to form a tight group by an RFLP study of the total *16S rRNA* gene (Lyra et al. 1997). More *16S rRNA* gene variability has been observed among the planktic *Anabaena* and *Nostoc* strains (Rudi et al. 1997). RFLP probing with short tandemly repeated DNA sequences found at many locations in the genome of *Anabaena* sp. PCC7120 helped to distinguish hepatotoxic *Anabaena* isolates from neurotoxic *Anabaena* strains and from *Nostoc* sp. (Rouhianen et al. 1995). However, the taxonomic value of Southern hybridization by using the randomly repeated sequences as a phylogenetic tool may be restricted to the identification of identical strains (Mazel et al. 1990).

All planktic cyanobacteria contain restriction endonucleases and Type II restriction endonucleases are widely distributed among cyanobacteria. To date, more than 200 cyanobacterial restriction endonucleases have been reported (Lyra et al. 2001). Iteman et al. (2002) observed that the sum of the band sizes obtained by restriction enzyme analyses considerably exceeded the length of amplicon (1450 bp) and concluded that this could result from sequence variations in the multiple *16Sr RNA* genes within individual strains. Most organisms have multiple copies of *rRNA* genes. Ligon et al. (1991) reported they had identified four rRNA operons in *Anabaena* strain PCC7120 but provided no method to support their statement. The *rRNA* operon copy number reflects the ecological strategy of the organism as there seems to be a correlation between the response rate of the organism to changing conditions and the rRNA operon copy number (Klappenbach et al. 2000).

The phylogenetic tree of the *16S rDNA* sequences obtained using fD1 and rD1 primers and a set of twenty eight *Anabaena* strains, revealed a tight grouping among all *Anabaena* strains. Microcystin synthetase genes are very useful in the detection of putative toxin producers in cyanobacteria and provide the possibility to relate the microcystin producing phenotype to the genotype independently of the taxonomic characterization of the organisms (Dittmann and Borner 2005). Microcystin biosynthesis genes have been extensively used to predict toxicity in laboratory strains of e.g. *Anabaena*, *Microcystis*, and *Planktothrix* (Rantala 2007). Hisbergues et al. (2003) designed specific primers towards condensation domain of microcystin synthetase gene (*mcyA*) for rapid identification of microcystin producing cyanobacterial genera of *Anabaena*, *Microcystis* and *Planktothrix*. Using the same set of primers, in our investigation, we have made an attempt to characterize these Indian *Anabaena* strains exhibiting fungicidal activity in terms of microcystin production, 16S rDNA and internal transcribed spacer (ITS) sequences (Gupta et al. 2012a). A similar amplified product of 297 bp was obtained only in six *Anabaena* strains belonging to five different species- *A. iyengarii* (RPAN6), *A. aphanizomenoides* (RPAN68), *A. oscillarioides* (RPAN69), *A. laxa* (RPAN8) and *A. spiroides* (RPAN50) and *Anabaena* sp. A7 (RPAN71). All the *mcyA*$^+$ *Anabaena* species studied under this investigation formed a separate subgroup in the phylogram and were found to be closely related with *Microcystis* species. This is an interesting finding which has tremendous significance in understanding evolutionary significance of *mcyA*, especially in relation to the global spread and diversity among cyanobacterial species (Gupta et al. 2012a).

The phylogenetic tree was also constructed using ITS (L) sequences which showed that most of the *mcyA*$^+$ *Anabaena* strains were grouped together in a separate Cluster, indicating again a close relation among toxic strains and confirmed the results based on *16S rRNA* gene sequences (Fig. 2). In general, a patchy distribution of toxigenecity is known to exist among strains belonging to the genera *Microcystis*, *Anabaena*, *Nodularia* and *Aphanizomenon* (Halinen et al. 2007, 2008). Gugger et al. (2002a) showed that *Anabaena* strains that produce different types of toxins clustered separately in the 16S rDNA, ITS1 and *rbc-LX* trees. It can be hypothesized that phylogeny may be correlated with the toxicity (in terms of microcystin synthesis) in the genus *Anabaena*, as recorded in our study (Gupta et al. 2012a). However, microcystin-producing *Anabaena* strains were found to be genetically heterogeneous and grouped together with few nontoxic strains, when distinguished by *16S rRNA* genes (Lyra et al. 2001; Rajaniemi et al. 2005; Halinen et al. 2007, 2008), internal transcribed spacer sequence (ITS-S) (Gulledge et al. 2002), *rboC1* (Halinen et al. 2008), *rboB* and *rbc-LX* trees (Rajaniemi et al. 2005). Similar findings were observed in the *16S rRNA* gene tree when phylogenetic tree was constructed using known microcystin-producing *Anabaena* strains of previous studies (Gugger et al. 2002a; Lyra et al. 2001; Rajaniemi et al. 2005; Halinen et al. 2007, 2008). Most of the known microcystin producing *Anabaena* strains fall into one cluster, but the Indian *Anabaena* strains were found genetically more diverse in the tree (Gupta et al. 2012a). Rantasarkka (2008) identified three different *16S rRNA Aphanizomenon* gene sequences in the Gulf of Finland, and Iteman et al. (2002) investigated multiple *rrn* operons within *Aphanizomenon* strains. *Anabaena* and *Aphanizomenon* are genetically closely related (Lyra et al. 2001; Gugger et al. 2002a; Rajaniemi et al. 2005) and therefore, the diversity in *16S rRNA* gene sequences in *Anabaena* may due to intragenomic variation. This can be further supported from the findings of Janson and Graneli (2002) who revealed four *Anabaena* strains, each of which had a unique *hetR* (heterocyst differentiation) genotype, suggesting genetically diverse *Anabaena* populations in the Baltic Sea.

Although the evolution of *mcy* has remained vertical, recombination events have occurred in *mcy* genes (Fewer et al. 2007). This polymorphism may reflect in the diversity of microcystin isoforms in *Anabaena* as produced by other cyanobacteria in earlier reports (Fewer et al. 2007). However, *mcyA*$^-$ *A. aphanizomenoides* RPAN25 was found group together with the microcystin producing *Anabaena* strains and was closely linked with the toxic *A. aphanizomenoides* RPAN68 (Fig. 2). Similar results were obtained in the earlier investigations (Lyra et al. 2001; Rajaniemi et al. 2005; Halinen et al. 2007, 2008). The clustering of few other *mcyA*$^-$ *Anabaena* strains such as *A. spiroides* RPAN58, *A. sphaerica* RPAN12, *A. oscillarioides* RPAN4 and *A. doliolum* RPAN53 was also found incoherent in relation to the*16S rRNA* gene tree (Fig. 2).

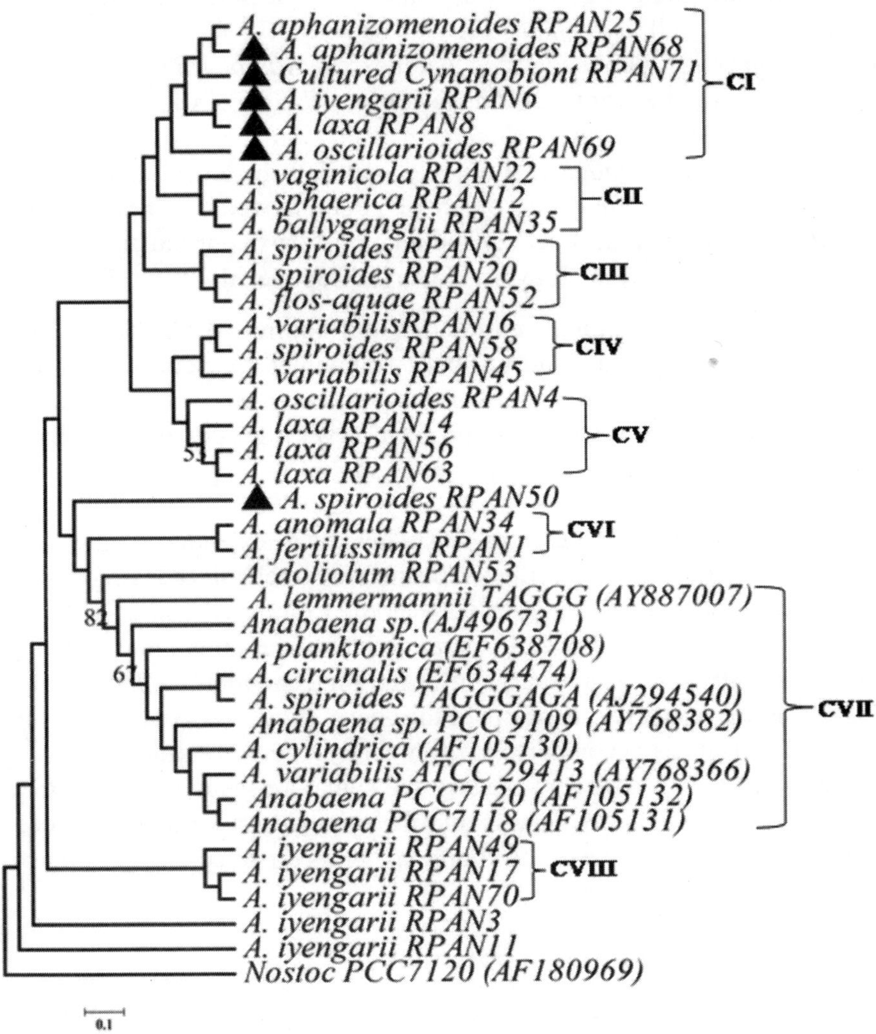

Fig. 2: Phylogenetic tree based on long internal transcribed spacer sequences (ITS-L) of all the *Anabaena* strains and available NCBI sequences, constructed by NJ method. Filled diamonds indicate *mcyA*$^+$ *Anabaena* strains. *A. variabilis* ATCC29413 (GenBank accession no. AY768366) was used as an outgroup.

This inconsistency may be due to the high variability in terms of number and composition of nucleotides in 16S-23S ITS region in these *Anabaena* strains. In addition, the Indian *Anabaena* strains exhibiting fungicidal activity could be differentiated in terms of presence of *mcyA* using proxies, such as *16S rRNA*, for identifying microcystin producers. The specific clustering based on *rpoC1* and *16S rRNA* gene has been used to identify Australian neurotoxic *Anabaena circinalis* strains (Baker et al. 2002). Similarly, hepatotoxic and nontoxic *Nodularia* strains have been distinguished by *16S rRNA* genes as well as by *hetR*, ITS1-S, and PC-IGS loci (Laamanen et al. 2001). Janse et al. (2004) suggested that the toxic *Microcystis* strains could be separated from the nontoxic ones, using *rRNA ITS* sequences. In our investigation, the Indian *mcyA*$^{+}$ *Anabaena* strains were genetically very diverse as compared to the *Anabaena* strains deposited at the NCBI data base. This can be attributed to the fungicidal nature of the strains studied. The phylogenetic tree clearly illustrated the diversity and relationships of our Indian strains with the available strains from all over the world (Gupta et al. 2012a; Fig. 2).

Analyses of morphological, molecular and combined (morphological + molecular) clearly indicated the utility of a combined dataset in discerning the genetic relationships among the strains of *Anabaena*. The importance of polyphasic approach is well recognized in several groups of cyanobacteria (Rajaniemi et al. 2005) and our previous studies (Nayak et al. 2007; 2009) further emphasized their significance, as it yielded a realistic relationship of the diversity within this genus. Similar studies can be extended to other genera, especially in germplasm collections/repositories for reliable indexing of strains. Our results undoubtedly suggested that the combined use of traditional phenotypic characterization and genotypic identification is more meaningful than their individual usage. The information generated to date, globally, emphasises the polyphyletic nature of the genus *Anabaena* and the uniqueness of the Indian isolates as compared to the globally recorded strains of this genus. Further work is in progress to identify the signature sequences for toxic and nontoxic Indian *Anabaena* strains and the microcystin variant produced by them, as a prelude to developing simple methods for screening germplasm or ecological samples for toxicity.

Deciphering the antifungal potential of *Anabaena* strains

Agriculture requires the extensive use of chemical pesticides to protect crops against pests and diseases, which urges the need for alternatives. One of these is the use of biological control agents, which include microorganisms, in live or preserved state or their products, suitably packaged in carriers, that can protect the plants against diseases (Akkopru and Demir 2005). The antifungal property of cyanobacterial metabolites has immense potential in agriculture. Endoglucanases/chitinases/chitosanases are identified as key enzymes which can be employed for biocontrol and restrict the growth of phytopathogens (Adams 2004). The successful use of *Pseudomonas* sp., *Serratia marcescens*, *Trichoderma* sp. *Bacillus* sp. and *Streptomyces* sp. producing chitinases/ endoglucanases has already been demonstrated (El-Mougy et al. 2011).

Patterson et al. (1991) obtained extracts from nearly 1000 strains of cyanobacteria from various habitats and screened them for possible antineoplastic activity, and identified cytotoxic antibiotics which were unique to cyanobacteria like acutiphycins, indolecarbazoles, mirabilene isonitriles, paracyclophanes, scytophycins, tantazoles, tolytoxin, toyocamycin and tubercidin. Laxaphycin A and B, for example, first isolated from the terrestrial cyanobacteria *Anabaena laxa*, and also found in *L. majuscula*, exert a synergistic effect against *Candida albicans* and inhibit lymphoblastic cell lines (Burja et al. 2001). Several other secondary metabolites produced by cyanobacteria such as fischerellin A, calophycin, scytonemin A, tanikolide, tolytoxin and schizotrin A, majusculamide C, AK-3 and benzoic acid derivative (Natarajan et al. 2012) have been reported for their fungicidal activity.

In our recent investigation, *Anabaena laxa* was selected for optimization, identification and characterization of biosynthesis of a fungicidal compound. Optimization of different growth conditions revealed that stationary phase (28d) culture grown under continuous light (CL) exhibited highest fungicidal activity against *Pythium debaryanum,* which in turn was regulated by different environmental/nutritional conditions. The fungicidal activity was enhanced significantly when high pH was given and phosphorous concentration was doubled (+344 μM P) Thin layer chromatography (TLC) and reverse phase high-performance liquid chromatography (HPLC) of the partially purified methanolic extracts of *A. laxa* (grown under optimized conditions) indicated that the fungicidal compound is a peptide in nature, with R_f value of 0.13 and retention time of 14.7, respectively. Fourier transform infrared (FT-IR), mass fragmentation pattern using GC-MS, ^{1}H-NMR and ^{13}C-NMR analysis revealed structural similarity of the isolated compound with majusculamide C from an alga, *Lyngbya majuscula*. The isolated majusculamide C compound was found to be soluble in solvents such as methanol, dichloromethane and acetone, but insoluble in water (Gupta et al. 2013a, manuscript submitted).

In order to identify the genes responsible for the production of a fungicidal compound by the same cyanobacterial strain- *A. laxa*, a genomic library was with 2000 clones (3-5 kb) was generated. Colony PCR using primers for known fungicidal compounds from *Pseudomonas* sp. revealed the presence of phenazine and thiquinolobactin homologs in this strain. The sequence of the biosynthetic locus (Accession number DQ119108) containing 11 open reading frames (ORFs) involved for the fungicidal compound production in *Pseudomonas* sp. was compared with the *Anabaena* genome. The corresponding ORFs (3-oxyacyl- acyl carrier-protein, beta-ketoacyl synthase, acetylorinithine aminotransferase etc.), particularly rich in PKS and NRPS genes with similar functions as in *Pseudomonas* sp. were also found in the *Anabaena* genome. Thus, it can be surmised that such a locus may be involved in the biosynthesis of fungicidal compound in *A. laxa*, although this needs in depth analyses (Gupta et al. 2013b, manuscript submitted).

Besides serving as allelochemicals and regulators of the succession of algal and bacterial populations in aquatic bodies, such chemicals can also serve as herbicide or bio-control agents (Jaiswal et al. 2008; Manjunath et al. 2010; Dukare et al. 2011). Our recent study showed that amendment of compost with *Anabaena laxa* RPAN8 or *Anabaena variabilis* RPAN59 or *Anabaena oscillarioides* RPAN69 in association with *Bacillus subtilis* strain provided effective biocontrol in tomato against plant pathogenic fungi. The increased PAL (phenylalanine ammonia lyase) and PPO (polyphenol oxidase) activity in shoots and roots of the plant, using composts amended with RPAN58 and 59 led to a significant increase in plant growth parameters, enhanced vigor index and suppression of the disease severity (Chaudhary et al. 2012) Cyanobacteria amended composts were found to provide bioprotection against *Fusarium* wilt in tomato, involving a strategy combining hydrolytic enzymes, elicitation of defense and pathogenesis related enzymes, plant growth promoting factors (including IAA production), which stimulated plant growth and yields, besides efficiently colonising roots and reducing fungal load (Prasanna et al. 2013b).

Potential role of hydrolytic enzymes in *Anabaena* strains and prospecting for genes responsible

Prasanna et al. (2008) for the first time evaluated the potential role of hydrolytic enzymes towards fungicidal activity in 35 strains of *Anabaena*. Significant differences were also observed among the strains in terms of their FPase, Chitosanase and xylanase activity, while low and relatively similar values of CMCase, cellobiase and protease activity were recorded in the strains analyzed. IAA production was also observed in all the strains. The high values of chitinase and xylanase activity in selected cultures may contribute to the fungicidal activity of the cyanobacterial strains, besides other bioactive compounds, including IAA, which are established promising traits for biocontrol agents. Prasanna et al. (2010b) undertook an investigation to mine for genes responsible for fungicidal activity in the selected cyanobacteria showing fungicidal activity (Fig. 3a-b). Microscopic examination of antifungal activity revealed abnormalities in the hyphal structure and severe disintegration of the mycelia in *Pythium aphanidermatum* (Fig. 3c-d). An amplified product of 172 bp obtained using specific primers derived from the coding region of chitinase (*chi IS*) gene in *Streptomyces* sp., showed 100% similarity with hydrogenbyrinic acid a, c-diamide cobaltochelatase gene in *Anabaena*, and significant similarity with *chi IS* gene under less stringent conditions. The 663 bp sequence obtained by employing specific primers for chitosanase (*cho* A) derived from *Mitsuaria chitosanitabida* 3001 strain, showed 100% similarity with glycoside hydrolase family 3 domain like protein(s). Gupta et al. (2010) characterized a novel *cho* gene encoding chitosanase associated with GH3-like family and showed that it is responsible for the antifungal activity in *A. fertilissima*. The chitosanase (*cho*) gene from two *Anabaena* species- *A. fertilissima* (exhibiting antifungal activity) and *A. sphaerica* (negative control) were amplified using chitosanase specific primers (as designed by Yun et al. 2005) in order to characterize the *cho* gene responsible for antifungal activity. Sequence analyses indicated amplified products of 1.086 and 1.101 kb, an open reading frame of 362 and 367 amino acids with a predicted molecular mass of 40 kDa and 40.6 kDa in *A. sphaerica* and *A. fertilissima*, respectively. The computational analyses of the corresponding amino acid sequences of both *cho* showed that they belong to glycoside hydrolase 3 (GH3) like family of *A. variabilis* ATCC 29413. The HPLC analyses of the purified recombinant Cho protein from *A. fertilissima* revealed its ability to degrade $(GlcN)_5$ into $(GlcN)_2$ and $(GlcN)_3$, thus confirmed the presence of a functional chitosanase in *A. fertilissima*. Site directed mutagenesis further revealed that critical Glu-121 and Glu-141 residues, as essential for the chitosanase/antifungal activity in *A. fertilissima* (Gupta et al. 2010).

The expression profile of *cho* was also analyzed under increasing dark period by quantitative real-time RT-PCR (qRT-PCR) which revealed increase in transcriptional leval under the condition of long dark phase in *A. fertilissima*. Besides L:D-8:16, the highest chitosanase/antifungal activity under the environmental and nutritional conditions of 43 µM P, pH 9.0 and a temperature 27°C (Gupta et al. 2012b). Further, the enzyme was purified under the optimized conditions which exhibited a specific activity of 822 U/mg. HPLC profile of the products of enzyme activity with different chitosan oligosaccharides revealed in the production of dimer units $(GlcN)_2$ or more confirming the endo-type nature of the purified chitosanase. The optimum pH and temperature of the purified enzyme was 7.5 and 27 °C, respectively. Further, the enzyme was stable in the pH range of 5.5-9.0 up to 12 h and temperature between 27-50 °C up to 3 h. The enzyme was strongly inhibited by Ag^+, Fe^{3+} and Hg^{2+} and stimulated by Cu^{+2} and Zn^{2+}. The investigation revealed significant features regarding the stability of the chitosanase enzyme from *A. fertilissima* under a broad range of pH and temperature and its potential in biocontrol. This study is a first time report on the presence of homologues of chitosanase in cyanobacteria which opens up new avenues to search for the detailed mechanism responsible for the antifungal activity in cyanobacteria.

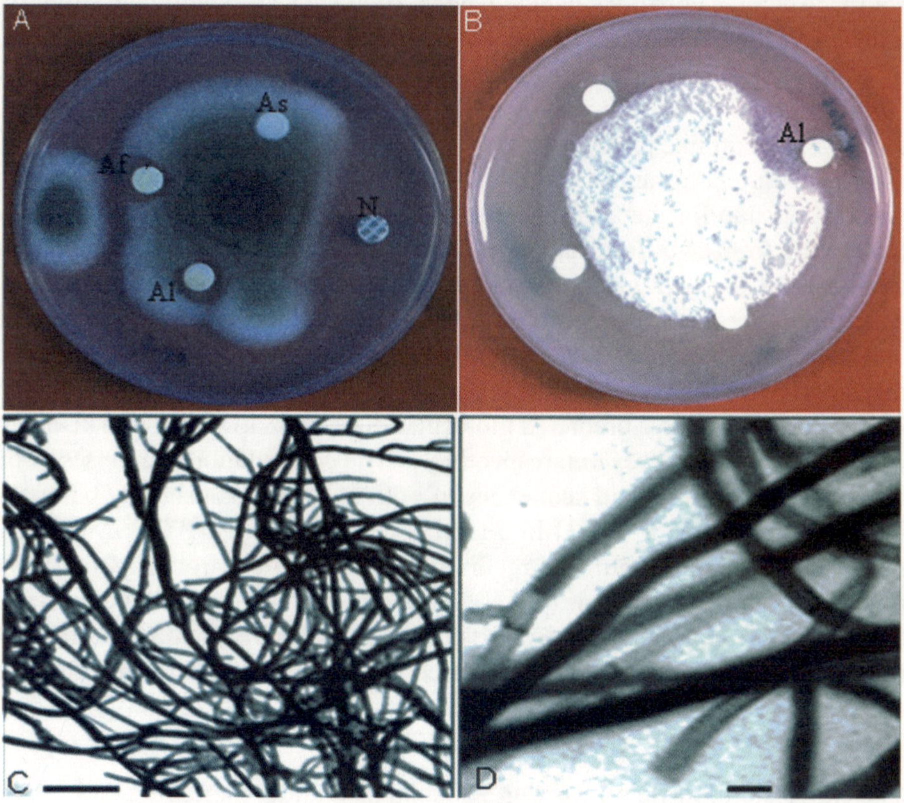

Fig. 3: A-B Fungicidal activity of *Anabaena* strains on the lawn of selected phytopathogenic fungi. **(A)** *Aspergillus candida* **(B)** *Fusarium solani*. Af, *A. fertilissima*; As, *A. sphaerica*; Al, *A. laxa* and N, nystatin. **(C-D)** Microscopic analyses of fungicidal activity. **(C)** indicates control and **(D)** shows fungicidal activity of *A. laxa*. The horizontal bars represent 50 μm.

Two antifungal endoglucanases- End1 (38 kDa) and End2 (74 kDa) have been purified and characterized from the cyanobacterial strain - *Anabaena laxa* RPAN8 (Gupta et al 2011; 2012c). Thin layer chromatography (TLC) analyses of the hydrolyzed products using different substrates indicated endo and exo/endo-type nature of End1 and End2, respectively. Both the purified End proteins showed the highest endoglucanase activities at pH 6.0 and temperature 30 °C. However, End1 and End2 were found to be stable over a wide range of pH (5.0-7.0 and 5.0-9.0, respectively) up to 12 h incubation; and temperature of

40 and 50 °C, up to 3h incubation. The three dimensional structures of both End1 and End2 were predicted using computational analyses and found to be distorted beta-barrel and beta-barrel shaped, respectively. The predicted active sites were identified and the computational models designed indicated that substrate - cellotetraose completely spans over the active site in the three dimensional cartoon and surface models for End2, confirming the biochemical and kinetic analyses. The active site residues were identified using site-directed mutagenesis which indicates Glu-23 and Cys-11 (present at signal peptide regions) as catalytic residues critical to antifungal activity in End1 and End2, respectively (Gupta et al. 2012c). To our knowledge, this study represents a first time in-depth investigation into the structural/functional aspects of purified endoglucanases from cyanobacteria.

Toxin production in *Anabaena*

Toxins, produced by bacterial algae/cyanobacteria can be considered as a sub group of allelochemicals, which serve to inhibit one or more groups of organisms, and help to provide a competitive advantage to the producer. These toxins are considered a serious toxicological health risk that may affect humans, animals, and ecosystems worldwide (Kaas and Henriksen 2000). Srivastava (1972) reported antagonism between species of *Anabaena* and *Hapalosiphon intricatus*. Several species of cyanobacteria have been shown to be poisonous and toxic to other algae (algicidal). Landsberg (2002) reported that *Anabaena flos-aquae* in freshwater environments produce toxins. During the course of our screening of the 70 *Anabaena* strains, we observed algicidal and fungicidal activities in 35 isolates obtained from diverse habitats (Prasanna et al. 2008), measured in terms of zone of inhibition (Table 2). Most of the cyanotoxins are non-ribosomal peptides and synthesized by a mechanism parallel to protein synthesis. On the basis of their chemical structures, cyanotoxins are classified as cyclic peptides (microcystins and nodularins), alkaloids [anatoxin-a, anatoxin-a(S)], saxitoxin, cylindrospermopsin, aplysiatoxins and lyngbyatoxin-a and lipopolysaccharides. *Anabaena* genera has ability to produce multiple toxins such as microcystins, saxitoxins, anatoxin-a, anatoxin-a(s) and ctylindrospermopsin (Dittmann and Wiegand 2006).

Several cyanobacterial taxa (*Microcystis*, *Planktothrix* and *Anabaena*) are capable of producing microcystins (Rouhiainen et al. 2004). Microcystins are potent inhibitors of two major eukaryotic signal transduction, serine phosphatase 1 and 2A enzymes. Saxitoxin or paralytic shellfish poisons (PSPs) are highly potent neurotoxins produced by several freshwater species of cyanobacteria- *Anabaena circinalis*, *Aphanizomenon* sp., *Aphanizomenon gracile, Cylindrospermopsis raciborskii* and *Lyngbya wollei* and marine dinoflagellates genera *Alexandrium*, *Pyrodinium* and *Gymnodinium* (Ballot et al. 2010). Eight cyanobacterial species have so far been identified as cylindrospermopsin producers: *Cylindrospermopsis raciborskii, Aphanizomenon ovalisporum, Aphanizomenon flosaquae, Umezakia natans, Rhaphdiopsis curvata, Anabaena bergii, Anabaena lapponica* and *Lygnbya wollei* (Seifert et al. 2007). Cylindrospermopsin has hepatotoxic, general cytotoxic and neurotoxic effects and is a potential carcinogen (Kiss et al. 2002). Its toxicity is due to the inhibition of glutathione and protein synthesis as well as the inhibition of cytochrome P450. Anatoxins are a group of alkaloids produced among a number of cyanobacterial species including *Anabaena flos–aquae, Aphanizomenon flos–aquae, Cylindrospermum*, *Raphidiopsis* and *Planktothrix* sp. (formerly *Oscillatoria* sp.) (Carmichael 1992). The biosynthetic pathway for anatoxin-a was recently described based on the genome sequence of *Oscillatoria* PCC 6506 (Mejean et al. 2009). Anatoxin-a is synthesized via a PKS pathway, but so far the genes have not been explored in other anatoxin producing strains.

Table 2: Allelopathic activity of *Anabaena* strains.

Strain number	*Anabaena* species	Allelopathic towards		Antifungal activity against		
		Synechocystis	*Nostoc*	*F. solani*	*A. solani*	*F. moniliforme*
RPAN57	*A. spiroides*	-	+	-	-	+
RPAN4	*A. oscillarioides*	+	+	-	+	+
RPAN20	*A. spiroides*	-	+	-	-	+
RPAN35	*A. ballyganglii*	+	-	-	+	-
RPAN34	*A. anomala*	-	+	-	-	+
RPAN1	*A. fertilissima*	-	+	+	+	+
RPAN11	*A. iyengarii*	+	-	-	+	-
RPAN49	*A. iyengarii*	+	+	-	+	+
RPAN17	*A. iyengarii*	-	+	+	-	-
RPAN70	*A. iyengarii*	+	-	+	-	-
RPAN3	*A. iyengarii*	-	+	-	+	+
RPAN68	*A. aphanizomenoides*	+	+	+	-	+
A7 (RPAN71)	*Anabaena* sp.	+	-	+	-	+
RPAN69	*A. oscillarioides*	+	+	+	-	+
RPAN6	*A. iyengarii*	+	+	+	+	+
RPAN8	*A. laxa*	+	-	+	+	+
RPAN50	*A. spiroides*	+	+	-	+	+
RPAN22	*A. vaginicola*	+	-	+	-	-
RPAN52	*A. flos- aquae*	-	+	-	+	+
RPAN56	*A. laxa*	+	-	-	-	+
RPAN25	*A. aphanizomenoides*	-	+	+	-	-
RPAN63	*A. laxa*	+	+	+	-	+
RPAN14	*A. laxa*	+	-	+	-	+
RPAN53	*A. doliolum*	-	+	+	+	+
RPAN45	*A. variabilis*	+	+	+	-	+
RPAN16	*A. variabilis*	+	-	+	+	-
RPAN12	*A. sphaerica*	-	+	-	-	+

Includes information from Prasanna et al. 2008 and Nayak et al. 2009

Conclusions

The genus *Anabaena* shows tremendous diversity in its morphology, qualitative and quantitative distribution of pigments, nitrogen -fixing potential, genetic aspects and secondary metabolites. Concerted efforts are required to develop databases on such important cyanobacterial genera, for their possible exploitation in environmental conservation, agriculture and industry.

References

Adams DJ (2004) Fungal cell wall chitinases and glucanases. Microbiology 150: 2029-2035.

Adams DG (1992) Multicellularity in Cyanobacteria. In: Prokaryotic structure and function: A New Perspective. Symp Soc Gen Microbiol 47: 341-384.

Akkopru A and Demir S (2005) Biological control of Fusarium wilts in tomato caused by *Fusarium oxysporum f. sp. lycopersici* by AMF Glomus intraradices and same rhizobacteria. J Phytopathol 153: 544-550.

Baker JA, Entsch B, Neilan BA and McKay DB (2002) Monitoring changing toxigenecity of a cyanobacterial bloom by molecular methods. Appl Environ Microbiol 68: 6070-6076.

Ballot A, Fastner J, Wiedner C (2010) Paralytic shellfish poisoning toxin-producing cyanobacterium *Aphanizomenon gracile* in northeast Germany. Appl Environ Microbiol 76: 1173-1180.

Ballot A, Dadheech P, Krienitz L (2004) Phylogenetic relationship of *Arthrospira, Phormidium,* and *Spirulina* strains from Kenyan and Indian water bodies. Arch Hydrobiol/Algol Stud 113: 37-56.

Bandaranayake WM (1998) Mycosporines: are they nature's sunscreens? Nat Prod Rep 15: 159-172.

Bornet E and Flahaut C (1886-1888) Revision des Nostocacees heterocystees. Ann Sci Nat Bot 7, Ser 3: 323-381; 4: 343-373; 5: 52-129; 7: 178-262.

Borowitzka LJ (1994) Commercial pigment production from algae. In: *Algal Biotechnology in the Asia Pacific Region* (Ed. Phang CJ), University of Malaya, Kuala Lumpur, Malaysia. pp. 82-84.

Bourrelly P (1985) Les Algues d,eau Douce. III, 2nd edn. Paris: N. Boubee and Cie.

Burja AM, Banaigs B, Abou-Mansour E, Burgess JG, Wright PC (2001) Marine cyanobacteria-a prolific source of natural products. Tetrahedron 57: 9347–9377.

Carmichael WW (1992) Cyanobacteria secondary metabolites-the cyanotoxins. J App Bacteriol 72: 445-459.

Castenholz RW (2001) Phylum BX. Cyanobacteria - Oxygenic photosynthetic bacteria. In: *Bergey's manual of Systematic Bacteriology*, 2nd Edn (Eds. Boone DR and Castenholz WR), Springer, Berlin.

Chapman DJ (1973) Biliproteins and bile pigments. In: *The Biology of blue green algae* (Eds. Carr NG and Whitton BA), University of California Press, Botanical Monographs No. 9, pp. 162 - 185.

Chaudhary V, Prasanna R, Nain L, Dubey SC, Gupta V, Singh R, Jaggi S, Bhatnagar, AK (2012) Bioefficacy of novel cyanobacteria-amended formulations in suppressing damping off disease in tomato seedlings. World J Microbiol Biotechnol 28: 3301-3310.

Chen F, Wang K, Kan J, Suzuki MT and Wommack KE (2006) Diverse and unique picocyanobacteria in Chesapeake Bay, reveled by 16S-23S rRNA internal transcribed spacer sequences. Appl Environ Microbiol 72: 2239-2243.

Desikachary TV and Anand N (1974) Preliminary studies on the genus *Anabaena* Bory. In:

Proceedings of the second international symposium in Phycology, Taxonomy of algae. Madras, University of Madras.

Desikachary TV (1970) Taxonomy of blue-green algae: Problems and prospects. *Schweiz.* Z Hydrol 32: 490-494.

Desikachary TV (1959) Cyanophyta – In: *ICAR Monographs on Algae*, New Delhi. pp. 686.

Dittmann E and Wiegand C (2006). Cyanobacterial toxins-occurrence, biosynthesis and impact on human affairs. Mol Nutr Food Res 50: 7-17.

Dittmann E and Borner T (2005) Genetic contributions to the risk assessment of microcystin in the environment. Toxicol Appl Pharmacol 203:192-200

Drouet F (1968) Revision of the classification of the Oscillatoriaceae Monograph. In: *Acad Nature Sci.* Philadelphia, Fulton Press, Lancaster, Pennsylvania.

Dukare AS, Prasanna R, Dubey SC, Chaudhary V, Nain L, Singh R and Saxena AK (2011) Evaluating novel microbe amended composts as biocontrol agents in tomato. Crop Protect 30: 436-442.

El-Mougy NS, Abdel-Kader MM and Alhabeb RS (2011) *In vitro* antifungal activity of chitinolytic enzymes produced by bioagents against root rot pathogenic fungi. Arch. Phytopathol. Plant Protect 44: 613-622.

Ezhilarasi A and Anand N (2009) Characterization of *Anabaena* species by RFLP of the 16S rRNA gene. J Pure Appl Microbiol 3: 273-278.

Fewer, DP, Rouhiainen L, Jokela J et al (2007) Recurrent adenylation domain replacement in the microcystin synthetase gene Cluster. BMC Evol Biol 7: 183-193.

Flechtner VR, Boyer SL, Johansen JR and DeNoble ML (2002) *Spirirestis rafaelensis* gen. et sp. nov. (Cyanophyceae), a new cyanobacterial genus from arid soils. Nova Hedwigia 74: 1-24.

Fox GE, Wisotzkey JD and Jurtshuk P (1992) How close is close: 16S rRNA sequence identity may not be sufficient to guarantee species identity. Int J Syst Bacteriol 42: 166-170.

Fritsch FE (1949) The genus *Anabaena* with special reference to the species recorded from the areas adjacent to Asiatic mainland. J Indian Bot Soc 28: 135-161.

Geitler L (1932) Cyanophyceae. In: *Kryptogamen flora van Deutschland, Osterrich under Schweiz (*Ed. Rabenhorsts), Vol. XIV. Academische Verlags, Leipzig, Reprinted 1971, Johnson Reprint Co., New York, London.

Gugger M, Lyra C, Suominen I, Tsitko I, Humbert JF, Salkinoja-Salonen M and Sivonen K (2002b) Cellular fatty acids as chemotaxonomic markers of the genera *Anabaena, Aphanizomenon, Microcystis, Nostoc* and *Planktothrix* (Cyanobacteria). Int J Syst Evol Microbiol 52: 1007–1015.

Gugger M, Lyra L, Henriksen P, Coute A, Humbert JF and Sivonen K (2002a) Phylogenetic comparison of the cyanobacterial genera *Anabaena* and *Aphanizomenon*. Int J Syst Evol Microbiol 52: 1867-1880.

Gulledge BM, Aggen JB, Huang HB et al (2002) The microcystins and nodularins: cyclic polypeptide inhibitors of PP1 and PP2A. Curr Med Chem 9: 1991-2003.

Gupta V, Prasanna R, Cameotra SS, Dureja P, Singh RN and Sharma J (2013a) Enhancing the production of an antifungal compound from *Anabaena laxa* through modulation of environmental conditions and its characterization. (manuscript submitted to Process Biochem).

Gupta V, Prasanna R and Sharma J (2013b) Allele mining of the genes involved for the biosynthesis of fungicidal compounds in *A. laxa*. (manuscript submitted to Folia Microbiol).

Gupta V, Natarajan C, Chaudhary V, Kumar A, Sharma E, Sharma J, Bhatnagar AK and Prasanna R (2012a) Analyses of diversity among fungicidal *Anabaena* strains. J Appl Phycol 24: 1395-1405.

Gupta V, Prasanna R, Srivastava AK and Sharma J (2012b) Purification and characterization of a novel antifungal endo-type chitosanase from *Anabaena fertilissima*. Ann Microbiol 62: 1089–1098.

Gupta V, Prasanna R, Chaudhary V, Nain L (2012c) Biochemical, structural and functional characterization of two novel antifungal endoglucanases from *Anabaena laxa*. Biocat Agr Biotechnol 1: 338–347.

Gupta V, Natarajan C, Kumar K and Prasanna R (2011) Identification and characterization of endoglucanases for fungicidal activity in *Anabaena laxa*. J Appl Phycol 23: 73-81.

Gupta V, Prasanna R, Natrajan C, Srivastava AK and Sharma J (2010) Identification, characterization and regulation of a novel antifungal chitosanase gene (*cho*) in *Anabaena* sp. Appl Environ Microbiol 76: 769-2777.

Halinen K, Fewer DP, Sihvonen LM, Lyra C, Eronen E and Sivonen K (2008) Genetic diversity in strains of the genus Anabaena isolated from planktonic and benthic habitats of the Gulf of Finland (Baltic Sea). FEMS Microbiol Ecol 64: 199-208.

Halinen K, Jokela J, Fewer DP, Wahlsten M and Sivonen M (2007) Direct evidence for production of microcystins by *Anabaena* strains from the Baltic Sea. Appl Environ Microbiol 73: 6543-6550.

Henson BJ, Hesselbrock SM, Watson LE and Barnum SR (2004) Molecular phylogeny of the heterocystous cyanobacteria (subsections IV and V) based on *nifD*. Int J Syst Evol Microbiol 54: 493–497.

Henson BJ, Watson LE and Barnum SR (2002) Molecular differentiation of the heterocystous cyanobacteria, *Nostoc* and *Anabaena*, based on complete *nifD* sequence. Curr Microbiol 45: 161-164.

Herdman M, Janvier M, Rippka R and Stanier RY (1979b) Genome size of cyanobacteria. J Gen Microbiol 111: 73-85.

Herdman M, Janvier M, Waterbury JB, Rippka R and Stanier RY (1979a) Deoxyribonucleic acid base composition of cyanobacteria. J Gen Microbiol 1i1: 63-71.

Hiroki M, Shimizu ALR, Watanabe M and Watanabe MM (1998) Development of database system useful for identification of *Anabaena* spp. (Cyanobacteria). Phycol Res 46: 85-93.

Hisbergues M, Christiansen G, Rouhiainen L, Sivonen K and Borner T (2003) PCR based identification of microcystin-producing genotypes of different cyanobacterial genera. Arch Microbiol 180: 402-410.

Iteman I, Rippka R, Tandeu de Marsac N and Herdman M (2002) rDNA analysis of planktonic heterocystous cyanobacteria, including numbers of the genera *Anabeanopsis* and *Cyanospira*. Microbiology 148: 481-496.

Jaiswal P, Singh PK and Prasanna R (2008) Cyanobacterial bioactive molecules-an overview of their toxic properties. Can J Microbiol 54: 701-717.

Janse I, Kardinaal WEA, Meima M, Fastner J, Visser PM and Zwart G (2004) Toxic and nontoxic *Microcystis* colonies in natural populations can be differentiated on the basis of rRNA gene internal transcribed spacer diversity. Appl Environ Microbiol 70: 3979-3987.

Janson S and Graneli E (2002) Phylogenetic analyses of nitrogen-fixing cyanobacteria from the Baltic Sea reveal sequence anomalies in the phycocyanin operon. Int J Syst Evol Microbiol 52: 1397-1404.

Kaas H and Henriksen P (2000) Saxitoxin (PSP toxins) in Danish lakes. Water Res 34: 2089-2097.

Kantz T and Bold HC (1969) Morphological and taxonomic investigations of *Nostoc* and *Anabaena* in culture. In: *Phycol. Stud.* Univ. of Texas Press, Austin.

Karthikeyan N, Prasanna R, Sood A, Jaiswal P, Nayak S and Kaushik BD (2009) Physiological characterization and electron microscopic investigations of cyanobacteria associated with wheat rhizosphere. Folia Microbiol 54: 43-51.

Karthikeyan N, Prasanna R, Lata and Kaushik BD (2007) Evaluating the potential of plant growth promoting cyanobacteria as inoculants for wheat. Europ J Soil Biol 43:23-30.

Kaushik BD (2000) Screening of cyanobacterial strains for natural colours. Indian Hydrobiol 3: 81-84.

Kiss T, Vehovsky A, Hiripi L, Kovacs M and Voros L (2002) Membrane effects of toxins isolated from a cyanobacterium, *Cylindrospermopsis raciborskii*, on identiûed molluscan neurones. Comp Biochem Physiol 131: 167-176.

Klappenbach JA, Dunabar JM and Schmidt TM (2000) rRNA operon copy number reflects ecological strategies of bacteria. Appl Environ Microbiol 66: 1328-1333.

Komarek J (2005b) The modern classification of cyanoprokaryotes (cyanobacteria). In: *Oceanological and Hydrobiological Studies* Vol. XXXIV, Supplement 3 Institute of Oceanography, University of Gdañsk, pp 5-17.

Komarek J (2005a) Studies on the cyanophytes (Cyanobacteria, Cyanoprokaryota) of Cuba 11. Freshwater *Anabaena* species. Preslia Praha 77: 211-234.

Komarek J (1992) Studies on the Cyanophytes of Cuba 7-9. Folia Geobot Phytotax 24: 171-206.

Komarek J and Kovacik L (1989) Trichome structure of four *Aphanizomenon* taxa (Cyanophyceae) from Czechoslovakia, with notes on the taxonomy and delimitation of the genus. Plant Syst Evol 164: 47-64.

Komarek J and Anagnostidis K (1989) Modern approach to the classification system of cyanophytes. 4 Nostocales. Arch Hydrobiol Suppl. 83. (Algol. Stud.) 56: 247-345.

Laamanen MJ, Gugger MF, Lehtimaki JM, Haukka K and Sivonen K (2001) Diversity of toxic and nontoxic *Nodularia* isolates (cyanobacteria) and filaments from the Baltic Sea. Appl Environ Microbiol 67: 4638-4647.

Landsberg G (2002). Discovering New Physics in the Decays of Black Holes. Phys Rev Lett 88: 81801-81804.

Larsson SC, Bergkvist L, Naslund I, Rutegard J and Wolk A (2007) Vitamin A, retinal, and carotenoids and the risk of gastric cancer: a prospective cohart study. Am J Clin Nutr 85: 497-503.

Li R (1998) Taxonomic studies on planktonic species of *Anabaena* (Cyanobacteria), Ph.D. thesis, Tsukuba University, Tsukuba.

Li RH and Watanabe MM (2002). DNA base composition of planktonic species of *Anabaena* (Cyanobacteria) and its taxonomic value. J Gen Appl Microbiol 48: 77-82.

Ligon PJB, Meyer KG, Martin JA and Curtis SE (1991) Nucleotide sequence of a *16S rRNA* gene from *Anabaena* sp. strain PCC 7120. Nucl Acids Res 19: 444-453.

Lyra C, Suomelainen S, Gugger M, Vezie C, Sundman P, Paulin L and Sivonen K (2001) Molecular characterization of planktic cyanobacteria of *Anabaena*, *Aphanizomenon*, *Microcystis* and *Planktothrix* genera. Int J Syst Evol Microbiol 51: 513-526.

Lyra G, Hantula J, Vainio E, Rapala J, Rouhiainen L and Sivonen K (1997) Characterization of cyanobacteria by SDS-PAGE of whole cell proteins and PCR/RFLP of the *16S rRNA* gene. Arch Microbiol 168: 176-184.

Manjunath M, Prasanna R, Sharma P, Nain L and Singh R (2011) Developing PGPR consortia using novel genera- *Providencia* and *Alcaligenes* along with cyanobacteria for wheat. Arch Agr Soil Sci 57: 873-887.

Manjunath M, Prasanna R, Lata, Dureja P, Singh R, Kumar A, Jaggi S and Kaushik BD (2010) Biocontrol potential of cyanobacterial metabolites against damping off disease caused by *Pythium aphanidermatum* in solanaceous vegetables. Arch Phytopathol Plant Protect 43: 666-677.

Marquardt J and Palinska KA (2007) Genotypic and phenotypic diversity of cyanobacteria assigned to the genus *Phormidium* (Oscillatoriales) from different habitats and geographical sites. Arch Microbiol 187: 397-413.

Mazel D, Houmard J, Castets AM and Tandeau de Marsac N (1990) Highly repetitive DNA sequences in cyanobacterial genomes. J Bacteriol 172: 2755-2761.

McGuire RF (1984) A numerical taxonomic study of *Nostoc* and *Anabaena*. J Phycol 20: 454-460.

Mejean A, Mann S, Maldiney T, Vassiliadis G, Lequin O and Ploux O (2009) Evidence that biosynthesis of the neurotoxic alkaloids anatoxin-a and homoanatoxin-a in the cyanobacterium *Oscillatoria* PCC 6506 occurs on a modular polyketide synthase initiated by L-proline. J Am Chem Soc 131: 7512-7513.

Mekonnen A, Prasanna R and Kaushik BD (2002) Cyanobacterial nitrogen fixation in the presence of nitrogen fertilizers. Ind J Exp Biol 40: 854–857.

Mulligan ME and Haselkorn R (1989) Nitrogen fixation (*nif*) genes of the cyanobacterium *Anabaena* species strain PCC 7120. J Biol Chem 264: 19200-19207.

Muralitharan G and Thajuddin N (2010) M13-based genotyping of marine cyanobacterial strains from the Indian subcontinent and maintained in the NFMC germplasm collection. J Appl Phycol 22: 709-716.

Nain L, Rana A, Joshi M, Jadhav SD, Kumar D, Shivay YS, Paul S and Prasanna R (2010) Evaluation of synergistic effects of bacterial and cyanobacterial strains as biofertilizers for wheat. Plant and Soil 331: 217-230.

Natarajan C, Prasanna R, Gupta V, Dureja P and Nain L (2012) Dissecting the fungicidal activity of *Calothrix elenkinii* using chemical analyses and microscopy. Appl Biochem Microbiol 48: 51–57.

Nayak S and Prasanna R (2007) Sol pH and its role in cyanobacterial abundance and diversity in rice field soils. Appl Ecol Environ Res 5: 103-113.

Nayak S, Prasanna R, Prasanna BM and Sahoo DB (2009) Genotypic and phenotypic diversity of *Anabaena* isolates from diverse rice agro-ecologies of India. J Basic Microbiol 49: 165-177.

Nayak S, Prasanna R, Prasanna BM and Sahoo DB (2007) Analysing diversity among Indian isolates of *Anabaena* (Nostocales, Cyanophyta) using morphological, physiological and biochemical characters. World J Microbiol Biotechnol 23: 1575-1584.

Nayak S, Prasanna R, Pabby A, Dominic TK and Singh PK (2004) Effect of BGA- *Azolla* biofertilizers on nitrogen fixation and chlorophyll accumulation at different depths in soil cores. Biol Fertil Soils 40: 67-72.

Palinska KA, Liesack W, Rhiel E and Kumbein WE (1996) Phenotype variability of identical genotypes: the need for a combined approach in cyanobacterial taxonomy demonstrated on merismopedia-like isolates. Arch Microbiol 166: 224-233.

Patterson GML, Baldwin CL, Bolis CM, Caplan FR, Kao H, Larsen LK, Levine IA, Moore RE, Nelson CS, Tschappat KD, Tuang GD, Furusaa E, Furusawa S, Norton TR and Raybourne RB (1991) Anti-neoplastic activity of cultured blue-green algae (cyanophyta). J Phycol 27: 530-536.

Pfeifer F, Betlach M, Martienssen R, Friedman J and Boyer HW (1983) Transposable elements of *Halobacterium halobium*. Mol Gen Genet 191: 182-188.

Prasanna R and Nayak S (2007) Influence of diverse rice soil ecologies on cyanobacterial diversity and abundance. Wetlands Ecol Manage 15: 127-134.

Prasanna R, Sharma E, Sharma P, Kumar A, Kumar R, Gupta V, Pal RK, Shivay YS and Nain L (2013a) Soil fertility and establishment potential of inoculated cyanobacteria in rice crop grown under non-flooded conditions. Paddy and Water Environ 11: 175-183.

Prasanna R, Chaudhary V, Gupta V, Babu S, Kumar A, Singh R, Shivay YS, Nain L (2013b) Cyanobacteria mediated plant growth promotion and bioprotection against *Fusarium* wilt in tomato. Eur J Plant Pathol DOI 10.1007/s10658-013-0167-x.

Prasanna R, Jaiswal P, Shrikrishna J, Joshi M, Nain L, Rana A, Shivay, YS (2012c). Evaluating the potential of rhizo-cyanobacteria as inoculants for rice and wheat. J Agric Technol 8: 157-171.

Prasanna R, Nain L, Pandey AK and Saxena AK (2012b) Microbial diversity and multidimensional interactions in the rice ecosystem. Arch Agr Soil Sci. 58;723-744.

Prasanna R, Joshi M, Rana A, Shivay YS and Nain L (2012a) Influence of co-inoculation of bacteria- cyanobacteria on crop yield and C- N sequestration in soil under rice crop. World J Microbiol Biotechnol 28: 1223-1235.

Prasanna R, Singh RN, Joshi M, Madhan K, Pal RK and Lata (2011b) Monitoring the biofertilizing potential and establishment of inoculated cyanobacteria in soil using physiological and molecular markers. J Appl Phycol 23: 301-308.

Prasanna R, Pattnayak S, Sugitha TCK, Nain L and Saxena AK (2011a) Development of cyanobacterium based biofilms and their in vitro evaluation for agriculturally useful traits. Folia Microbiol 56: 49-58.

Prasanna R, Joshi M, Rana A and Nain L (2010a) Modulation of IAA production in cyanobacteria by tryptophan and light. Polish J Microbiol (formerly *Acta Microbiologica Polonica)* 59: 99-105.

Prasanna R, Gupta V, Natarajan C and Chaudhary V (2010b) Bioprospecting for genes involved in the production of chitosanases and microcystin-like compounds in *Anabaena* strains. World J Microbiol Biotechnol 26: 717-724.

Prasanna R, Madhan K, Singh RN, Chauhan AK and Nain L (2010c) Developing biochemical and molecular markers for cyanobacterial inoculants. Folia Microbiol 55: 474-480.

Prasanna R, Jaiswal P, Nayak S, Sood A and Kaushik BD (2009a) Cyanobacterial diversity in the rhizosphere of rice and its ecological significance. Ind J Microbiol 49: 89-97.

Prasanna R, Nain L, Ancha R, Shrikrishna J, Joshi M and Kaushik BD (2009b) Rhizosphere dynamics of inoculated cyanobacteria and their growth-promoting role in rice crop. Egypt J Biol 11: 26-36.

Prasanna R, Lata N, Tripathi R, Gupta V, Middha S, Joshi M, Ancha R and Kaushik BD (2008) Evaluation of fungicidal activity of extracellular filtrates of cyanobacteria - possible role of hydrolytic enzymes. J Basic Microbiol 48: 186-194.

Prasanna R, Kumar R, Sood A, Prasanna BM and Singh PK (2006) Morphological, physiochemical and molecular characterization of *Anabaena* strains. Microbiol Res 161: 187-202.

Prasanna R, Pabby A and Singh PK (2004) Effect of glucose and light/dark environment on pigmentation profiles in *Calothrix elenkenii.* Folia Microbiol 49: 26-30.

Prasanna R, Prasanna BM, Mohammadi SA and Singh PK (2003b) Analysis of phycobiliprotein content in *Tolypothrix* germplasm. Folia Microbiol 48: 59-64.

Prasanna R, Tripathi U, Dominic TK, Singh AK, Yadav AK and Singh PK (2003a) An improvised technique for measurement of nitrogen fixation by blue green algae and *Azolla* using intact soil cores. Expl Agric 39: 145–150.

Premanandh J, Priya B, Teneva I, Dzhambazov B, Prabaharan D and Uma L (2006) Molecular characterization of marine cyanobacteria from the Indian subcontinent deduced from sequence analysis of the phycocyanin operon (*cpcB*-IGS-*cpcA*) and 16S-23S ITS region. J Microbiol 44: 607-616.

Rajaniemi P, Hrouzek P, Kastovska K, Willame R, Rantala A, Hoffmann L, Komarek J and Sivonen K (2005) Phylogenetic and morphological evaluation of genera *Anabaena*, *Aphanizomenon*, *Trichormus* and *Nostoc* (Nostocales, Cyanobacteria). Int J Syst Evol Microbiol 55: 11-26.

Rantala A (2007) Evolution and detection of cyanobacterial hepatotoxin synthetase genes. In: Dissertationes Biocentri Viikki Universitatis Helsingiensis, 29/2007. Edita Prima Oy, Helsinki, Finland.

Rantasarkka K (2008) Cyanobacteria and other bacteria in the plankton diversity, and vertical and horizontal distribution in the Gulf of Finland and northern Baltic Proper. M.Sc. Thesis, University of Helsinki, Finland.

Rastogi RP and Sinha RP (2009) Biotechnological and industrial significance of cyanobacterial secondary metabolites. Biotechnol Adv 27: 521–539.

Rezanka T and Dembitsky VM (2006) Metabolites produced by cyanobacteria belonging to several species of the family Nostocaceae. Folia Microbiol 51: 159-82.

Rippka R, Castentholtz RW, Iteman I and Herdman M (2001) Form - genus I. *Anabaena* Bory In: *Bergey's Manual of Systematic Bacteriology* (Eds. Boone DR, Castenholtz WR), 2nd edn., Springer, Berlin.

Rippka R, Deruelles J, Waterbury JB, Herdman M and Stanier RY (1979) Generic assignments, strain histories and properties of pure cultures of cyanobacteria. J Gen Microbiol 111: 1-61.

Rodriguez H, Rivas J, Guerrero MG and Losada M (1989) Nitrogen-fixing cyanobacterium with a high phycoerythrin content. Appl Environ Microbiol 59: 758-760.

Roger PA and Kulasooriya SA (1980) Blue green algae and rice. International Rice Research Institute, PO Box 933. Manila, Philippines.

Rouhiainen L, Vakkilainen T, Siemer BL, Buikema W, Haselkorn R and Sivonen K (2004) Genes coding for hepatotoxic heptapeptides (microcystins) in the cyanobacterium *Anabaena* strain 90. Appl Environ Microbiol 70: 686-692.

Rudi K, Skulberg OM, Larsen F and Jakobsen KS (1997) Strain characterization and classification of oxyphotobacteria in clone cultures on the basis of 16S rRNA sequences from the variable regions V6, V7, and V8. Appl Environ Microbiol 63: 2593-2599.

Samsonoff WA and MacColl R (2001) Biliproteins and phycobilisomes from cyanobacteria and red algae at extreme habitat. Arch Microbiol 176: 400-405.

Seifert M, McGregor G, Eaglesham G, Wickamasinghe W and Shaw G (2007) First evidence for the production of cylindrospermopsine and deoxy-cylyndrospermopsin by the freshwater benthic cyanobacterium, *Lyngbia wollei* (Fallow ex Gomont) Spezialeand Dick. Harmful Algae 6: 73-80.

Sidler WA (1994) Phycobilisome and phycobiliprotein structures. In: *The Molecular Biology of Cyanobacteria* (Ed. Bryant DA, Kluwer Academic Publ., Netherlands. pp. 139-216.

Sinha RP, Klisch M, Helbling EW and Hader D (2001) Induction of mycosporine-like amino acids (MAAs) in cyanobacteria by solar ultraviolet-B radiation. J Photochem Photobiol Biol 60: 129-135.

Smith JK, Parry JD, Day JG and Smith RJ (1998) A PCR technique based on Hip1 interspersed repetitive sequence distinguishes cyanobacterial species and strains. Microbiology 144: 2791-2801.

Sood A, Prasanna R and Singh PK (2007) Utilization of SDS-PAGE of whole cell proteins for characterization of *Azolla* species. Ann Bot Fennici 44: 283-286.

Srivastava PN (1972) Antagonism between two species of blue green algae. In: *Taxonomy and Biology of Blue green Algae* (Ed. Desikachary TV), University of Madras, Madras. pp. 391-392.

Stackebrandt E and Goebel BM (1994) Taxonomic note: a place for DNA-DNA reassociation and 16S rRNA sequence analysis in the present species definition in Bacteriology. Int J Syst Bacteriol 44: 846–849.

Starmach K (1966) Cyanophyta - sinice - Flora slodkow. Polski Warszawa, 2: 753 pp.

Stulp BK and Stam WT (1984) Genotypic relationships between strains of *Anabaena* (Cyanophyceae) and their correlation with morphological affinities. Br Phycol J 19: 287–301.

Stulp BK (1982) Morphological variability of *Anabaena* strains (Cyanophyceae) under different culture conditions. Arch Hydrobiol 63: 165-176.

Stulp BK and Stam WT (1982) General morphology and akinete germination of a number of *Anabaena* strains (Cyanophyceae) in culture. Arch Hydrobiol Suppl 63: (Algol, Stud. 30) 35-52.

Venkataraman GS (1972) Algal biofertilizer and rice cultivation. Today and Tomorrow printers, Faridabad, Haryana, India.

Wiegand C and Pflugmacher S (2005) Ecotoxicological effects of selected cyanobacterial secondary metabolites: a short review. Toxicol Appl Pharmacol 203: 201-218.

Wilmotte A and Herdman M (2001) Phylogenetic relationships among the cyanobacteria based on 16S rRNA sequences. In: *Bergey's Manual of Systematic Bacteriology* (Boone DR and Castenholz), 2nd edn, Springer, New York. pp. 487-493.

Wilmotte A (1994) Molecular evolution and taxonomy of the cyanobacteria. In: *The molecular biology of cyanobacteria.* (Ed. Bryant D and Dordrecht A), Kluwer Academic Publishers, The Netherlands. pp. 1-25.

Wolk CP, Ethai J, Kuritz T and Holland D (1993) Amplified expression of a transcriptional pattern formed during development of *Anabaena.* Mol Microbiol 7: 441-445.

Yoshida T and Ancajas RR (1971) Nitrogen fixation by bacteria in the root zone of rice. Soil Sci Soc America Proc 35: 156-157.

2

Long-term Stability of Mycosporine-like Amino Acid-producing Cyanobacteria Isolated in Syria

Jaoudat Faddoul, Shailendra P. Singh, Azizullah Azizullah, Adeel Nasir, Roland Richter, Peter R. Richter, Donat-P. Häder and Michael Lebert

Abstract

Twelve different cyanobacteria strains from different locations in Syria were collected and analyzed. They were identified as strains from *Anabaena*, *Nostoc*, *Phormidium* and *Chroococcus*. All strains produce mycosporine-like amino acids (MAAs) as UV-screens upon UV-radiation. Composition of the MAAs and amount of UV-induced MAA production differ between the strains. In order to evaluate the long-term stability of completely encapsulated cyanobacteria aliquots were transferred into hermetically closed vessels. The survival was non-invasively determined by means of pulse-amplitude modulated fluorescence and by delayed fluorescence detection. It was found that 4 strains (one *Chroococcus*, two *Anabaena* and one *Nostoc* species) show photosynthetic performance after 4 years of encapsulation. In addition, many cyanobacteria are desiccation tolerant and are able to fix atmospheric nitrogen. They may play an important role in biological life support systems e.g. in space research.

Keywords: Cyanobacteria; Long-term stability; Photosynthetic performance; Delayed fluorescence; Closed biological system

Introduction

In about 2 billion years of evolution cyanobacteria have adapted to almost all habitats. In plant cells they live as endosymbionts since almost 1.5 billion years (McFadden 2001). As they invented oxygenic photosynthesis, cyanobacteria are prerequesite of all higher life on Earth (Dismukes et al. 2001). Cyanobacteria have adapted to different environmental conditions including extreme habitats such as hot springs (Miller and Castenholz 2000), terrestrial habitiats (Sinha et al. 1999a), and permafrost (Erokhina et al. 1998; Erokhina et al. 2000). Many strains are capable of fixing nitrogen from the air and play an vital role e.g. in rice fields as biofertilizers (Vaishampayan et al. 2001). On the other hand cyanobacteria often produce toxins, which are of relevance e.g. in aquaculture, because these toxins may accumulate in the food web and end up in seafood (Ibelings and Chorus 2007).Cyanobacteria have evolved strategies to cope with solar radiation and deleterious UV, by a production of UV-screening pigments, downregulation of photosynthesis by means of D1 degradation, state transition of the LHC complexes or implementation of alternative electron transport pathways (Kirilovsky 2007; Mullineaux and Allen 1990; Sinha and Häder 2003).

Some cyanobacteria produce scytonemins (Carreto et al. 1990; Garcia-Pichel et al. 1993a; Garcia-Pichel and Castenholz 1991; Proteau et al. 1993; Squier et al. 2004), others mycosporine-like amino acids (MAAs) as screening pigments (Cockell and Knowland 1999; Kräbs et al. 2002; 2004; Singh et al. 2008). Generally cyanobacteria produce MAAs only in response to UV (Franklin et al. 1999; Kräbs et al. 2004; Sinha et al. 2002). Under moderate culture room conditions (photosynthetic active radiation (PAR) only) cultures contain only traces of MAAs. Production of MAAs can be induced by exposure to artificial or solar radiation. Many valuable data about this phenomenon can be collected by using different light regimes or wavelengths, such as determination of effective wavelengths (action spectra), duration of synthesis, protection effects of UV-screens etc. (Garcia-Pichel et al. 1993b; Sinha et al. 1999b; Sinha et al. 2001; Sinha et al. 2002; Sinha et al. 2003a; Sinha et al. 2003b). Recently, the molecular biological mechanism of MAA synthesis has been elucidated (Balskus and Walsh 2010; Singh et al. 2009).

In the course of research with cyanobacteria from rice paddy fields and with Syrian cyanobacteria we observed that a variety of strains are very undemanding. Without any change of medium the cultures were stable for at least 5 years in flasks (3 l Erlenmeyer). After desiccation some cyanobacteria can be revitalized by addition of fresh medium or water (Richter et al. 2006). In closed containers (about 500 ml) *Anabaena oryzae* and another *Anabaena* species from India still show photosynthetic performance after 3 years. In contrast, *Scytonema* sp.

is less resistant under these conditions and bleaches after about 1 year. In the recent years, extremophile and resistant organisms gained interest in space research, because they may play a role in manned space flight as integral part of biological life support systems, in which organic matter is recycled (uptake of carbon dioxide and other waste products, supply of oxygen and biomass). In space experiments biological samples were even exposed to free space without loss of vitality (Horneck et al. 1999). Among others these experiments were perfomed to contribute to the discussion about the panspermia hypothesis (Horneck et al. 2001).

In this work several MAA-producing cyanobacteria isolated from different habitats in Syria are described and their durability was tested in the course of a long-term experiment.

Materials and Methods

Cyanobacteria and collection sites

Species were collected at different sites in Syria (Table 1). The samples were photographed using a inverse microsope with digital camera (Biozero, Keyence). Fluorescence images were recorded using a filter set measureing phycocyanin fluorescence (exc. 590 nm, em. 620 nm) sollte bei 620 absorbieren und bei 640 nm emittieren. The different strains are cultured in the Cell Biology Division, Erlangen, Germany.

Induction of MAAs and HPLC analysis

Samples were exposed under a light field with radiation in PAR, UV-A and UV-B. The light intensity was 175 W/m^2 PAR, 35 W/m^2 UV-B and 0.4 W/m^2 UV-B. Certain short wavelength bands were removed from the light by covering the samples with suitable filters: Montagefolie (Nr. 10155099, Folex, Dreieich, Germany) to remove UV-B (radiation <320 nm)or Ultraphan (UV Opak, Difrega, Munich, Germany) to remove UV-B and UB-A (radiation <400 nm).

Two-ml aliquots were drawn from each sample tube after homogeneously distributing the cell suspension by gentle mixing of the tubes. The cells were harvested by centrifugation (10 min at 5000 x g) and extracted with 100 % MeOH (12 h at 4°C in darkness). After centrifugation at 5000 x g for 10 min, the absorption spectra of the supernatants were measured with a photometer (model und Firma, Ort) between 250 and 750 nm. The different MAAs were analyzed by means of HPLC as described earlier (Klisch et al. 2001). In brief, after measurement of the absorbance MeOH was removed by means of a heat plate (45° C) and a gentle air current. The residue was dissolved in 1 mL 0.02 %

acetic acid. Analysis and purification of MAAs were done using a HPLC system (Merck Hitachi; Interface D-7000, UV-Detector L-7400, Pump L-7100, Darmstadt, Germany) equipped with a LiCrospher RP 18 column and guard (5 µm packing; 250 x 4 mm I.D.).

Long-term closed container experiments

About 20 ml of cyanobacteria were located in heremetically sealed containers of 50 ml total volume. The samples were placed for 3 years at an east-facing window and encountered temperatures between 18°C to 35°C. During the fourth year the samples were kept in a temperature controlled chamber at 22 °C under low-light conditions (25 W m^{-2}).

PAM analysis

Photosynthetic efficiency of the cells was determined by means of a PAM (pulse amplitude modulated) fluorometer (PAM-2000, Walz, Effeltrich Germany). The measurements were performed with the instrument probe through the transparent wall of the container (without opening them). The photosynthetic yield was determined according to the following equation (Genty et al. 1989; Weis and Berry 1987):

Yield = (Fm'-Ft)/Fm',

with Fm' representing the maximal fluorescence measured after saturation of the photosynthetic electron transport chain (light pulse) and Ft as temporary fluorescence.

Eight measurements were performed for each sample. Data are presented as mean values with standard deviation.

Determination of delayed fluorecence

In order to monitor whether photosynthetic charge separation takes place, the samples were investigated by means of a sinlge-photon camera (C-2400-07ER; Hamamatsu Photonics K.K., Hamamatsu, Japan). The samples were irradiated with a short pulse (10 s) of white light, and the resulting delayed fluorescence was observed with the single photon camera. The analogous signal was transferred to a computer equipped with a Matrox frame grabber card and a recent version of the software Wintrack 2000 (Lebert and Häder 1999). The resulting appearing light flashes were counted over 5-s time intervals. Taking in account the sampling rate of the system of 40 ms, the dihedral angle observed by the camera (given by the distance of the cell container from the aperture of the camera) it was possible to calculate in rough approximation DF-photons orginating from recombination of energetic photons via PS II with the water splitting complex. As the signal decays exponentially only the average of the delayed photoens of the first 15 s after excitiation was used for calculation.

Results and Discussion

Determination of the strains

Using appropiate determination keys (e.g. Desikachary 1959; Desikachary 1970) the cyanobacteria samples were identified as follows (Table 1):

Sample 4 is *Chroococcus* sp., samples 3, 8 and 11 are are species of the genus *Nostoc*. As it can be seen from the fluorecence images the cells possess heterocysts, modified cells with a capacity of nitrogen reduction. Samples 2 and 5 are *Phormidium* sp. strains. Samples 1, 6, 7, 9, 22 and 33 are *Anabaena* species.

Table 1: Determination of the collected Syrian cyanobacteria, indication of sampling location, and presence of heterocysts (specialized N-fixing cells).

Sample	Species	Sampling location	Heterocysts
Sample #1	*Anabaena* sp.	Hims	+
Sample #2	*Phormidium* sp.	Duma	-
Sample #3	*Nostoc* sp.	Hannah	+
Sample #4	*Chroococcus* sp.	Al Buraykah	-
Sample #5	*Phormidium* sp.	Aleppo	-
Sample #6	*Anabaena* sp.	Duma	+
Sample #7	*Anabaena* sp.	Lattakia	+
Sample #8	*Nostoc* sp.	Lattakia	+
Sample #9	*Anabaena* sp.	Lattakia	+
Sample #11	*Nostoc* sp.	Daraa	+
Sample #22	*Anabaena* sp.	Duma	+
Sample #33	*Anabaena* sp.	Duma	+

MAA induction

It was found that all strains produce mycosporine-like amino acids upon irradiation with UV radiation. Grown at low light conditions the algae contained only minimal amounts of MAAs. HPLC analysis relvealed that shinorine and prophyra 334 were the most prominent MAA species (Table 2).

Table 2: Different mycosporine-like amino acids (MAAs) identified in the cyanobacteria samples. Plus in bold indicates high amounts of the corresponding MAA.

Retention time	2.1 min Shinorine	2.8 min Porphyra 334	3 min	4.1 min Palythine	5.2 min Mycosporine-glycine	8.1 min Palythinol
Sample #1	+	+	+			
Sample #2	+	+	+		+	
Sample #3	+	+	+	+	+	
Sample #5	+	+				
Sample #6	+					
Sample #7	+	+	+			
Sample #8	+					
Sample #22	+	+				
Sample #33	+	+	+			+

*Data for sample #4, #9 and #11 not available

Photosynthetic performance of the cyanobacteria after 4 years encapsulationAfter 4 years two Anabaena sp., one Chroococcus and one Nostoc-strain (Fig. 1) showed considerable photosynthetic efficiency as measured with PAM fluorimetry and delayed fluorescence analysis. Photosynthetic yield was between 0.2 in *Anabaena* sp. # 7 and *Nostoc* sp. #8 and about 0.4 in *Anabaena* sp. # 6 and *Chroococcus* sp. #4 (Fig.2). These samples emitted a considerably high number of delayed fluorescence photons with typical decay kinetic (Fig. 3 and 4). Also the remaining samples showed low intensities of delayed fluorescence, indicating that some of the cells are still capable of performing light driven charge separation, but no clear photosynthetic efficiency could be measured by means of PAM fluorimetry.

Several new strains of cyanobacteria were isolated from different habitats in Syria. Due to the geographic location and the climatic conditions the cyanobacteria are exposed to intensive solar radiation as well as high temperatures. All strains were shown to produce mycosporine-like amino acids as screening substances. In addition, these algae posses the property to fix atmospheric nitrogen by means of heterocysts and all the strains are desiccation tolerant. After desiccation some cyanobacteria can be revitalized upon addition of fresh medium or water, respectively. Among others desiccation tolerance was investigated in the terrestrial cyanobacteria Nostoc flagelliforme which survives long time in a dehydrated state (Gao and Ye 2007).

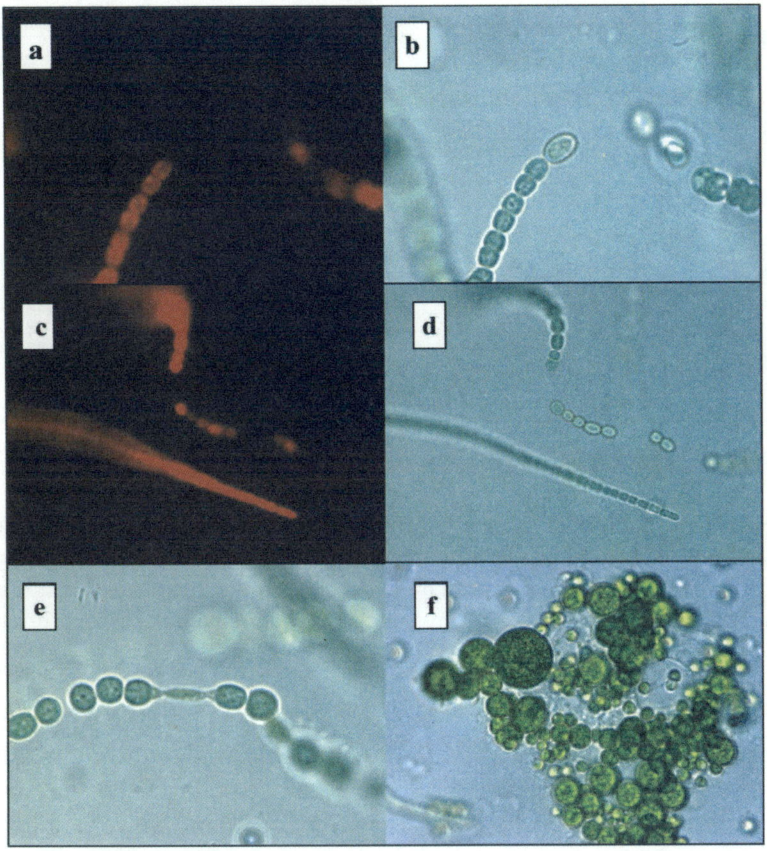

Fig. 1: Photographs of the cyanobacteria strains, which survived 4 years of encapsulation in hermetically closed containers. a, *Anabaena* sp. # 6 (isolated near Duma) fluorescence image, b: same species in bright field ; c, d *Anabaena* sp. # 8 (isolated near Lattakia)-fluorecence image and bright field image, respectively; e: *Nostoc* sp. # 8 (isolated near Lattakia). f: *Chroococcus* sp. # 4 (isolated near Al Buraykah).

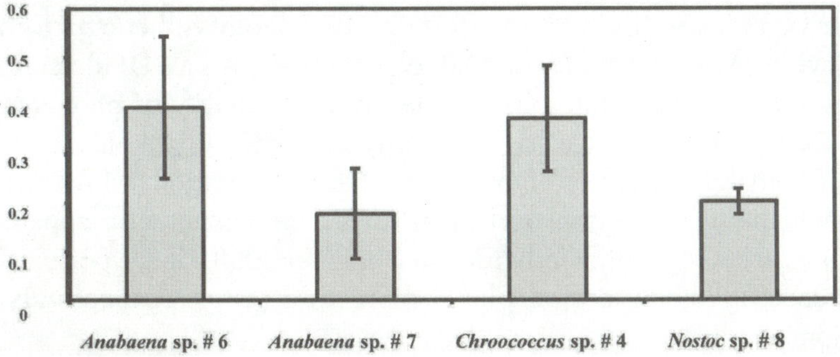

Fig. 2: Photosynthetic efficiency of four cyanobacteria samples after 4 years in hermetically closed containers as determined with a PAM fluorimeter.

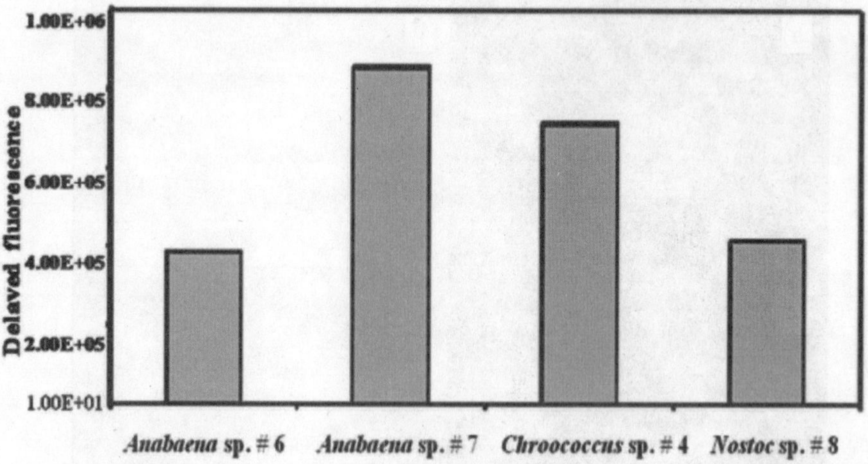

Fig. 3: Approximate number of delayed fluorescence (DF) photons originating from cyanobacteria after 4 years in closed containers. DF (generated by recombination of separated charges at PS II) was induced by 10-s flashes of white light and in subsequent darkness. The photons were recorded by means of a single photon camera. For details see text.

The durability and resistance of cyanobacteria is supported by formation of stable resting states, akinetes and hormogonia (Agrawal and Singh 2000). In the course of research with cyanobacteria from rice paddy fields and with the Syrian cyanobacteria we observed that a variety of strains are very undemanding. Without any change of medium the cultures are stable for at least 5 years in flasks (3 l Erlenmeyer). Inclosed containers (about 500 ml) Anabaena oryzae and another Anabaena species from India are still living after 3 years. *Scytonema* sp. in contrast is less resistant under these conditions and bleaches after about 1 year (unpublished results). Most of the cyanobacteria from Syria have been shown to be extremely stable for a long time period.

After four years four of the strains still show good photosynthetic performance as tested with PAM fluorimetry and DF detection. Above all DF detection has been shown to be a valuable tool in order to show vitality of photosynthetic samples. Delayed fluorescence originates only from intact thylakoid membranes, because it is mainly based on light driven charge separation, which can only work, when the electron transport chain of the photosynthetic apparatus is intact (Bodemer et al. 2000; Gerhardt and Bodemer 2000). It is a perfect tool to see the recovery of dhydrated plants or resting states, respectively after rehydration.

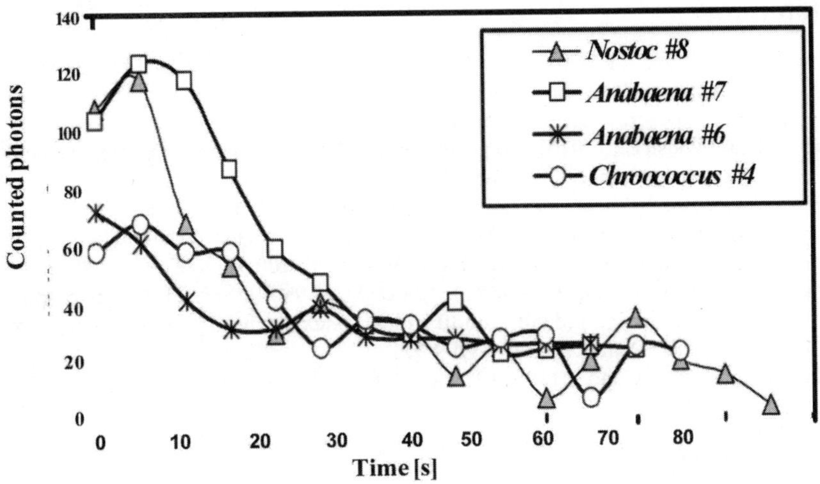

Fig. 4: Decay of the delayed fluorescence (DF) in darkness following a 10-s excitation light flash. The photons were counted by means of a single photon camera. For details see text

Conclusion

The results of the experiments in combination with other scientific results suggests that cyanobacteria are possibly very interesting organisms to be employed in biological regenerative life support systems, because they possess different valuable properties:

- They produce oxygen such as higher plants
- Fast growth rate can be induced.
- They transfer N_2 into biological accessible forms of N (primary in the form of NH_3).
- Many cyanobacteria are dessiccation tolerant and can be stored in dehydrated state.
- Due to the ability to form screening pigments, they are tolerant against excessive solar radiation.

Acknowledegment

The work was financially supported by the DAAD (Award for a Research Stay A06/11/964).

References

Agrawal SC and Singh V (2000) Vegetative survival, akinete formation and germination in three blue-green algae and one green alga in relation to light intensity, temperature, heat shock and UV exposure. Folia Microbiol 45: 439-446.

Balskus EP and Walsh CT (2010) The genetic and molecular basis for sunscreen biosynthesis in cyanobacteria. Science 329: 1653-1656.

Bodemer U, Gerhardt V, Yacobi YZ, Zohary T, Friedrich G and Pohlmann M (2000) Phytoplankton abundance and composition in freshwater systems determined by DF excitation spetroscopy and conventional methods. Arch Hydrobiol 55: 87-100.

Carreto JI, Carignan MO, Daleo G and De Marco SG (1990) Occurrence of mycosporine-like amino acids in the red tide dinoflagellate Alexandrium excavatum: UV-protective compounds? J Plankton Res 12: 909-921.

Cockell CS and Knowland J (1999) Ultraviolet radiation screening compounds. Biol Rev 74: 311-345.

Desikachary TV (1959) Cyanophyta. Indian Council of Agriculture Research, New Delhi, India. pp 222-223.

Desikachary TV (1970) Taxonomy of blue-green algae - problems and prospects. Schweiz Z Hydrol 32: 490-494.

Dismukes GC, Klimov VV, Baranov SV, Kozlov YuN, DasGupta J and Tyryshkin A (2001) The origin of atmospheric oxygen on earth: the innovation of oxygenic photosynthesis. Proc Nat Acad Sci USA 98: 2170-2175.

Erokhina LG, Spirina EV, Shatilovich AV and Gilichinskii DA (2000) Chromatic adaptation of viable ancient cyanobacteria from Arctic permafrost. Microbiol Engl Tr 69: 730-731.

Erokhina LG, Vishnivetskaya TA and Gilichinskii DA (1998) The content and composition of phycobilisome pigments in cells of ancient viable cyanobacteria from Arctic permafrost. Microbiol Engl Tr 67: 682-687.

Franklin LA, Yakovleva I, Karsten U and Lüning K (1999) Synthesis of mycosporine-like amino acids in Chondrus crispus (Florideophyceae) and the consequences for sensitivity to ultraviolet B radiation. J Phycol 35: 682-693.

Gao K and Ye C (2007) Photosynthetic insensitivity of the terrestrial cyanobacterium Nostoc flagelliforme to solar UV radiation while rehydrated or dessiccated. J Phycol 43: 628-635.

Garcia-Pichel F and Castenholz RW (1991) Characterization and biological implications of scytonemin, a cyanobacterial sheath pigment. J Phycol 27: 395-409.

Garcia-Pichel F, Wingard AE and Castenholz RW (1993a) Evidence regarding the UV sunscreen role of a mycosporine-like compound in the cyanobacterium *Gloeocapsa* sp. Appl Environ Microbiol 59: 170-176.

Garcia-Pichel F, Wingard CE and Castenholz RW (1993b) Evidence regarding the UV sunscreen role of a mycosporine-like compound in the cyanobacterium *Gloeocapsa* sp. Appl Environ Microbiol 59: 170-176.

Genty BE, Briantais JM and Baker NR (1989) Relative quantum efficiencies of the two photosystems of leaves in photorespiratory and non-photorespiratory conditions. Plant Physiol Biochem 28:1-10

Gerhardt V and Bodemer U (2000) Delayed fluorescence excitation spectroscopy: a method for determining phytoplankton composition. Arch Hydrobiol 55: 101-120.

Horneck G, Rettberg P, Reitz G, Wehner J, Eschweiler U, Strauch K, Panitz C, Starke V and Baumstark-Khan C (2001) Protection of bacterial spores in space, a contribution to the discussion on Panspermia. Origins Life Evol Biosphere 31: 527-547.

Horneck G, Wynn-Williams DD, Mancinelli RL, Cadet J, Munakata N, Ronto G, Edwards HGM, Hock B, Wänke H, Reitz G, Dachev T, Häder D-P and Brioullett C (1999) Biological experiments on the expose facility of the International Space Station. Proceedings of the 2nd European Symposium on the Utilisation of the International Space Station, ESTEC, Noordwijk, The Netherlands, 16-18 Nov. 1998 (ESA SP-433). pp 459-468.

Ibelings BW and Chorus I (2007) Accumulation of cyanobacterial toxins in freshwater "seafood" and its consequences for public health: A review. Environ Pollut 150: 177-192.

Kirilovsky D (2007) Photoprotection in cyanobacteria: the orange carotenoid protein (OCP)-related non-photochemical-quenching mechanism. Photosynth Res 93: 7-16.

Klisch M, Sinha RP, Richter PR and Häder D-P (2001) Mycosporine-like amino acids (MAAs) protect against UV-B-induced damage in *Gyrodinium dorsum* Kofoid. J Plant Physiol 158: 1449-1454.

Kräbs G, Bischof K, Hanelt D, Karsten U and Wiencke C (2002) Wavelength-dependent induction of UV-absorbing mycosporine-like amino acids in the red alga *Chondrus crispus* under natural solar radiation. J Exp Mar Biol Ecol 268: 69-82.

Kräbs G, Watanabe M and Wiencke C (2004) A monochromatic action spectrum for the photoinduction of the UV-absorbing mycosporine-like amino acid shinorine in the red alga *Chondrus crispus*. Photochem Photobiol 79: 515-519.

Lebert M and Häder D-P (1999) Image analysis: A versatile tool for numerous applications. G I T Imaging Microsc 1: 5-6.

McFadden GI (2001) Primary and secondary endosymbiosis and the origin of plastids. J Phycol 37: 951-959.

Miller SR and Castenholz RW (2000) Evolution of thermotolerance in hot spring cyanobacteria of the genus *Synechococcus*. Appl Environ Microbiol 66: 4222-4229.

Mullineaux CW and Allen JF (1990) State 1-state 2 transitions in the cyanobacterium *Synechococcus* 6301 are controlled by the redox state of electron carriers between photosystem I and II. Photosynth Res 23: 297-311.

Proteau PJ, Gerwick WH, Garcia-Pichel F and Castenholz R (1993) The structure of scytonemin, an ultraviolet sunscreen pigment from the sheaths of cyanobacteria. Experientia 49: 825-829.

Richter PR, Sinha RP and Häder D-P (2006) Scytonemin-rich epilithic cyanobacteria survive acetone treatment. Curr Trends Microbiol 2: 13-19.

Singh SP, Klisch M, Sinha RP and Häder D-P (2009) Genome mining of mycosporine-like amino acid (MAA) synthesizing and non-synthesizing cyanobacteria: A bioinformatics study. Genomics 95: 120-128.

Singh SP, Kumari S, Rastogi RP, Singh KL and Sinha RP (2008) Mycosporine-like amino acids (MAAs): Chemical structure, biosynthesis and significance as UV-absorbing/screening compounds. Ind J Exp Biol 46: 7-17.

Sinha RP, Ambasht NK, Sinha JP and Häder D-P (2003a) Wavelength-dependent induction of a mycosporine-like amino acid in a rice-field cyanobacterium, *Nostoc commune*: role of inhibitors and salt stress. Photochem Photobiol Sci 2: 171-176.

Sinha RP, Ambasht NK, Sinha JP, Klisch M and Häder D-P (2003b) UV-B-induced synthesis of mycosporine-like amino acids in three strains of *Nodularia* (cyanobacteria). J Photochem Photobiol B Biol 71: 51-58.

Sinha RP and Häder D-P (2003) Photoprotection in cyanobacteria. Recent Res Devel Biochem 4: 915-924.

Sinha RP, Klisch M, Vaishampayan A and Häder D-P (1999a) Biochemical and spectroscopic characterization of the cyanobacterium *Lyngbya* sp. inhabiting mango (Mangifera indica) trees: presence of an ultraviolet-absorbing pigment, scytonemin. Acta Protozool 38: 291-298.

Sinha RP, Klisch M and Häder D-P (1999b) Induction of a mycosporine-like amino acid (MAA) in the rice-field cyanobacterium *Anabaena* sp. by UV irradiation. J Photochem Photobiol B Biol 52: 59-64.

Sinha RP, Klisch M, Helbling EW and Häder D-P (2001) Induction of mycosporine-like amino acids (MAAs) in cyanobacteria by solar ultraviolet-B radiation. J Photochem Photobiol B Biol 60: 129-135.

Sinha RP, Sinha JP, Gröniger A and Häder D-P (2002) Polychromatic action spectrum for the induction of a mycosporine-like amino acid in a rice-field cyanobacterium, *Anabaena* sp. J Photochem Photobiol B Biol 66: 47-53.

Squier AH, Hodgson DA and Keely BJ (2004) A crotocal assessment of the analysis and distribution of scytonemin and related UV screening pigments in sediments. Organic Geochem 35: 1221-1228.

Vaishampayan A, Sinha RP, Häder DP, Dey T, Gupta AK, Bhan U and Rao AL (2001) Cyanobacterial biofertilizers in rice agriculture. Bot Rev 67: 453-516.

Weis E and Berry A (1987) Quantum efficiency of photosystem II in relation to the energy dependent quenching of chlorophyll fluorescence. Biochim Biophys Acta 894: 198-208.

3

Liquid Chromatography-Mass Spectrometry Applications in Algal Biochemistry

Olimpio Montero and Marta Velasco

Abstract

Present implementation of mass spectrometers has allowed developing new approaches in algal research. Soft ionization techniques like atmospheric pressure chemical ionization (APCI), electrospray ionization (ESI), and matrix assisted laser desorption ionization (MALDI) extend the analytical possibilities of mass spectrometry (MS) beyond classical fatty acid analysis by gas chromatography (GC) with electron impact ionization (EI). New MS-based analytical methods have been developed over the last two decades for algal molecule research including membrane lipids, carotenoids, mycosporine-like amino acids and peptides. MS-based Metabololomics approaches have also been used to gain insights into algal biochemistry and physiology. This review aims at highlighting the achievements in algal research gained by using these recent research tools and their future prospects.

Keywords: Lipids; Carotenoids; Hopanoids; Mycosporine-like amino acids; LC-MS; ESI.

Introduction

Biochemical composition underlies most of physiological processes in any organism. Algal characterization from ecophysiological, biotechnological and phylogenetical viewpoints is dependent upon a precise and sensitive determination of different metabolites. This challenge needs of improved analytical techniques like mass spectrometry (MS), in particular for such compounds which are difficult to analyze by classical biochemical methods. Electron impact ionization (EI) has been used since its origins in mass spectrometers with magnetic sector analyzers for structural elucidation of isolated compounds. From an analytical viewpoint, fatty acids have been analyzed by gas chromatography coupled to mass spectrometry (GC-MS) with ionization by electron impact (EI) for decades. However, present implementation of mass spectrometric techniques provides new efficient tools to profile not only single molecular types but a broad range of compounds in a comprehensive manner.

Until Molecular Biology techniques have been available to any laboratory and in particular automated polymerase chain reaction (PCR) and sequencing, chlorophyll and carotenoid profile have been used with a phylogenetical aim for algal species identification (Falkowsky and Raven 1997; McIntyre et al. 2002; Mackey et al. 1996; Brotas and Plante-Cuny 2003). Even though sequencing of 16S cDNA for cyanobateria or 18S cDNA for eukaryote is required to date for a valuable phylogenetical characterization of any new algal species at present, photosynthetic pigment profile is still a suitable tool in phylogenetic studies. As well, membrane lipids and the so called mycosporine-like amino acids are of concern in phylogenetical and physiological studies.

In this chapter, mass spectrometry applications in the analysis of relevant compounds in algal research are revised, with special focus on methods that imply interfacing with liquid chromatography. Examples of liquid chromatography-mass spectrometry analysis are afforded for membrane lipids and carotenoids as measured in the authors' laboratory.

Mass Spectrometry

A brief introduction to mass spectrometry (MS) is afforded here, with particular attention being paid to the recent developments devoted to biochemical analysis using the so called hyphenated techniques, especially liquid chromatography (LC). MS is suited for the experimental measurement of the mass of gas-phase ions generated from molecules of an analyte. This microanalytical technique can be used selectively to detect and ascertain the amount of a given analyte (Watson and Sparkman 2007; Hoffmann and Stroobant 2007). The first requirement for mass spectrometry measurements is the production of gas-phase ions of the substances of interest, which therefore require having low vapor pressure, the ions are subsequently separated according to their mass-to-charge ratio (m/z) in the mass analyzer. Initially, prior to 1970, ions were obtained by impinging the molecule mixture with a 70 V electron flow, which gave rise to a set of positively charged ions stemming from molecule fragments. This ionization technique, named as electron impact (EI), was efficient and produced reproducible spectra for the majority of substances, which allows mass spectra to be compiled in databases; however, the molecular ion was generally absent due to the highly energetic ionization process breaks completely the molecule. This feature has limited substantially the types of molecules that were amenable to mass spectrometric analysis for long time.

EI imposes as well an additional constrain to biochemical analysis through the so called hyphenated-techniques, reducing the options to gas chromatography-mass spectrometry (GC-MS). Over the 80s and 90s of the last century new ionization techniques were developed that circumvented this problem (Larsen 1998; Cody 2002; Siuzdak 2003; Watson and Sparkman 2007). The most popular ionization techniques are electrospray ionization (ESI), atmospheric pressure chemical ionization (APCI) and matrix-assisted laser desorption ionization (MALDI). Both ESI and APCI can be interfaced to liquid chromatography, thus triggering the chance to analyze a large set of biomolecules, including proteins (Abian 1999; Cody 2002; Watson and Sparkman 2007). Molecules are ionized and transferred to the gas phase in ESI and APCI from droplets of the carrier solvent through high temperatures and a drying gas flux, although differences exist between these methods in regard to how the ions are formed. Protonated ($[M+H]^+$) or de-protonated ($[M-H]^-$) ions can be obtained in both positive and negative operating mode, respectively, along with a set of adducts such as $[M+Na]^+$ or $[M+CH_3COO]^-$, the exact mass of the molecule being thus accessible. In addition, specific fragmentation of these ions can be achieved through MS/MS and MS^n experiments, which is an invaluable tool in molecule structure elucidation.

Ionized molecules are subsequently separated in the mass spectrometer analyzer and their m/z determined. Current mass spectrometer analyzers are triple quadruple (TQ), ion trap (IT, linear and 3D), and time-of-flight (ToF) or quadrupole time-of-flight (QToF). Each analyzer has its own characteristics that enable it for a particular analysis type (Cody 2002; Niesen 2003; Douglas et al. 2005; Xing et al. 2007; Mirsaleh-Kohan 2008). ToF analyzers provide exact mass with up to four decimals and a high resolution capacity (20,000-40,000), ion traps allow successive fragmentations, while TQs can be operated in selected ion monitoring (SIM) and multiple reaction monitoring (MRM) besides they yield more reproducible analysis. Orbitrap instruments are also becoming of increasing interest because of they have a resolution capacity (>100,000) above that of any other instrument (Scigelova and Makarov 2006; Schuhmann et al. 2012), although their price make these mass spectrometers unavailable for many laboratories at present.

Analytical applications of mass spectrometers extend from proteins to single lipids and sugars (Siuzdak 2003). MS methodologies that allow acquisition of exact mass for both the intact ion and its fragments (MS^E) in the same run have been raised by some mass spectrometer manufacturers (Plumb et al. 2006). Recent applications of mass spectrometry have focused in the development of a whole small molecule (< 1500 Da) analysis in one, a new discipline arising which is called Metabolomics, with sub-disciplines as Lipidomics (Fiehn 2002; Dettmer et al. 2007; Welti et al. 2007; Schuhmann et al. 2012). This methodology involves the use of advance statistical methods for sample clustering and metabolite mining that could serve as biomarkers. Such an approach has been used recently by Lu et al. (2012) with the aim to detect biomarkers of the response to salt stress of the snow alga *Chlamydomonas nivalis*. The scope of new developments in mass spectrometry is increasing day-to-day and, hence, MS analysis is expected to contribute significantly to the understanding of algal biochemistry and physiology in the next years. Some of the most recent results of LC-MS and MALDI-MS analysis in microalgae research are presented in this chapter.

Lipids and Related Compounds

The term lipid comprises a large variety of hydrocarbon-related molecules with chemical structures that vary from fatty acids, carotenoids and prenyllipids to complex phospholipids, glycolipids, sphingolipids, cardiolipin or cholesterol derivatives. The lipid profile of algae and cyanobacteria accounts for a diversity of lipid classes that largely surpasses that of mammals and higher plants (Harwood 1998; Guschina and Harwood 2006). Separation of lipids into different classes is still accomplished by thin layer chromatography (TLC) in many laboratories, using reaction with iodine for visualization. Quantification can afterwards be conducted by the charring method, radiolabeling or phosphorus measurement for the case of phospholipids. Mass spectrometry has been used for longer in combination with gas chromatography (GC-MS) to quantify fatty acids (FAs) after hydrolysis and methylation, thus rendering fatty acid methyl esters (FAMEs), or other derivatizing methods. GC-MS determination of FAMEs is reproducible and highly sensitive though cumbersome and time consuming. Ascertaining FA profile and content in algae is demanded from diverse research areas. FA profile along with lipid class composition may help determining the phylogenetic position of an algal species (Goss and Wilhelm 2009). Biotechnological applications for antioxidants and polyunsaturated fatty acid (PUFA) production using algal cells as factory require sensitive and reproducible FA measurement (Guschina and Harwood 2006; De Angelis et al. 2005; Herrero et al. 2007). Recently, biofuel production from microalgae is of raising concern as a consequence of fossil combustible reserve is being depleted worldwide with concurrent price increase (Chisti 2007). However, at present not only the FA profile is of concern but the different glycerolipid classes in which the FA is esterifying the glycerol backbone is also relevant as algae have been shown to contain special lipid classes, like betaine lipids, which are not present in other organisms (Goss and Wilhelm 2009; Guschina and Harwood 2006). Results from diverse applications of MS and LC-MS in analysis of membrane lipids and carotenoids are shown here.

(a) Membrane lipids

While algal membrane lipids include almost all phospholipid classes, sphyngoid-like and phytosterols, cyanobacterial membrane lipids are composed by phosphatidylglycerol (PG), mono- and di-galactosyldiacylglycerol (MGDG and DGDG, respectively), and sulfoquinovosyldiacylglycerol (SQDG) (Harwood 1998; Goss and Wilhelm 2009; Sato and Wada 2009). In addition to PG, phospholipids of eukaryotic algae include phosphatidylcholine (PC), phosphatidylserine (PS), phosphatidylethanolamine (PE) and phosphatidic acid (PA). Phosphatidylinositol-phosphate (PIP) and phosphatidylinositol-4,5-bisphosphate (PIP_2) seem to act in intracellular signaling pathways in the model alga *Chlamydomonas reinhardtii* (Arisz et al. 2000). As well, betain lipids, which are not present in vascular plants, have been reported to replace PC in brown and green algal groups (Goss and Wilhelm 2009). Diacylglycerolhydroxymethylalanine (DGTA) in brown algae, and diacylglyceryltrimethylserine (DGTS) in green algae are the better known betaine lipids (Fig. 1). Hydroxy and oxygenated fatty acids, these compounds named oxylipins, have also been found to esterify the membrane glycerolipids, and they seem to act as components of defense system against the attack of bacteria and viruses. Most knowledge about algal lipids has been obtained from the green alga *Chlamydomonas reinhardtii* (Moellering et al. 2009), but an increasing number of organisms are being studied using new MS-based technique implementation, which is allowing to unveil up-to-date unknown lipids (Guschina and Harwood 2006).

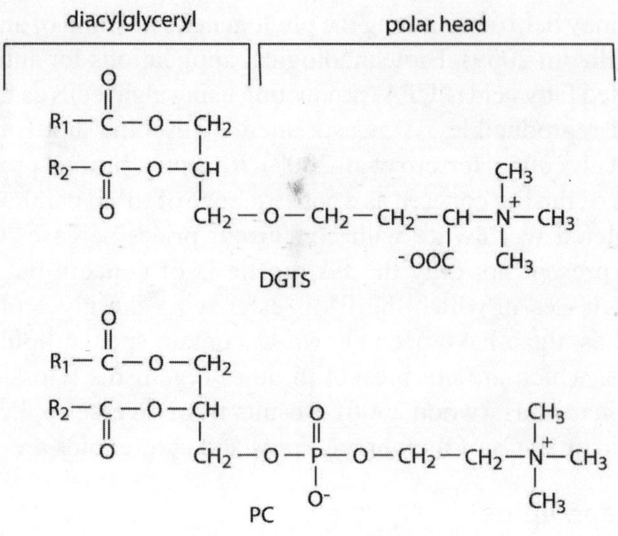

Fig. 1: Structure of two common membrane lipids in algae: DGTS, diacylglyceryltrimethylserine; and PC, diacylglycerylphosphatidylcholine. Both lipids have zwitterion properties but DGTS lacks phosphate group.

With the advent of soft ionization techniques, determination of lipids by liquid chromatography hyphenated to mass spectrometry (LC-MS) has started to be a current analytical methodology in a number of laboratories. These techniques have giving rise to the possibility of directly determining the whole lipid molecule as well as to obtain structural information by relatively controlled fragmentation through collision induced dissociation (CID) (Kim et al. 1997; Welti et al. 2003; Murphy et al. 2001; Pulfer and Murphy 2003; Han and Gross 2003; Guella et al. 2003; Montero 2011). Lipid determination by liquid chromatography-mass spectrometry (LC-MS) is a well suited methodology at present. Most knowledge about ionization and fragmentation processes of diverse lipid classes was accounted by fast-atom bombardment mass spectrometry (FAB-MS), the same mechanisms being later found to occur when atmospheric pressure ionization techniques (ESI and APCI) are used. Detailed description of MS analysis of a large variety of lipids is afforded in Murphy et al. (2001), Pulfer and Murphy (2003), Kim et al. (1999), Welti et al. (2003), and Seiwert et al. (2009). As well, protocols for new lipid analysis and their MS and MS/MS behavior is eliciting a survey of papers that will bring new insights into this field research.

Thus, monogalactosylmonoacylglycerides (MGMG) and fragmentation patterns of different glycolipids have been reported in the cyanobacterium *Spirulina* using extraction with supercritical fluids (SCFE) and ESI-ToF-MS/MS analysis

(Herrero et al. 2007). Diverse hexose-hexacosanediols and hexose-octacosanetriols were found in cyanobacterial heterocysts (Baversachs et al. 2009). More than 100 triacylglycerol (TG) species in which the oleic acid was predominant could be measured by MacDougall et al. (2011) in *Botryococcus braunii* and *Scenedesmus obliquus* using UHPLC-MS with an Orbitrap MS instrument to a lower limit of detection of 11-19 nmol/L; a survey of phospholipids and glycolipids were measured in the same run as TGs were measured. These authors detected TGs as the sodium and ammonium adducts ($[TG+Na]^+$ and $[TG+NH_4]^+$, respectively), the Na^+ adducts being more stable and resistant to fragmentation than NH_4^+ adducts. In this study, a 3D map of the LC-MS chromatogram for *B. braunii* lipid extract is depicted, which may provide other researchers a better understanding of chromatographic behavior of the different lipid classes (MacDougall et al. 2011). Very-long-chain fatty acids (VLCFAs) were measured as picolinyl esters in a dinoflagellate (*Amphidinium carterae*) and an antarctic cyanobacterium of the genus *Calothrix* by Rezanka et al. (2008; 2009) using LC-MS with either APCI or ESI sources. PUFAs were previously removed using Ag^+-TLC, and typical fragments of m/z 92, 108, 151 and 164 in LC-MS/MS analysis were reported. Fractions were collected from silica-column chromatography and analyzed afterwards by ESI/MS and ESI/MS/MS, they showing $[M+Na]^+$ ions of betaine lipids, namely DGTA and DGTS, which contained C20:5n-3 (Roche and Leblond 2010) and a minor new lipid, diacylglycerylcarboxyhydroxymethylcholine (DGCC), which had not been previously reported was also found. Chlorosulfolipids and sulfoglycolipids have also been analyzed using MALDI-ToF-MS; these studies showing C20:4 and C20:5 fatty acids esterify the *sn*-2 position of glycerol (Darsow et al. 2007; Naumann et al. 2007) in some of the glycerolipid species. The differential composition of glycerolipids in the diatom *Cyclotella meneghiniana* and the green alga *Chlamydomonas reinhardtii* was studied by Vieler et al. (2007) using MALDI-ToF-MS and TLC; this study showed that betaine lipids (GTMHS) and PE account for the major membrane lipid components in the green alga, whereas SQDG and PG are predominant in the diatom. A recent paper has open new frontiers in biochemical analysis through the combination of microspectroscopy techniques with MALDI-MS (Urban et al. 2011). This study points out a new methodology through with the arrangement of photosystem components and phospholipid plus pigment composition can be analyzed in single cells of the green alga *Euglena gracilis*. An interesting application of LC-MS analysis of lipids is that concerned with studies on biota distribution in sediments (Schubotz et al. 2009). Indeed, intact polar lipids (IPLs) including betaine lipids are considered to be distinctive biomarkers and a suitable tool that provides reliable data in such a type studies (Wakeham et al. 2007). Using GC-MS and HPLC-ESI-ion trap-MS^n analysis of extracts from sediments, main components of biota in oxic waters above the thermocline were found to be eukaryotic photosynthetic algae (Schubotz et al. 2009). A new approach to lipid analysis was tackled by Lu et al. (2012) taking advantage of multivariate statistical analysis of exact mass data generated by UPLC-QToF measurements (a lipidomics-based approach). Seven types and 35 kinds of polar lipid molecules were selected as biomarkers to track the response of the snow alga *Chlamydomonas nivalis* to salt stress in the study by Lu et al. (2012); within each lipid class, different glycerolipid species seem to behave according to different roles played by each species in regulation of the salt stress response, thus showing a species-specific behavior.

Regiochemistry of acyl chains at the *sn*-positions of the glycerol backbone is an issue of interest in glycerolipid identification (Zepke et al. 1978; Welti et al. 2003; Guella et al. 2003). C18 acyl chains are preferred at the *sn*-1 position and are C16 acyl chains at the *sn*-2 position in prokaryotic organism membranes, whereas the opposite occurs in eukaryotic organism membranes (Sato and Wada 2009). Structural information on such issue can easily be gained from MS analysis of glycerolipids; thus it has been shown that the *lyso*-glycerolipid formed after losing the acyl chain from the *sn*-1 position in MS/MS of $[M+Na]^+$ experiments is more abundant than the corresponding *lyso*-glycerolipid formed after losing the acyl chain from the *sn*-2 position (Guella et al. 2003; Seiwerts et al. 2009). Hence, this feature shows LC-MS/MS analysis can be a powerful tool in glycerolipid structure ascertaining.

The lipid profile of two strains of the marine cyanobacterium *Synechococcus* sp., along with the model strain *Synechococcus* sp. PCC7002, was analyzed using LC-ESI-ion trap-MS (Montero et al. 2011; Montero 2011). Different species of SQDG, PG, MGDG and DGDG were found to be present in a methanolic extract of these cyanobacterial species. SQDG species containing C17:1 and C17:3 fatty acyls were reported for the first time (Montero 2011). Two isomers of MGDG (34:3) could be chromatographically separated, and positions *sn*-1 and *sn*-2 of the glycerol backbone were shown to be esterified by C18:2 and C16:1 fatty acids in the MGDG species eluting earlier, while C18:3 and C16:0 fatty acids esterified the glycerol in the other MGDG species. The relative content of these two MGDG species varied with the illumination conditions (Montero et al. 2012). Samples were analyzed utilizing two chromatographic solvent gradients. In Montero (2011) an LC elution gradient based on a solvent system constituted by acetonitrile/methanol/aqueous ammonium acetate gave predominant $[M\text{-}H]^-$ ions for SQDG and PG, and $[M+CH_3COOH\text{-}H]^-$ and $[M+ClH\text{-}H]^-$ ions for MGDG and DGDG species, in negative mode, while $[M+NH_4]^+$ ions predominate in positive mode for all the glycerolipid species. UV-Vis detection was also conducted on-line in this analysis system. In Montero et al. (2011) a methanol/hexane/aqueous ammonium acetate-based elution gradient brought about $[M-H]^-$ ions for SQDG and PG and $[M+Na]^+$ ions for MGDG and DGDG by changing the polarity mode after 21 min of elution, after this time being when MGDG and DGDG species were expected to elute from the column; in this way, each glycerolipid species could be measured in one run through its optimal ionization mode. The first gradient system yielded a better species resolution and MS/MS fragmentation pattern, but two different chromatographic runs were necessary for positive and negative mode analysis. Hence, the LC-MS system of Montero (2011) is recommended for a first analysis of unknown samples, whereas the LC-MS system of Montero et al. (2011) is better suited for routine analysis instead. A typical chromatogram obtained for MS and UV-Vis detection using the gradient system of Montero (2011) is illustrated in Fig. 2, where the relative elution pattern of membrane lipids and carotenoids can be compared.

A methanolic extract of *Synechococcus* sp. was injected in a UPLC-QToF-MS system and analyzed in positive and negative mode using a MS^E method (Plumb et al. 2006; Xu et al. 2010). Because the same elution solvent system was used for both ionization modes (methanol/acetonitrile/aqueous formic acid, pH 3.8), SQDGs and PGs were poorly ionized in negative mode, which points out that basic pH is necessary to get efficient ionization of these anionic glycerolipids when analyzed in negative mode, and excess ammonium or sodium cations that allow formation of the corresponding adduct ions when they are analyzed in positive mode. By screening EIC of m/z 243.2131 the position of MGDGs and DGDGs in the chromatogram can be easily viewed (Fig. 3).

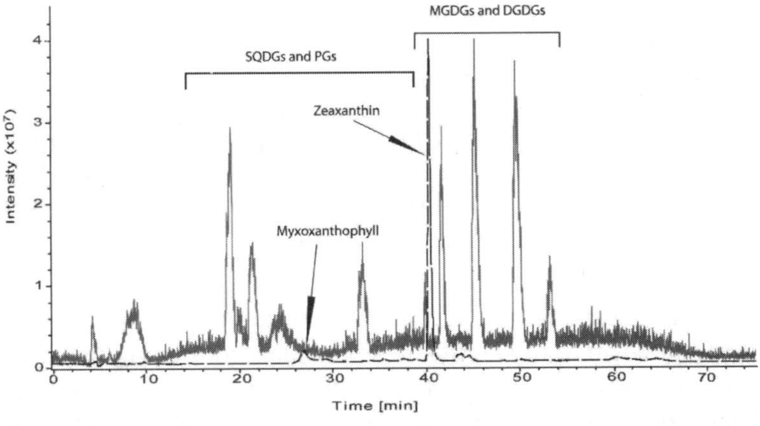

Fig. 2: A typical chromatogram obtained in positive mode for a methanolic extract of *Synechococcus* cells with LC-UV/Vis-ion trap-MS analysis. Major peaks correspond to glycerolipids. For details see Montero (2011). Dotted line represents the chromatogram from UV-Vis detection.

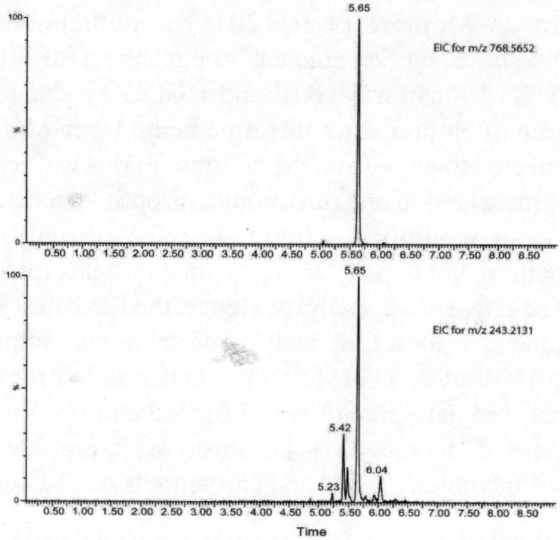

Fig. 3: Extracted ion chromatograms (EIC) of m/z 768.5652 ($[M+NH_4]^+$ of MGDG(18:3/16:1)) and m/z 243.2131, from positive ionization mode analysis of a methanolic extract of *Synechococcus* cells with UPLC-QToF-MS. The peak at 5.42 min proceeds from DGDG(18:3/16:1). (Unpublished results).

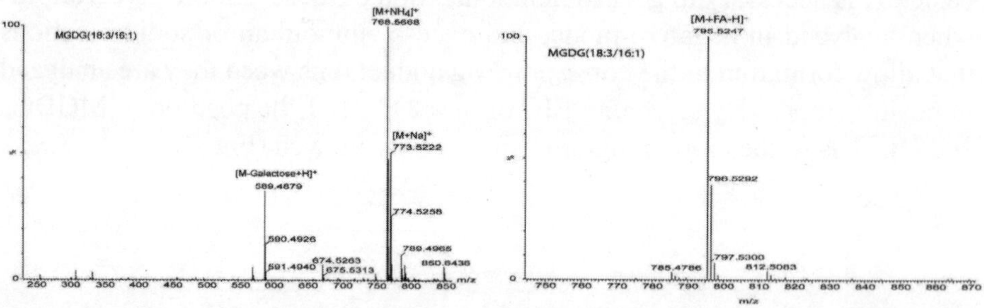

Fig. 4: Spectrum of m/z 768.5652 ($[M+NH_4]^+$) and 795.5247 ($[M+HCOOH-H]^-$) corresponding to MGDG(18:3/16:1) in positive and negative ionization mode, respectively, obtained by UPLC-QToF-MS analysis of a methanolic extract from *Synechococcus* cells. (Unpublished results)

In contrast to anionic glycerolipids, efficient ionization of MGDGs and DGDGs was obtained in positive mode, the most abundant ions being the ammonium and sodium adducts, and the formate adducts in negative mode (Fig. 4) (unpublished results). The measured exact masses for the $[M+NH_4]^+$ and $[M+Na]^+$ adducts of MGDG(18:3/16:1) were 768.5668 and 773.5222 Da, which mean they were measured with an error of 4.2 mDa (5.5 ppm). The formate adduct ion gave an error of 12.0 mDa (calculated exact mass 795.5259 Da).

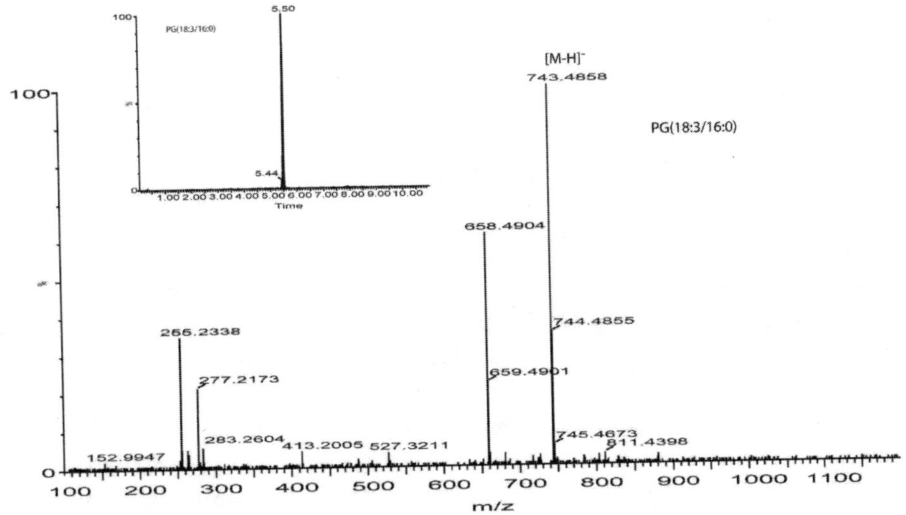

Fig. 5: Mass spectrum of m/z 743.4858 ([M-H]$^-$) from the high energy function (MS^E method), and which corresponds to PG(18:3/16:0) in negative ionization mode. The extracted ion chromatogram (EIC) from the low energy function, as obtained by UPLC-QToF-MS analysis of a methanolic extract from *Synechococcus* cells, is shown in the inset at 5.50 min. (Unpublished results).

In Fig. 5 the m/z corresponding to the two acyl chains generated by collision induced dissociation in the fragmentation of the negative ion of PG(18:3/16:0) ([M-H]$^-$, m/z 743.4858) are shown. The C16:0 chain gives m/z 255.2338, and the C18:3 chain gives m/z 277.2173. The fragment with m/z 152.9947 corresponds to the phosphoglyceryl moiety ([171.0051-H_2O]$^-$, where m/z 171.0051 is the m/z of deprotonated phosphoglycerol), but the fragment of m/z 227.02 which is expected to be produced in PG species fragmentation (Pulfer and Murphy 2003; Xu et al. 2010; Montero 2011) could not be detected. These experimental data show different features of glycerolipid analysis by LC-MS, and along with research revised above point out the usefulness of LC-MS in the analysis of glycerolipids in algae without the need for cumbersome sample preparation.

(b) Carotenoids

The most standing characteristic of carotenoids is the long polyene chain, which stems from their isoprenoid nature and may extend to 15 conjugated double bonds; in addition, up to seven different end groups contribute to increase diversity of the basic hydrocarbon structure. Hydroxyl and/or keto functions added to the hydrocarbon core give rise to the specific carotenoid group of xanthophylls (Fraser and Bramley 2004). Regarding algae, carotenoids have served useful taxonomical markers for decades (Mackey et al. 1996; Brotas and Plante-Cuny 2003), in addition to their feasible biotechnological applications (Das et al. 2007). A comprehensive review of analytical methods for carotenoid measurement in different matrices is afforded in a recent paper by Rivera and Canela-Garayoa (2012). Hence, this chapter section will primarily tackle with specific developments of mass spectrometry applications to carotenoid assessment in alga.

As in any organic compound identification, off-line mass spectrometry with electron impact ionization has been used for longer, and it is already used at present in structural studies regarding unknown carotenoid and related substances (Bjørnland et al. 2003). High tendency of carotenoids to form degradation products because they are thermally labile, as well as their low polarity and high vaporization pressure, entail a constraint for MS analysis of these substances. A first approach to the analysis of carotenoids by on-line LC-MS with ionization by electrospray was conducted by van Breemen (1995). A description of previous assays intended to measuring carotenoids by on-line mass spectrometry can be found in van Breemen (1997). This author used a C_{30} reversed-phased column for carotenoid and xanthophyll separation, which was interfaced to a UV-Vis diode-array and then to a single quadrupole mass spectrometer. In order to enhance ionization through solution-phase carotenoid oxidation, a solution of 2,2,3,3,4,4,4-heptafluoro-1-butanol (0.1%, v/v) in methanol/methyl *tert*-butyl ether was added post-column, the chromatographic eluent being also methanol/methyl *tert*-butyl ether but with 1.0 mM ammonium acetate. Carotenes and xanthophylls were detected as the radical ion ($M^{+}\cdot$) in positive ionization. Contrary to that shown in previous studies by other authors, van Breemen found that a protic solvent like methanol did not interfer ionization to a large extent while improving chromatographic separation. Indeed, the *cis* isomer of â-carotene could be detected in this study by mass spectrometry at a concentration 100-fold lower than it was detected by the UV-Vis spectrum because of a higher signal-to-noise ratio. Further work by van Breemen et al. showed differences in carotenoid ionization using atmospheric pressure chemical ionization (APCI) instead of electrospray (van Breemen et al. 1996; van Breemen 1997). Thus, positive APCI rendered the protonated ions $[M+H]^{+}$ for both carotenes and xanthophylls, the proton coming from the solvent as determined using deuterochloroform as unique solvent (van Breemen et al. 1996). Protic solvents like alcohols were found to increase the abundance of protonated ions. MS/MS of the pseudo-molecular ion by collision induced dissociation (CID) led to fragments $[M-56]^{+}\cdot$ (retro-Diels-Alder process) for á-carotene, and $[M-92]^{+}\cdot$ (typical loss of toluene) for á-carotene and lutein. In the study by van Breemen et al. (1996), negative ionization brought about the pseudo-molecular ion $[M-H]^-$ along with the radical ion $M^{-\cdot}$. Interestingly, the low limit of detection (LOD) was higher under positive ionization than under negative ionization for lutein (13 pmol vs. 1 pmol) whereas it remained the same for α-carotene (3 pmol); these LODs were below the detection limits obtained with UV-Vis. It was shown that a possible advantage of APCI against ESI is that the linearity of the dynamic range of carotenoid concentrations that can be measured is of at least three orders of magnitude higher for APCI than for ESI (van Breemen 1997). Later on, diverse authors have applied the methods developed by van Bremeen et al. (1997) to the analysis of carotenoids in several matrices, with particular focus upon food related ones. Because of the possible informative value, a list of radical or pseudo-molecular ions, along with characteristic fragment ions from LC-MS/MS is depicted in Table 1 as reported in different studies, irrespective the source of carotenoids that was used in the study.

Regarding algae, the majority of studies have been concerned with identification of astaxanthin and other xanthophylls esters and glucosides, for which mass spectrometry has showed particularly useful since the specific fatty acyls that esterify the carotenoid cannot be distinguished by the UV-Vis spectrum. Frassanito et al. (2005; 2006; 2008) have applied mass spectrometry analysis to the assessment of carotenoids from freshwater microalgae. These authors were able to elucidate co-elution of diadinoxanthin and diatoxanthin through differential mass of these carotenoids with close similar UV-Vis spectrum (Frassanito et al. 2005). In other study, palmitic and/or myristic esters, but not di-myristic or di-palmitic esters, of astaxanthin were found to be the most abundant carotenoids in the dinoflagellate *Tovellia sanguinea* (Frassanito et al. 2006). The fragmentation pattern of astaxanthin esters has also been investigated in lipid extracts from the alga *Haematococcus pluvialis* with ESI-MS (Frassanito et al. 2008) and APCI-MS (Breithaupt 2004; Miao et al. 2006; Holtin et al. 2009). APCI-MS led to fragmentation patterns dominated by the loss of the fatty acid and water, whereas the neutral losses of C_7H_8 and C_8H_{10} were also frequent in ESI-MS. Predominant fatty acids were found to be oleic (C18:1), linoleic (C18:2), and linolenic (C18:3), the astaxanthin monoesters accounting for up to > 90% of the astaxanthin content. Contrary to other studies, Breithaupt et al. (2004) used negative ionization mode, which provided $M^{-\cdot}$ ions predominantly and allowed them obtaining spectra clearer than using positive ionization.

Table 1: Values of m/z of common carotenoids and their current fragments ionized in positive mode. For more details see Airs et al. (2001), De Rosso and Mercadante (2007) and Mendiola et al. (2009).

Carotenoid	m/z ($[M+H]^+/M^{+\cdot}$)	Main fragments MS/MS (m/z)
Siphonaxanthin	545	528, 489
Diatoxanthin	567	549
Lutein	569/568	551
Zeaxanthin	569/568	551, 481/445, 429
Alloxanthin	565	
Monadoxanthin	567	549
Carotene diol	569	552, 496
Adonirubin	581	
Canthaxanthin	565	537, 505
Rhodoxanthin	563	545
Crocoxanthin	551	533, 494
Okenone	579	547
Echinenone	551	533
Myxoxanthophyll	751	733, 589
19'-butanoyloxyfucoxanthin	-/744 (M^+)	726, 638, 298, 283, 221, 195
Diadinoxanthin	582 (M^+)	502, 490, 221
Gyroxanthin	-/614 (M^+)	596, 578, 197
Astaxanthin	-/596	
Violaxanthin	601	583, 565, 509, 491, 221
β-Cryptoxanthin	553	535, 495, 461,
β-Carotene	537	444
β-zeacarotene	539	455, 402, 310

As well, opposite to previous studies (Renstrom and Liaaen-Jensen 1981) using thin layer chromatography and further GC-MS analysis of methyl esters of fatty acids (FAMEs), Breithaupt (2004) reported lack of polyunsaturated fatty acids (PUFAs) esterifying astaxanthin. In the alga *Scenesdemus obliquus*, a set of carotenoids were detected by LC-MS as the $M^{+\cdot}$ ion using ESI-ion trap-mass spectrometry, which included adonixanthin, astaxanthin, 3'-hydroxyechinenone, 3-hydroxy-β,β-caroten-4,4'-dione (adonirubin), canthaxanthin, echinenone and β-carotene (Qin et al. 2008). From the analysis of different intermediates which were produced under diverse culture conditions, the authors conclude that astaxanthin is synthesized from β-carotene through three possible pathways. Zeaxanthin and myxoxanthophyll were detected as the $[M+H]^+$ ions at m/z 569 and m/z 751, respectively, in extracts of the cyanobacterium *Spirulina platensis* using supercritical CO_2 with 10% ethanol as modifier. LC-APCI-MS and LC-ESI-MS/MS experiments were conducted in this study by Mendiola et al. (2009), obtaining characteristic fragments of m/z 445 and m/z 429 for zeaxanthin, and m/z 589 and m/z 733 for myxoxanthophyll. Several phaeophytin allomers could be also ascertained.

Mass spectrometry has also been used on-line with supercritical fluid chromatography (SCF-MS) for carotenoid measurement in the alga *Chlamydomonas reinhardtii* (Matsubara et al. 2009). Despite neoxanthin and violaxanthin could not be chromatographically resolved, they could be distinguished by the mass spectrum. More than 100 molecular species of astaxanthin glucoside diesters were shown by LC-APCI-MS analysis to constitute the pigment profile of the alga *Chlamydomonas nivalis* (Rezanka et al. 2008). Major fragments of the astaxanthin glucoside diesters were those in which the fatty acid or the fatty acyl-glucoside moieties were lost, although a 70 V cone voltage was necessary and Rezanka et al. (2008) suggested the fragment $[M+H-FA]^+$ could be used for quantification.

Even though not algae but plant cells and extracts being used, it warrants mention the work by Fraser et al. (2007), where carotenoid composition was determined using a matrix assisted laser desorption ionization time-of-flight mass spectrometry (MALDI-ToF-MS) with sensitivity at pmol level. Carotenoids were only detected using positive reflectron mode, obtaining in this mode strong signals with mass accuracies of 10 ppm. Sinapinic acid (SN) and 2,5-dihydroxybenoic acid (DHB) matrices rendered similar efficiencies when used at a concentration of 10-fold in excess that of the sample. By comparing MS analysis with HPLC-PAD analysis, intermediates (*pro*-lycopene, neurosporene, β-carotene, phytofluene, and phytoene) of the β-carotene synthesis pathway were identified by m/z in this study. A comparison of quantification relative to canthaxanthin between MALDI-ToF and HPLC-PAD did not show significant content differences, although linearity was not possible to be obtained for lycopene and zeaxanthin. Through multivariate methods and search in databases (METLIN, KEGG) Fraser et al. (2007) were able to identify the majority of metabolites and genotype differentiation. However, the isomers lutein/zeaxanthin and lycopene/β-carotene could not be distinguished. A metabolomics approach with HPLC-ESI-HRMS analysis was also used by Chu et al. (2011) to characterize the intermediates in the astaxanthin biosynthesis pathway in the alga *Scenedesmus* sp. submitted to high salt treatment. Compounds were identified by retention time, UV-Vis spectrum and accurate mass (< 2 ppm error). β-apo-8'-carotenal was utilized as reference for quantification in this study. Lutein and β-carotene elicited $M^{+\cdot}$ ions while other carotenoids ionized through the protonated molecular ion ($[M+H]^+$). Sodium acetate and sodium chloride enhanced ionization although sodium adducts were not abundant. The authors of this study conclude that astaxanthin is formed from â-carotene through sequential hydroxylations.

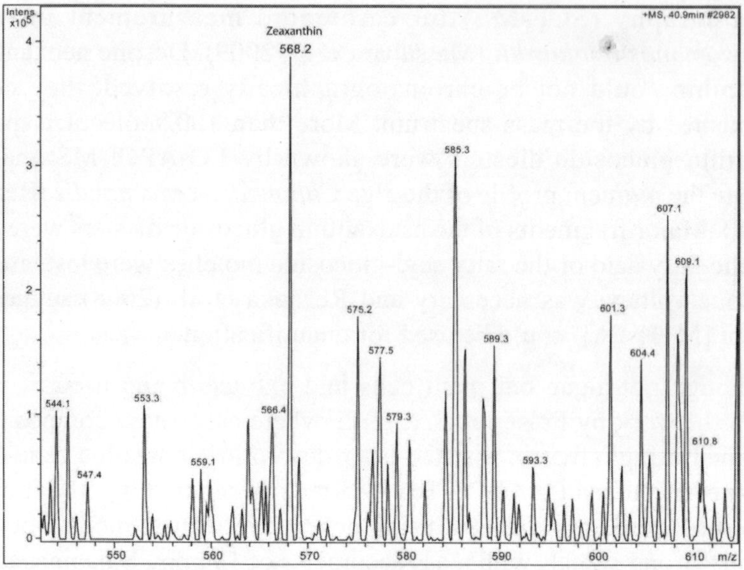

Fig. 6: Spectrum of the region where zeaxanthin eluted as shown by the UV-Vis chromatogram (see also Fig. 2 and Fig. 7). $M^{+\cdot}$ of zeaxanthin is m/z 568.2.

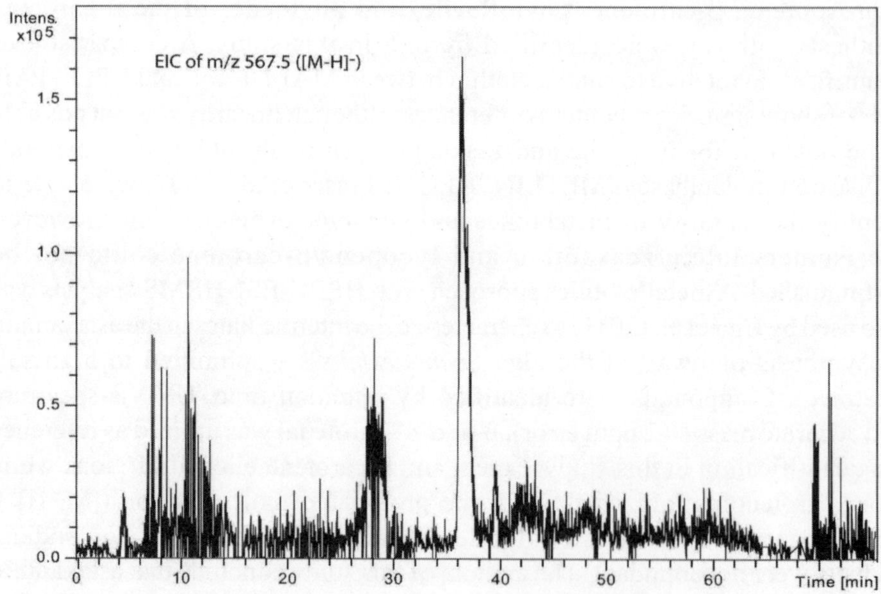

Fig. 7: Extracted ion chromatogram (EIC) of m/z 567.5 from the chromatogram obtained in the analysis in negative mode of a methanolic extract of *Synechococcus* sp. cells. Details on the experimental setup can be found in Montero (2011).

In the analysis of a methanolic extract of *Synechococcus* cells with HPLC-ESI-ion trap-MS (Bruker Daltonics GmbH) (see above) it was found that carotenoids did not render visible peaks in the base-peak-chromatogram (BPI) (Fig. 2), which might be due to an ionization efficiency and relative content much lower than those of glycerolipids. Nonetheless, the m/z 568.2 ($M^{+\cdot}$, where M is the mass of zeaxanthin) could be observed at relative high intensity (Fig. 6), in spite of the extracted ion chromatogram (EIC) of m/z 568.2 could not be obtained, which is likely due to the low signal/noise level. When the extract was analyzed in negative mode, the m/z 567 corresponding to zeaxanthin was detected and the EIC could be drawn (Fig. 7), it showing a small peak whose elution time was close similar to that of the zeaxanthin peak as detected by UV-Vis spectrophotometry (Fig. 2).

When the methanolic extract of *Synechococcus* sp. was injected in a UPLC-QToF-MS system (Acquity UPLC and SYNAPT HDMS G2 from WATERS) in the laboratory of this chapter's authors, narrow chromatographic peaks were obtained for zeaxanthin and echinenone in spite of glycerolipid peaks were also much more intense than those of carotenoids (Fig. 8), as shown by the respective extracted ion chromatograms (EICs).

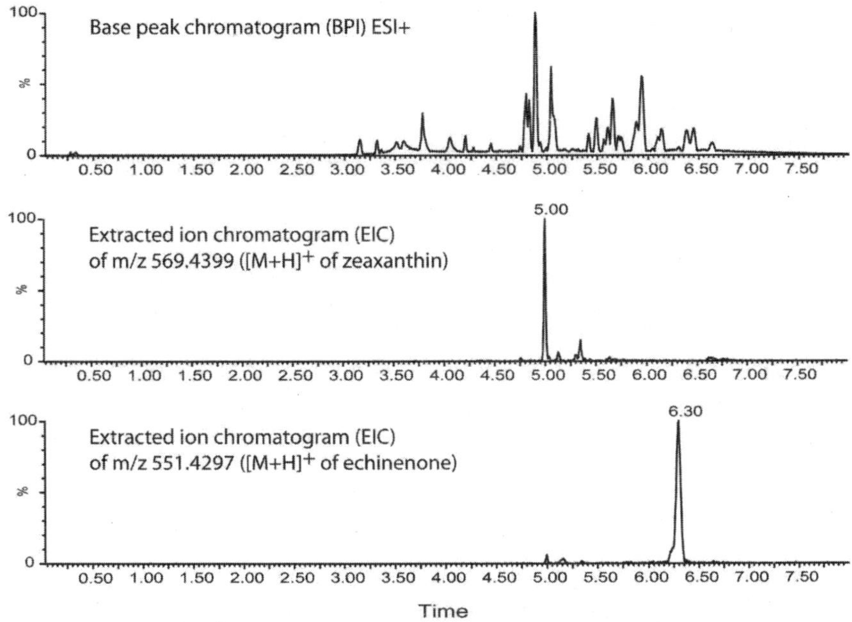

Fig. 8: Base peak chromatogram (BPI) obtained for a methanolic extract of *Synechococcus* sp. cells analyzed by UPLC-ESI-QToF-MS in positive mode, and extracted ion chromatograms (EIC) of zeaxanthin and echinenone.

Echinenone is currently not easy to detect by UV-Vis because it co-elute with b-cryptoxanthin and chlorophyll *a*. An extracted ion chromatogram of the m/z 569.4399 corresponding to zeaxanthin is depicted in Fig. 8. Zeaxanthin was found to elute at 5.00 min according to the mass spectrum obtained for such chromatographic peak (Fig. 9). The ions of m/z 568.4223 and m/z 569.4399 in Fig. 9 correspond to $M^{+\cdot}$ and $[M+H]^+$, respectively. These results are in agreement with the double ionization in positive mode of carotenoids as shown by other authors (see above).

In spite of MS analysis of carotenoids which is more selective and sensitive than UV-Vis detection (van Breemen et al. 1996; Breithaupt 2004), the majority of researchers still prefer quantification of carotenoids to be accomplished from the UV-Vis spectrum owing to the Beer-Lambert law provide a well-defined relation between concentration and absorbance. Because there is not a mathematical relationship of the concentration to the measured magnitude (eg. ion intensity) in mass spectrometry, quantification of carotenoids through mass spectrometry measurements requires previous experimental definition of the particular mass spectrometer response to a range of concentrations (dynamic range) using proper purified standards. Hence, further work is devoted in this substance analysis to establish a suitable methodology that allows validated quantitative measurement of carotenoids (and related substances like isoprenoids).

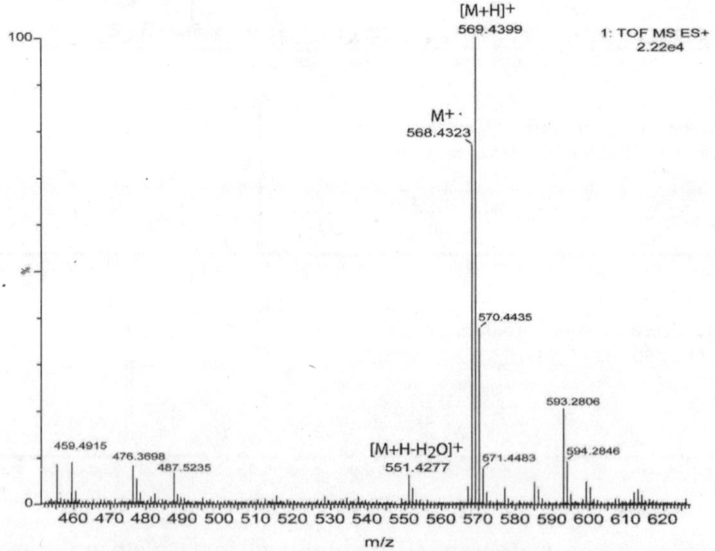

Fig. 9: Spectrum of the peak eluting at 5.00 min (see Fig. 8) in the analysis of a methanolic extract of *Synechococcus* sp. cells analyzed by UPLC-ESI- QToF-MS in positive mode. The ions of m/z 568.4323 ($M^{+\cdot}$) and m/z 569.4399 ($[M+H]^+$) correspond to zeaxanthin.

(c) Hopanoids

Hopanoids are a group of hydrocarbon-related compounds (Fig. 10), pentacyclic triterpenoids synthesized from squalene by some bacterial species, including cyanobacteria. Hopanoids are believed to play a role in membrane fluidity and stress tolerance, a role similar to that played by sterols in eukaryotic organisms (Jürgens et al. 1992). These compounds account for particular significance regarding paleochemistry studies because they have been found in rocks and sediments from early periods of Earth formation (Ourisson et al. 1979). On the basis of morphological studies, organic geochemical spots within stromatolites and microfossils pertaining to ancient rocks of up to 3,465 Myr-old have putatively been ascribed to a prokaryotic, maybe cyanobacterial, origin. Indeed, about two-thirds of the named microfossil taxa in the Apex Chert (Western Australia) exhibit 'cyanobacterium-like' morphology (Shopf 1993; 2006).

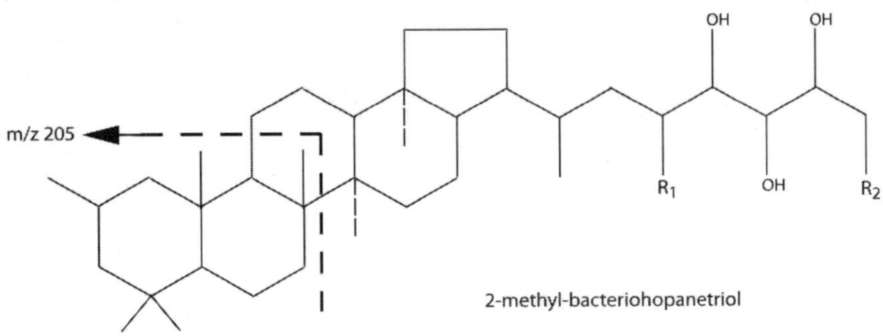

Fig. 10: Chemical structure of 2-methyl-bacteriohopanetriol where the typical fragment of m/z 205 obtained in collision induced dissociation (CID) is shown.

Because cyanobacteria are the evolutionary most advanced lineage within bacterial types, with photosynthetic oxygen production and respiratory oxygen consumption capabilities, their presence in Earth's history as early as 3500 Myr ago is of concern to establish the origin of both life and the oxygen-rich atmosphere (Buick 1992; Summons et al. 2006). However, the above contentions are guided by the assumption that similar morphology is indicative of similar biochemistry, which might have not been the case (Shopf 2006); therefore, finding molecules which could be biochemically ascribed to a given microbial origin is of relevance for the issues on concern related above.

Within hopanoids, 2-methyl-bacteriohopanepolyols (2-Me-BHP) have been found to be present in high proportion of cultured cyanobacteria and cyanobacterial mats. Even though other BHP methylated at C2 have been reported in some species of methylotrophs (eg. *Methylobacterium organophilum*) and nitrogen-fixing bacteria (Talbot et al. 2007; Welander et al. 2010), they are currently related to hopanoids of the diplopterol type (C_{31} skeleton instead of C_{32} hopanol in cyanobacteria), and account for a much lesser proportion than in cyanobacteria. Because hydrocarbon skeletons of BHP are extremely refractory and resist biodegradation during diagenesis, 2-Me-BHP have become a biomarker for cyanobacterial species within fossil records, they being expected to help in the quantification of the ecological relevance of cyanobacteria through geological time. Cyanobacterial strains accounting for hopanoids in their lipid profile include, amongst others, species of the genus *Synechococcus*, *Phormidium*, *Oscillatoria*, *Prochlorothrix*, *Gleobacter*, *Nostoc* and *Calothrix* (Summons et al. 1999). Marine cyanobaceria are poorly represented within the 2-Me-BHP producing strains (Summons 2006), although the majority of records found in sediments are believed to proceed from restricted marine settings like shallow-water continental shelves or saline coastal lakes (Summons and Jahnke 1992).

Hopanoids in general and BHP in particular are amenable to analysis by hyphenated techniques such as GC-MS and LC-MS (Ourisson et al. 1979; Talbot et al. 2001; Talbot et al. 2003a; 2003b; Summons et al. 2008). However, BHP need to be chemically transformed to the primary alcohol by cleavage of the vicinal alcohols with periodic acid and sodium borohydride, this followed by thin-layer chromatography (TLC) purification and further acetylation before GC-MS analysis can be performed (Rohmer 1993; Summons et al. 1999). This procedure cannot avoid loss of specific source-related information regarding the functional groups (Talbot et al. 2003b). Liquid chromatography coupled to mass spectrometry was introduced in hopanoid analysis by Fox et al. (1998). These authors detected intact bacteriohopanepolyols and composite hopanes as chlorine adducts ($[M+Cl]^-$) in negative mode using atmospheric pressure chemical ionization (APCI) and triple quadrupole analyzer (TQ-MS), and applied this methodology to sedimentary samples. Chlorine addition has the advantage of presenting a characteristic isotopic pattern since the heavier isotope of chloride (^{37}Cl) has an abundance one-third that of the lighter isotope (^{35}Cl). Hopanoids were extracted by using chloroform:methanol (2:1, v/v) and after the solvent was evaporated to dryness, hopanoids were solubilized in methanol:water (1:1, v/v) with further evaporation of methanol under a nitrogen gas flux. As in other molecule analysis, liquid chromatography is well suited for minimal manipulation of molecule, thus avoiding cumbersome work and artifact formation. Later, Talbot et al. (2001; 2003a; 2003b; 2007) used LC-MS but with positive ionization mode for the analysis of acetate derivatives of intact bacteriohopanepolyols. Normal phase columns were found by these authors to yield poorly defined peaks and strong retention of aminobacteriohopanepolyols, thus substantially lengthening the duration of the chromatographic run (Talbot et al. 2001). Hence, reversed-phased HPLC separation was conducted in further studies of these authors (Talbot et al. 2003a; b). Mass spectrometry offers the possibility of determining hopanoids in an unambiguous manner through their specific fragmentation pattern (Ourisson et al. 1979; Talbot et al. 2003a). Under APCI ionization, Talbot et al. (2003a) have shown that acetylated 2-Me-BHP renders a characteristic fragment of m/z 205 (Fig. 10), with neutral loss of a 206 Da fragment, this fragmentation being characteristic of the hopane ring system as it had been previously shown in GC-MS analysis with electron impact (EI) ionization (Ourisson et al. 1979). This fragmentation produces additional characteristic ions of m/z 489, 463, 403, 343 and 283, which allows distinguishing 2-Me-BHP from non-methylated hopanes because the fragment due to rings A and B cleavage would then have m/z 191 (Fig. 10). Composite BHP, namely 35-*O*-*β*-3,5-anhydro-galacturono-pyranolsylbacteriohopanetetrol and its C2 methylated homologue, have been detected by these authors in the prochlorophyte *Prochlorothrix hollandica* and cultures of the cyanobacterium *Cyanothece* strain RC. Additionaly, Talbot et al. (2003a) were able to find new composites, possibly amino-sugar derivatives, of tetra- and penta-hydroxylated hopanes in cultures of the cyanobacterium *Chlorogloeopsis* strain LA. These hopanoids had previously been observed in samples from Lake Druzhby in Antarctica (Talbot et al. 2003b).

As mentioned above, 2-Me-BHP (and composite hopanepolyols) determination by LC-MS might help quantifying cyanobacterium distribution, including prochlorophytes, with high specificity and sensitivity within different ecosystems and, in particular, these related to ocean, where current methods depend upon variable molecular signatures as chlorophyll species or carotenoid profile (Falkowsky and Raven 1997; McIntyre et al. 2002). Even though LC-MS methods are likely not as sensitive as GC-MS methods, sensitivity of MS detection can largely be augmented using MS methodologies as selected ion monitoring (SIM) or multiple reaction monitoring (MRM) (Talbot et al. 2001). Recently, the authors of this chapter have collaborated with RAMAN's group of Prof. F. Rull (Valladolid, Spain) to look for possible biomolecules present in RAMAN-detected carbonaceous spots inside rock samples from Barberton (South Africa) by UPLC-QToF-MS analysis; three m/z values from positive ionization and one from negative ionization were tentatively suggested to be compatible with different m/z hopanoid ions according to search in METLIN database (Rull et al., unpublished). Hence, it is expected that MS methodology give rise to new insights into hopanoid chemistry and biology that help unveiling history of life in our planet.

Mycosporine-like amino acids (MAAs)

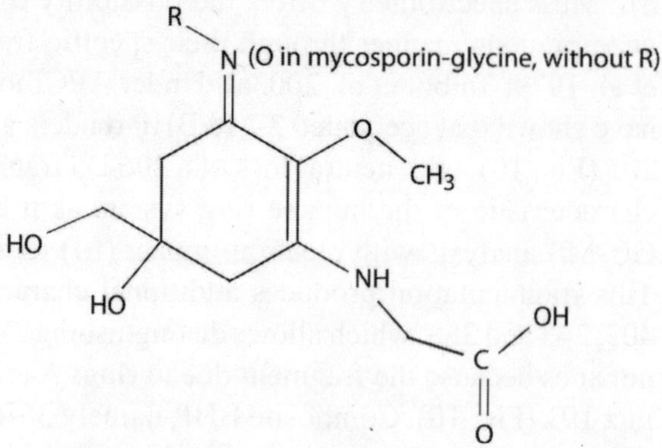

Fig. 11: Core structure of mycosporine-like amino acids. Group R determines the different MAA molecular species (Whitehead and Hedges 2002).

The occurrence of a group of UV-absorbing compounds with a similar structure related to amino-cyclohexenone or amino-cycloheximine rings (Fig. 11) was firstly described in aquatic organisms during the late 1960s. These compounds, whose UV absorption maxima fall within the range 310 to 360 nm, were named mycosporine-like amino acids (MAAs) and were shown to be capable of dissipating absorbed radiation. Hence, a possible role as sunscreens against UV radiation in cyanobacteria, algae, fungi and phytoplankton was ascribed to MAAs (Shick and Dunlap 2002; Bandaranayake 1998; Sinha et al. 1998; Banaszak et al. 1998). Bound to the ring system through C3 there may be an amino acid or an amino alcohol, which determines the particular absorption maxima of each MAA (Bandaranayake 1998).

Regarding daily analysis of MAAs beyond structural elucidation of new molecules, it has currently been accomplished by high performance liquid chromatography (HPLC) with UV or photodiode array (PDA) detection, using the differential signature of each MAA absorption spectrum to characterize the MAA content in a given sample (Nakamura et al. 1982; Whitehead and Hedges 2002; Montero and Lubián 2003). However, identification of MAAs by the UV spectrum is imprecise given that there are similarities in the absorption maxima and chromatographic retention time between different MAAs, as for example between shinorine and porphyra-334 (λ_{max}=334) (Whitehead and Hedges 2002; Kamio et al. 2011). Furthermore, there are not standards for MAAs available commercially. MAAs are amenable to direct analysis by LC-MS with electrospray ionization (ESI) because of their acidic and amino groups can be ionized through proton acquisition or proton release, apart from the formation of typical adducts with negative or positive ions (Na^+, Cl^-, NH_4^+, CH_3COO^-, etc). In addition to identification by mass, it is assumed at present that the particular fragmentation pattern of each MAA may render on-line structural information and avoid misidentification (Cardozo et al. 2006; Carreto and Carignan 2011; Llewellyn and Airs 2010). As well, a common fragment to all MAAs may serve for direct, fast and unequivocal screening of the whole MAA profile in a given sample through MS approaches like SIM and MRM. Indeed, Whitehead et al. (2001) shown LC-MS analysis could help detecting MAA misidentification by UV spectrum.

Since Whitehead et al. (2001) reported a first paper on MAA analysis in plankton by liquid chromatography coupled to mass spectrometry with electrospray ionization, there has been a number of papers dealing with such issue in algae and related organisms (Whitehead and Hedges 2002; 2003; Montero and Lubián 2003; Cardozo et al. 2006; 2008; 2009; Volkmann and Borgushina 2006; Sommaruga et al. 2006; Ingalls et al. 2010; Cardozo et al. 2011). LC-MS and LC-MS/MS analyses of an extract from the halophilic unicellular cyanobacterium *Euhalothece* sp., it including exact mass determination by using a ToF mass spectrometer, led to the identification of a novel MAA, euhalothece-362, with absorption maximum at 362 nm, and whose chemical structure could not have been determined previously by using nuclear magnetic resonance (NMR) after purification because of the interference of iron cations in the extract, whose presence was suggested by the authors to be likely related to a chelation effect by the hydroxyl groups of the MAA (Volkmann et al. 2006). A method for extraction and analysis of MAAs from different matrices by HPLC-PAD-MS/MS, which is expected to be of general application, was proposed by Volkmann and Gorbushina (2006). This method encompasses lyophylization and solid-liquid extraction with 0.2% aqueous acetic acid plus 0.5% methanol for 12 h at 4 °C, and further chromatographic separation using reversed-phase C_{18} chromatography with isocratic elution utilizing the same solvent as for extraction.

A first attempt to characterize the fragmentation pattern of different MAAs was accomplished by Whitehead and Hedges (2003) using a triple quadrupole (TQ) instrument. These authors showed that GC-MS analysis with electron impact (EI) ionization, which provides more reproducible spectra, was not an optimal choice for MAA analysis with mass spectrometry because MAAs tend to form multiple tetra-methyl-silyl derivatives, thus extreme cleaning of MAA extract becoming essential. As well, these authors failed to get negative ions through deprotonation of the carboxylic group. More insights into the MAA fragmentation processes including a scheme for it in base to high resolution experimental data and theoretical calculations has been provided in further works (Cardozo et al. 2008; 2009).

Table 2: Mass spectrometry data of mycosporine-like amino acids (MAAs) as reported by different authors (Whitehead and Hedges 2002, 2003, Volkmann and Gorbushina 2006, Carignan et al. 2009). Most significant fragments are indicated in bold. More data on MAAs can be found in Sinha et al. (1998, 2007).

Substance	m/z for $[M+H]^+$	m/z of the main fragments from MS/MS
Shinorine	333	137, 168, 185, **186**, 197, 211, 212, 230, 241, 256, 255, 265, 274, 287, 300, 303, 318
Palythine	245	137, 150, 155, 162, 184, **186**, 197, 199, 209, 230
Asterina 330	289	137, 150, 168, **186**, 197, 199, 212, 213, 230, 243, 271, 273, 289
Porphyra 334	347	137, 151, 164,168, 185, **186**, 191, 197, 200, 209, **227**, 243, 288, 303
Mycosporine-serinol	262	**244**, 216, 194
Mycosporine-glutaminol	303	285, 267, **235**
Palythinol	303	186, 226, 230, 244, 285, 288
Mycosporine-glutaminol-glucoside	465	447, 429, **303**, 285, 267
Mycosporine-alanine	260	242, **171**, 147
Mycosporine-serine	276	**258**, 228, 212
Mycosporine-glutamicol	304	286, 268, **236**
Mycosporine-glutamicol-glucoside	466	448, 430, **304**, 286, 268, 236
Euhalothece 362	331	316, 272, 242, 228, 213
Mycosporine-glutamicol-glucoside	465	447, 429, 365, **303**, 285, 250, 235, 218
Palythine-threonine	289	**245**, 169, 151, 134, (106)
Palythine-serine	275	122, 154, 172, 224, 245, 260
Mycosporine-methylamine-serine	289	131, 175, 186, 197, 212, 230, 259, 274
Mycosporine-glycine	246	153, 172, 188, 200, 210
Mycosporine-methylamine-threonine	303	106, 165, 183, 224, 259
Palythene	285	197, 226, 241, 267, 270
Palythenic acid	329	226, 233, 252, 261, 270, 283, 296, 311, 314

A current concern regarding LC-MS/MS analysis of MAAs with ESI ionization in ion-trap mass spectrometers is the apparent formation of a $[M+H-15]^+$ ion in the collision induced dissociation (CID) process owing to the loss of a methyl group (Whitehead and Hedges 2002; Cardozo et al. 2008). This feature difficult quantitation because some MAAs did not follow this pattern, and undergo losses of rather 44 or 30 Da, which can be likely attributed to CO_2 and CH_2O, respectively. A neutral loss of 46 Da was observed in all MAA standards analyzed apart from palythinol and porphyra (Table 2).

However, a distinctive pattern of fragmentation for the different MAAs was found for fragment ions of less than 200 Da, with a common fragment to all MAA standards of m/z 197 (Whitehead and Hedges 2003). Sensitivity to ESI ionization to render intense $[M+H]^+$ ions was found to depend on polarity of the imine substituent, the higher polarity the higher sensitivity, with palythene being the more sensitive (Whitehead and Hedges 2002). Because of unavailability of appropriate standards for each MAA and the differential response to ionization, quantification by LC-MS from the full-scan using the $[M+H]^+$ ion is at present dependent on previous UV quantification through the extinction coefficient and absorbance to every MAA λ_{max} (Whitehead and Hedges 2002; Carreto and Carignan 2011). Nonetheless, tyramin hydrochloride (THC) has been proved to be a feasible compound to be used as internal standard for MAA quantification in LC-MS determinations, although a relative high concentration of it is required. In spite of these constraints, LC-MS quantitation was shown to be about 100-fold more sensitive than UV quantitation, with a low limit of detection of 2 pg injected and precision of less than 3% (Whitehead and Hedges 2002).

Next step expected to be carried out in MAA analysis by mass spectrometry seems to be the development of methods for fast on-line screening. This can be accomplished through selected ion monitoring (SIM) or precursor ion measurements. The apparently common fragment to all MAAs of m/z 197 might be a suitable candidate for such measurements, although this fragment has not been further reported for other MAAs; hence, it is necessary to establish MAA exceptions whose fragmentation pattern does not account for such fragment. The work by Whitehead and Hedges (2002; 2003) and Cardozo et al. (2008) is of high value for such a purpose.

Concluding Remarks

New insights are being gained in algal research with the suitability of liquid chromatography-mass spectrometry (LC-MS) for molecule analysis. Mass spectrometry can be considered a separation technique that enforces liquid chromatography separation of biomolecules, as well as allowing their unequivocal identification through exact mass. These advantages of MS are promoting the introduction of LC-MS platforms for routing analysis in many laboratories worldwide. New molecules are expected to be discovered in the near future, as well as the involvement of these new molecules and already known ones in physiological processes is expected to be unveiled. LC-MS-based metabolomics (lipidomics) approaches with targeted- or untargeted analysis are being developed for a better understanding of different physiological pathways and their interplay, contributing to the development of the so called System Biology (Frassanito et al. 2005; Lu et al. 2012). This is promising research field for algal research which may be used in combination with genomics approaches to search for new findings on processes like UV effects on algal cell physiology. Combination of different instrumental techniques can also be a productive experimental setup as shown by Urban et al. (2011).

Acknowledgement

Financial support from the Spanish Ministry of Education and Science (project reference AGL2004-08215-C03-02) is greatly appreciated.

References

Abian J (1999) The coupling of gas and liquid chromatography with mass spectrometry. J Mass Spectrom 34: 157-168.

Arisz SA, van Himbergen JAJ, Musgrave A, van del Ende H and Munnik T (2000) Polar glycerolipids of *Chlamydomonas moewusii*. Phytochemistry 53: 265-270.

Banaszak AT, Lesser MP, Kuffner IB and Ondrusek M (1998) Relationship between ultraviolet (UV) radiation and mycosporine-like amino acids (MAAs) in marine organisms. Bull Mar Sci 63: 617-628.

Bandaranayake WM (1998) Mycosporines: are they nature's sunscreens? Nat Prod Rep 15: 159-172.

Bjørnland T, Haxo FT and Liaaen-Jensen S (2003) Carotenoids of the Florida red tide dinoflagellate *Karenia brevis*. Biochem Syst Ecol 31: 1147-1162.

Breithaupt DE (2004) Identification of astaxanthin esters in shrimp (*Pandalus borealis*) and in a microalga (*Haematococcus pluvialis*) by liquid chromatography-mass spectrometry using negative ion atmospheric pressure chemical ionization. J Agric Food Chem 52: 3870-3875.

Brotas V and Plante-Cuny M-R (2003) The use of HPLC pigment analysis to study microphytobenthos communities. Acta Oecol 24: S109-S115.

Buick R (1992) The antiquity of oxygenic photosynthesis: Evidence from stramatolites in sulfate-deficient Archean lakes. Science 255: 74-77.

Cardozo KHM, Carvalho VM, Pinto E and Colepicolo P (2006) Fragmentation of mycosporine-like amino acids by hydrogen/deuterium exchange and electrospray ionisation tandem mass spectrometry. Rapid Commun. Mass Spectrom 20: 253-258.

Cardozo KHM, Marques LG, Carvalho VM, et al. (2011) Analyses of photoprotective compounds in red algae from the Brazilian coast. Braz J Pharmacogn 21: 202-208.

Cardozo KHM, Vessecchi R, Carvalho VM, et al. (2008) A theoretical and mass spectrometry study of the fragmentation of mycosporine-like amino acids. Int J Mass Spectrom 237: 11-19.

Cardozo KHM, Vessecchi R, Galembeck SE, et al. (2009) A fragmentation study of di-acidic mycosporine-like amino acids in electrospray and nanospray mass spectrometry. J Braz Chem Soc 20: 1625-1631.

Carreto JI and Carignan MO (2011) Mycosporine-like amino acids: relevant secondary metabolites. Chemical and ecological aspects. Mar Drugs 9: 387-446.

Chisti Y (2007) Biodiesel from microalgae. Biotechnol Adv 25: 294-306.

Chu FL, Pirastru L, Popovic R and Sleno L (2011) Carotenogenesis up-regulation in *Scenedesmus* sp. using a targeted metabolomics approach by liquid chromatography-high resolution mass spectrometry. J Agric Food Chem 59: 3004-3013.

Cody RB (2002) Electrospray ionization mass spectrometry. In: Pramanik BN (ed.) Applied electrospray mass spectrometry. Marcel Dekker Inc, NY, USA, pp 433.

Darsow KH, Lange HA, Resch M, Walter C and Buchholz R (2007) Analysis of a chlorosulfolipid form *Ochromonas danica* by matrix-assisted laser desorption/ionization quadrupole ion trap time-of-flight mass spectrometry. Rapid Commun Mass Spectrom 21: 2188-2194.

Das A, Yoon S-H, Lee S-H, Kim J-Y, Oh D-K, Kim S-W (2007) An update on microbial carotenoid production: application of recent metabolic engineering tools. Appl Microbiol Biotechnol 77: 505-512.

De Angelis L, Rise P, Giavarini F, Galli C, Bolis CL and Colombo ML (2005) Marine microalgae analyzed by mass spectrometry are rich sources of polyunsaturated fatty acids. J Mass Spectrom 40: 1605-1608.

De Hoffmann E and Stroobant V (2007) *Mass spectrometry. Principles and applications*. John Wiley & Sons, Ltd, England. pp 819.

De Rosso VV and Mercadante A (2007) Identification and quantification of carotenoids, by HPLC-PDA-MS/MS, from amazonian fruits. J Agric Food Chem 55: 5062-5072.

Dettmer K, Aronov PA and Hammock (2007) Mass spectrometry-based metabolomics. Mass Spectrom Rev 26: 51-78.

Douglas DJ, Rank AJ and Mao D (2005) Linear ion traps in mass spectrometry. Mass Spectrom Rev 24: 1-29.

Falkowsky PG and Raven JA (1997) *Aquatic Photosynthesis* (Malden, Massachusetts: Blackwell Science).

Fiehn O (2002) Metabolomics – the link between genotypes and phenotypes. Plant Mol Biol 48: 155-171.

Fox PA, Carter J and Farrimond P (1998) Analysis of bacteriohopanepolyols in sediment and bacterial extracts by high performance liquid chromatography/atmospheric pressure chemical ionization mass spectrometry. Rapid Commun Mass Spectrom 12: 609-612.

Fraser PD and Bramley PM (2004) The biosynthesis and nutritional uses of carotenoids. Prog Lipid Res 43: 228-265.

Fraser PD, Enfissi EMA, Goodfellow M, Eguchi T and Bramley PM (2007) Metabolite profiling of plant carotenoids using the matrix-assisted laser desorption ionization time-of-flight mass spectrometry. Plant J 49: 552-564.

Frassanito R, Cantonati M, Flaim G, Mancini I and Guella G (2008) A new method for the identification and the structural characterization of carotenoid esters in freshwater microorganisms by liquid chromatography/electrospray ionization tandem mass spectrometry. Rapid Commun Mass Spectrom 22: 3531-3539.

Frassanito R, Flaim G, Mancini I, Guella G (2006) High production of unexpected carotenoids in Dinophyceae. Astaxanthin esters from the freshwater dinoflagellate *Tovellia sanguinea*. Biochem Syst Ecol 34: 843-853.

Goss R and Wilhelm C (2009) Lipids in algae, lichens and mosses. In: Wada H and Murata N (eds.) *Lipids in photosynthesis: essential and regulatory functions*, Springer science + business media BV, pp. 117-135.

Guella G, Frassanito R and Mancini I (2003) A new solution for an old problem: the regiochemical distribution of the acyl chains in galactolipids can be established by electrospray ionization tandem mass spectrometry. Rapid Commun Mass Spectrom 17: 1982-1994.

Guschina IA and Harwood JL (2006) Lipids and lipid metabolism in eukaryotic algae. Prog. Lipid Res 45: 160-186.

Guschina IA and Harwood JL (2006) Lipids and lipid metabolism in eukaryotic algae. Prog. Lipid Res 45: 160-186.

Han X and Gross RW (2003) Global analyses of cellular lipidomes directly from crude extracts of biological samples by ESI/mass spectrometry: a bridge to lipidomics. J Lipid Res 44: 1071-1079.

Harwood JL (1998) Membrane lipids in algae. In: Siegentaler PA and Murata N (eds.) *Lipids in photosynthesis*. Kluwer, Dordrecht, pp 53-64.

Herrero M, Vicente MJ, Cifuentes A and Ibañez E (2007) Characterization by high-performance liquid chromatography/electrospray ionization quadrupole time-of-flight mass spectrometry of the lipid fraction of *Spirulina platensis* pressurized ethanol extract. Rapid Commun Mass Spectrom 21: 1729-1738.

Holtin K, Kuehnle M, Rehbein J, Schuler P, Nicholson G and Albert K (2009) Determination of astaxanthin and astaxanthin esters in the microalga *Haematococcus pluvialis* by LC-(APCI)MS and characterization of predominant carotenoid isomers by NMR spectroscopy. Anal Bioanal Chem 395: 1613-1622.

Ingalls AE, Whitehead K and Bridoux MC (2010) Tinted windows: the presence of the UV-absorbing compounds called mycosporine-like amino acids embedded in the frustules of marine diatoms. Geochim Cosmochim Acta 74: 104-115.

Jürgens UJ, Simonin P, Rohmer M (1992) Localisation and distribution of hopanoids in membrane systems of the cyanobacterium *Synechocystis* PCC 6714. FEMS Microbiol Lett 92: 285-288.

Kamio M, Kicklighter CE, Nguyen L, Germann MW and Derby CD (2011) Isolation and structural elucidation of novel mycosporine-like amino acids as alarm cues in the defensive ink secretion of the sea hare *Aplysia californica.* Helv Chim Acta 94: 1012-1018.

Kim YH, Choi J-S, Yoo JS, Park Y-M and Kim MS (1999) Structural identification of glycerolipid molecular species isolated from cyanobacterium *Synechocystis* sp. PCC 6803 using fast-atom bombardment tandem mass spectrometry. Anal Biochem 267: 260-270.

Larsen MCN (1998) Mass spectrometry of biological materials. B.S., Marcel Dekker Inc., New York, USA.

Llewellyn CA and Airs RL (2010) Distribution and abundance of MAAs in 33 species of microalgae across 13 classes. Mar. Drugs 8: 1273-1291.

Lu N, Wei D, Chen F and Yang ST (2012) Lipidomic profiling and discovery of lipid biomarkers in snow alga *Chlamydomonas nivalis* under salt stress. Eur J Lipid Sci Technol 114: 252-265.

MacDougall KM, McNichol J, McGinn PJ, O'Leary SJB and Melanson JE (2011) Triacylglycerol profiling of microalgae strains for biofuel feedstock by liquid chromatography-high-resolution mass spectrometry. Anal Bioanal Chem 401: 2609-2616.

Mackey MD, Mackey DJ, Higgins HW and Wright SW (1996) CHEMTAX – a program for estimating class abundances from chemical markers: application to HPLC measurements of phytoplankton. Mar Ecol Prog Ser 144: 265-283.

Matsubara A, Bamba T, Ishida H, Fukusaki E and Hirata K (2009) Highly sensitive and accurate profiling of carotenoids by supercritical fluid chromatography coupled with mass spectrometry. J Sep Sci 32: 1459-1464.

McIntyre HL, Kana TM, Anning T and Geider RJ (2002) Photoacclimation of photosynthesis response curves and photosynthetic pigments in microalgae and cyanobacteria. J Phycol 38: 17-38.

Mendiola JA, Marín FR, Hernández SF, Arredondo BO, Señoráns FJ, Ibañez E and Reglero G (2009) Characterization via liquid Chromatography coupled to diode array detector and tandem mass spectrometry of supercritical fluid antioxidant extracts of *Spirulina platensis* microalgae. J Sep Sci 28: 1031-1038.

Miao F, Lu D, Li Y and Zeng M (2006) Characterization of astaxanthin esters in Haematococcus pluvialis by liquid chromatography – atmospheric pressure chemical ionization mass spectrometry. Anal Biochem 352: 176-181.

Mirsaleh-Kohan N, Robertson WD and Compton RN (2008) Electron ionization time-of-flight mass spectrometry: historical review and current applications. Mass Spectrom Rev 27: 237-285.

Moellering ER, Miller R and Benning C (2009) Molecular genetics of lipid metabolism in the model green alga *Chlamydomonas reinhardtii*. In: Wada H and Murata N (eds.) *Lipids in photosynthesis: essential and regulatory functions*, Springer science + business media BV, pp. 139-155.

Montero O (2011) Lipid Profiling of *Synechococcus* sp. PCC7002 and Two Related Strains by HPLC Coupled to ESI–(Ion Trap)–MS/MS. Z. Naturforsch 66c: 149-158.

Montero O and Lubián LM (2003) Mycosporine-like amino acid production by *Heterocapasa* (*sic*) sp. (Dinophyceae) in indoor cultures. Biomol Eng 20: 183-189.

Montero O, Porta JM, Porta J, Martínez G and Lubián LM (2011) Characterization of two *Synechococcus* sp. PCC7002-related cyanobacterial strains in relation to 16S rDNA, *crtR* gene, lipids and pigments. Phycol Res 59: 147-155.

Montero O, Sánchez-Guijo A, Lubián LM and Martínez-Rodríguez G (2012) Changes in membrane lipids and carotenoids during light acclimation in a marine cyanobacterium *Synechococcus* sp. J Biosci 37(4): 635-645.

Murphy RC, Fiedler J and Hevko J (2001) Analysis of nonvolatile lipids by mass spectrometry. Chem Rev 101: 479-526.

Nakamura H, Kobayashi J and Hirata Y (1982) Separation of mycosporine-like amino acids in marine organisms using reverse-phase high performance liquid chromatography. J Chromatogr 250: 113-118.

Naumann I, Darsow KH, Walter C, Lange HA and Buchholz R (2007) Identification of sulfoglycolipids from the alga *Porphyridium purpureum* by matrix-assisted laser desorption/ionization quadrupole ion trap time-of-flight mass spectrometry. Rapid Commun Mass Spectrom 21: 3185-3192.

Niesen WMA (2003) Progress in liquid chromatography-mass spectrometry instrumentation and its impact on high throughput screening. J Chromatogr A 1000: 413-436.

Plumb RS, Johnson KA, Rainville P, Smith BW, Wilson ID, Castro-Pérez JM and Nicholson JK (2006) UPLC-MS^E; a new approach for generating molecular fragment information for biomarker structure elucidation. Rapid Commun Mass Spectrom 20: 1989-1994.

Pulfer M and Murphy RC (2003) Electrospray mass spectrometry of phospholipids. Mass Spectrom Rev 22: 332-364.

Qin S, Liu G-X and Hu Z-Y (2008) The accumulation and metabolism of astaxanthin in *Scenesdemus obliquus* (Chlorophyceae). Process Biochem 43: 795-802.

Renstrom B and Liaaen-Jensen S (1981) Fatty acid composition of some esterified carotenols. Comp. Biochem Physiol 69B: 625-627.

Rezanka T, Nedbalova L, Elster J and Sigler K (2008) Identification of very-long-chain polyunsaturated fatty acids from *Amphidinium carterae* by atmospheric pressure chemical ionization liquid chromatography-mass spectroscopy. Phytochemistry 69: 2391-2399.

Rezanka T, Nedbalova L, Elster J, Cajthaml T and Sigler K (2009) Very-long-chain iso and anteiso branched fatty acids in N-acylphosphatidylethanolamines from a natural cyanobacterial mat of *Calothrix* sp. Phytochemistry 70: 655-663.

Rezanka T, Nedvalová L, Sogler K and Cepák V (2008) Identification of astaxanthin diglucoside diesters from snow alga *Chlamydomonas nivalis* by liquid chromatography-atmospheric pressure chemical ionization mass spectrometry. Phytochemistry 69: 479-490.

Rivera SM and Canela-Garayoa R (2012) Analytical tools for the analysis of carotenoids in diverse materials. J Chromatogr A 1224: 1-10.

Roche SA and Leblond JD (2010) Betaine lipids in chlorarachniophytes. Phycol Res 58: 298-305.

Rohmer M (1993) The biosynthesis of triterpenoids of the hopane series in the eubacteria: a mine of new enzyme reactions. Pure Appl Chem 65: 1293-1298.

Rull F, Venegas G, Montero O and Medina J (unpublished) Study of carbonaceous material in Archean samples from Barberton (South Africa) and possible astrobiological implications.

Sato N and Wada H (2009) Lipid biosynthesis and its regulation in Cyanobacteria. In: Wada H and Murata N (eds.) *Lipids in photosynthesis: essential and regulatory functions*, Springer science + business media BV, pp. 157-177.

Schopf JW (1993) Microfossils of the Early Archean Apex chert: new evidence of the antiquity of life. Science 260: 640-646.

Schopf JW (2006) Fossil evidence of Archaean life. Phil Trans R Soc B 361: 869-885.

Schubotz F, Wakeham SG, Lipp JS, Fredricks HF and Hinrichs KU (2009) Detection of microbial biomass by intact polar membrane lipid analysis in the water column and surface sediments of the Black Sea. Environ Microbiol 11: 2720-2734.

Schuhmann K, Almeida R, Baumert M, Herzog R, Bornstein SR and Shevchenco A (2012) Shotgun lipidomics on a LTQ Orbitrap mass spectrometer by successive switching between acquisition polarity modes. J Mass Spectrom 47: 96-104.

Scigelova M and Makarov A (2006) Orbitrap mass analyzer – overview and applications in proteomics. Proteomics 6: 16-21.

Seiwert B, Giavalisco P and Willmitzer L (2009) Advanced mass spectrometry methods for analysis of lipids from photosynthetic organisms. In: Wada H and Murata N (eds.) *Lipids in photosynthesis: essential and regulatory functions*, Springer science + business media BV, pp. 445-461.

Shick JM and Dunlap WC (2002) Mycosporine-like amino acids and related gadusols: biosynthesis, accumulation, and UV-protective functions in aquatic organisms. Annu Rev Physiol 64: 223-262.

Sinha RP, Klisch M, Gröniger A and Häder DP (1998) Ultraviolet-absorbing/screening substances in cyanobacteria, phytoplankton and macroalgae. J Photochem Photobiol B Biol. 47: 83-94.

Sinha RP, Singh SP and Häder DP (2007) Database on mycosporine and mycosporine-like amino acids (MAAs) in fungi, cyanobacteria, macroalgae, phytoplankton and animals. J Photochem Photobiol B: Biol 89: 29-35.

Siuzdak G (2003) The expanding role of mass spectrometry in Biotechnology. MCC Press, San Diego, USA. 272 pp.

Sommaruga R, Whitehead K, Shick JM and Lobban CS (2006) Mycosporine-like amino acids in the zooxanthelia-ciliate symbiosis *Maristentor dinoferus*. Protist 57: 185-191.

Summons RE and Jahnke LL (1992) Hopenes and hopanes methylated in ring-A: correlation of the hopanoids of extant methylotrophic bacteria with their fossil analogues. In: Moldowan JM, Albrecht P and Philip RP (eds.) *Biomarkers in Sediments and Petroleum*. Prentice Hall, Englewood Cliffs, New Jersey. Pp. 182-200.

Summons RE, Albrecht P, McDonald G and Moldowan JM (2008) Molecular Biosignatures. Space Sci Rev 135: 133-159.

Summons RE, Bradley AS, Jahnke LL and Waldbauer JR (2006) Steroids, triterpenoids and molecular oxygen. Phil Trans R Soc B 361: 951-968.

Summons RE, Jahnke LL, Hope JM and Logan GA (1999) 2-Methylhopanoids as biomarkers for cyanobacterial oxygenic photosynthesis. Nature 400: 554-557.

Talbot HM, Rohmer M and Farrimond P (2007) Rapid structural elucidation of composite bacterial hopanoids by atmospheric pressure chemical ionization liquid chromatography/ion trap mass spectrometry. Rapid Commun Mass Spectrom 21: 880-892.

Talbot HM, Squier AH, Keely BJ and Farrimod P (2003b) Atmospheric pressure chemical ionization reversed-phase liquid chromatography/ion trap mass spectrometry of intact bacteriohopanepolyols. Rapid Commun Mass Spectrom 17: 728-737.

Talbot HM, Summons R, Jahnke L and Farrimond P (2003a) Characteristics fragmentation of bacteriohopanepolyols during atmospheric pressure chemical ionization liquid chromatography/ion trap mass spectrometry. Rapid Commun Mass Spectrom 17: 2788-2796.

Talbot HM, Watson DF, Murrell JC, Carter JF and Farrimond P (2001) Analysis of intact bacteriohopanepolyols from methanotrophic bacteria by reversed-phased high-performance liquid chromatography – atmospheric pressure chemical ionisation mass spectrometry. J Chromatogr A 921: 175-185.

Urban PL, Schmid T, Amantonico A and Zenobi R (2011) Multidimensional analysis of single algal cells by integrating microspectroscopy with mass spectrometry. Anal Chem 83: 1843-1849.

Van Breemen RB (1995) Electrospray liquid chromatography-mass spectrometry of carotenoids. Anal Chem 67: 2004-2009.

Van Breemen RB (1997) Liquid chromatography/mass spectrometry of carotenoids. Pure Appl Chem 69: 2061-2066.

Van Breemen RB, Huang CR, Tan Y, Sander LC and Schilling AB (1996) Liquid chromatography/mass spectrometry of carotenoids using atmospheric pressure chemical ionization. J Mass Spectrom 31: 875-981.

Vieler A, Wilhelm C, Goss R, Sub R and Schiller J (2007) The lipid composition of the unicellular green alga *Chlamydomonas reinhardtii* and the diatom *Cyclotella meneghiniana* investigated by MALDI-TOF and TLC. Chem Phys Lipids 150: 143-155.

Volkmann M and Gorbushina AA (2006) A broadly applicable method for extraction and characterization of mycosporines and mycosporine-like amino acids of terrestrial, marine and freshwater origin. FEMS Microbiol Lett 255: 286-295.

Volkmann M, Gorbushina AA, Kedar L and Oren A (2006) Structure of euhalothece-362, a novel red-shifted mycosporine-like amino acid, from a halophylic cyanobacterium (*Euhalothece* sp.). FEMS Microbiol Lett 258: 50-54.

Watson JT and Sparkman OD (2007) *Introduction to Mass Spectrometry. Instrumentation, applications and strategies for data interpretation* (Fourth edition). John Wiley & Sons, Ltd, England. 819 pp.

Welander PV, Coleman ML, Sessions AL, Summons RE and Mewman DK (2010) Identification of a methylase required for 2-methylhopanoid production and implications for the interpretation of sedimentary hopanes. Proc Natl Acad Sci USA 107: 8537-8542.

Welti R, Shah J, Li W, Li M, Chen J, Burke JJ, Fauconnier M-L, Chapman K, Chye M-L and Wang X (2007) Plant lipidomics: discerning biological function by profiling plant complex lipids using mass spectrometry. Front Biosci 12: 2494-2506.

Welti R, Wang X and Williams TD (2003) Electrospray ionization tandem mass spectrometry scan modes for plant chloroplast lipids. Anal Biochem 314: 149-152.

Whitehead K and Hedges JI (2002) Analysis of mycosporine-like amino acids in plankton by liquid chromatography electrospray ionization mass spectrometry. Mar. Chem. 80: 27-39.

Whitehead K and Hedges JI (2003) Electrospray ionization tandem mass spectrometric and electron impact mass spectrometric characterization of mycosporine-like amino acids. Rapid Commun Mass Spectrom 17: 2133-2138.

Whitehead K, Karentz D and Hedges JI (2001) Mycosporine-like amino acids (MAAs) in phytoplankton, a herbivorous pteropod (*Limacina helicina*), and its pteropod predator (*Clione antarctica*) in McMurdo Bay, Antarctica. Mar Biol 139: 1013-1019.

Xing J, Xie C and Lou H (2007) Recent applications of liquid chromatography-mass spectrometry in natural products bioanalysis. J Pharm Biomed Anal 44: 368-378.

Xu JL, Chen DY, Yan XJ, Chen JJ and Zhou CX (2010) Global Characterization of the photosynthetic glycerolipids from a marine diatom *Stephanodiscus* sp by ultra performance liquid chromatography coupled with electrospray ionization-quadrupole-time of flight mass. Anal Chim Acta 663: 60-68.

Zepke HD, Heinz E, Radunz A, Linscheid M and Pesch R (1978) Combination and positional distribution of fatty acids in lipids from blue-green algae. Arch Microbiol 119: 157-62.

4

UV-induced Oxidative Stress in Cyanobacteria

Rajesh P. Rastogi, Rajeshwar P. Sinha and Aran Incharoensakdi

Abstract

Being an obligate photoautotroph, cyanobacteria must cope with a wide range of fluctuating environmental factors, including high light and ultraviolet radiation (UVR) that may lead to oxidative stress by upsetting the cellular redox status. Photo-induced oxidative stress by means of oxidizing agent such as reactive oxygen species (ROS) may damage several biological processes. In response to several adverse effects of solar UVR, cyanobacteria have developed different strategies such as heat dissipation mechanisms, synthesis of UV-absorbing/screening compounds and antioxidant systems to overcome the detrimental effects of intracellular oxidant molecules. This review deals with UV-induced oxidative stress and their counteracting mechanisms operating in cyanobacteria.

Keywords: Cyanobacteria; UV radiation; ROS; Oxidative stress; Heat dissipation; Antioxidants.

Introduction

Cyanobacteria are a group of photosynthesis mediated oxygen evolving, Gram-negative, primitive prokaryotes that appeared on the Earth between 2.8 and 3.5 x 10^9 years ago during the Precambrian era (Fischer 2008). Cyanobacteria are cosmopolitan in distribution ranging from hot springs to the Arctic and Antarctic regions (Stanier and Cohen-Bazire 1977) and are major biomass producers both in aquatic as well as terrestrial ecosystems (Häder et al. 2011). Besides atmospheric nitrogen fixation, these prokaryotes are imperative sources of various natural products of medicinal and industrial importance (Cardozo et al. 2007; Rastogi and Sinha 2009; Rastogi et al. 2010a; Rastogi et al. 2011a). It has been shown that cyanobacteria may be employed as an efficient source of energy production (Khetkorn et al. 2010).

In the past few decades, incidence of harmful ultraviolet radiation (UVR) on Earth's atmosphere has increased due to depletion of stratospheric ozone layer caused by increase in anthropogenically released chlorofluorocarbons (CFCs), chlorocarbons (CCs) and organobromides (OBs). It has been shown that substantial amounts of reactive nitrogen species (RNS) such as nitric oxide, nitrous oxide and peroxynitrite, produced naturally or from anthropogenic sources such as biomass or fuel burnings, may also contribute to the depletion of the ozone layer (Kramlich and Linak 1994). Increased solar UVR (280–400 nm), particularly UV-B radiation (280–315 nm) on the Earth's surface is absorbed by important biomolecules such as nucleic acids and proteins leading to several abnormalities on the biological systems. The high-energetic UV-B radiation may affect normal state of life either through direct effects on cellular DNA and proteins or indirectly by the production of reactive oxygen species (ROS) (Vincent and Neale 2000; Rastogi et al. 2010b). In contrast, UV-A (315-400 nm) radiation which is not absorbed directly by the DNA, can still induce DNA damage either by producing a secondary photoreaction of existing DNA photoproducts or via indirect photosensitizing reactions (Hargreaves et al. 2007).

Several environmental stress conditions such as high light intensity, temperature, desiccation, high salinity, chemical toxicity, often result in oxidative damage. UV-induced production of ROS has been reported in cyanobacteria (Rastogi et al. 2010c; He and Häder 2002a). DNA, proteins and lipids are prime targets of ROS, however, ROS-mediated damage to photosynthetic apparatus followed by inhibition of photosynthesis has also been observed in cyanobacteria (He and Häder 2002b). Moreover, UVR may affect morphology, cell differentiation, survival, growth, pigmentation, motility and orientation, N_2 metabolism, and several other metabolic activities of cyanobacterial cells (Gao et al. 2007; Lesser 2008; Sinha et al. 2008a). ROS are unavoidably generated as intermediates of O_2 reduction, or by its energization. ROS, including singlet oxygen ($^1O_2$), superoxide anion (O_2^-), hydrogen peroxide (H_2O_2) and the hydroxyl radical (OHÿ) are potent oxidizing agents. $^1O_2$ is produced by energy input to oxygen; it is highly reactive, has a short half life in cells (Gorman and Rodgers 1992) and reacts with target molecules (DNA, proteins, pigments, and lipids) in the immediate neighborhood.

Nevertheless, living organisms have developed various strategies to protect themselves against oxidative damage (Imlay 2003). Cyanobacteria developed a number of defense mechanisms to mitigate the direct or indirect damaging effects of UVR (Fig. 1).

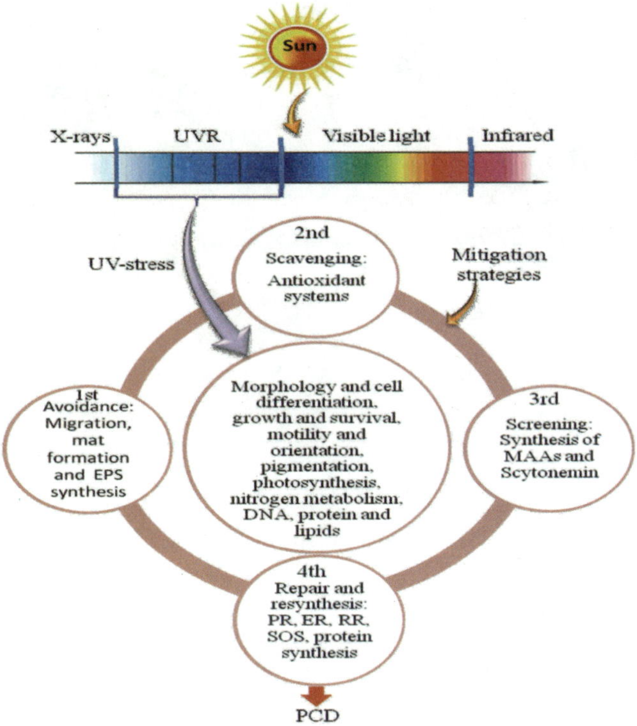

Fig. 1: Effects of solar UVR on normal physiology and biochemistry of cyanobacteria and adoption of possible defense strategies/mechanism. In case of failure of repair mechanisms the cells may undergo for apoptosis or PCD.

Several reviews have addressed the oxidative stress responses in plants and heterotrophic bacteria, and only few literatures have been reviewed on cyanobacteria (Nishiyama et al. 2006; Latifi et al. 2009). In the subsequent chapter effects of UV-induced oxidative stress and probable role of defense systems against ROS are briefly reviewed in cyanobacteria.

Cellular production of ROS

ROS are generally produced in the plants as byproducts of several metabolic activities, such as photosynthesis, photorespiration and respiration. Moreover, in photosynthetic organisms including cyanobacteria, ROS are generated by means of photosynthetic electron transport chain (Fig. 2) (Latifi et al. 2009).

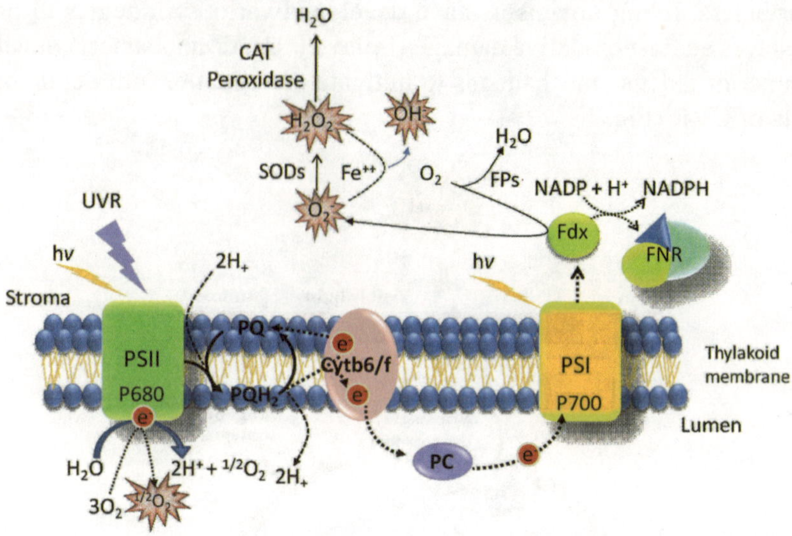

Fig. 2. A simplified schematic view of ROS generation by means of photosynthetic electron transport chain. Energy input to oxygen from photosensitized chlorophyll molecules in PSII, leads to formation of $^1O_2$. Univalent reduction of O_2 by means of electrons relocating from PSII results in production of O_2^- radicals in PSI. In presence of SOD the O_2^- radical forms H_2O_2 and O_2 disproportionately, and the formed H_2O_2 is reduced to form H_2O through CAT and peroxidises. The reduction of H_2O_2 by metal catalysed Fenton reaction produce very toxic OH^- radicals. Reduction of O_2 by FPs leads to generation of H_2O molecule. FNR (Fdx + NADP + reductase) catalyzes one terminal step of the conversion of light energy into chemical energy during photosynthesis. (For details see Latifi et al., 2009).

ROS are generally found in almost all intracellular organelles or compartments as a result of normal metabolic activity. Each organelle or compartment has potential targets for oxidative damage, as well as mechanisms for the elimination of excess ROS accumulation. $^1O_2$ produced by an energy input to oxygen (O_2) from photosensitized chlorophyll, is believed to inhibit the repair of photosystem II (PSII). It has been shown that high light intensity beyond the normal capacity of the photosynthetic electron flow may cause production of other ROS along with $^1O_2$ leading to inactivation of photosystems. In this case, oxygen rather than ferredoxin (Fdx) can be used as an electron acceptor, which generates a superoxide anion (O_2^-) as a primary product and the phenomenon on the donor side of the photosystems I (PSI) in chloroplasts was described as Mehler reaction (Mehler 1951). After Mehler's finding, it is assumed that reduction of oxygen by the PSI electron flow occurs in all oxygenic phototrophs, though, the extent of Mehler reaction in cyanobacteria is rather controversial.

It has been shown that in contrast to algae and higher plants, cyanobacteria undergo a high degree of O_2 reduction by consuming 50 % of the photosynthetic electrons instead of only 15 % for plants (Badger et al. 2000). However, Tichy and Vermaas (1999) found H_2O_2 production in *Synechocystis* PCC 6803 stands for 1% of the maximum rate of photosynthetic electron transport *in vivo*. It has been observed that the Mehler reaction in *Synechocystis* PCC 6803 does not produce ROS and that A-type flavoproteins (FPs) directly reduce O_2 to water (H_2O), a role similar to their function in obligate anaerobes (Vicente et al. 2002; Helman et al. 2003). Moreover, photosynthetic production of H_2O_2 has been recorded in *Anacystis nidulans* (Patterson and Myers 1973).

UV-B-induced production of ROS has been reported in a number of cyanobacteria. He and Häder (2002a) for the first time reported *in vivo* production of ROS in the cyanobacterium *Anabaena* sp. with the oxidative stress-sensitive probe 2', 7'-dichlorodihydrofluorescein diacetate (DCFH-DA). Recently, Rastogi et al. (2010c) also detected the formation of ROS under UVR in the cyanobacterium *Anabaena variabilis* PCC7937 by means of confocal image analysis with ROS sensing probe DCFH-DA (Fig. 3). Much attention has been focused on photosynthetic electron transport as the major source of ROS production and little is known about the contribution of the respiratory chain in ROS production. Much effort should be dedicated towards elucidating the biological effects and the specific targets of ROS and their specific physiological responses in cyanobacteria.

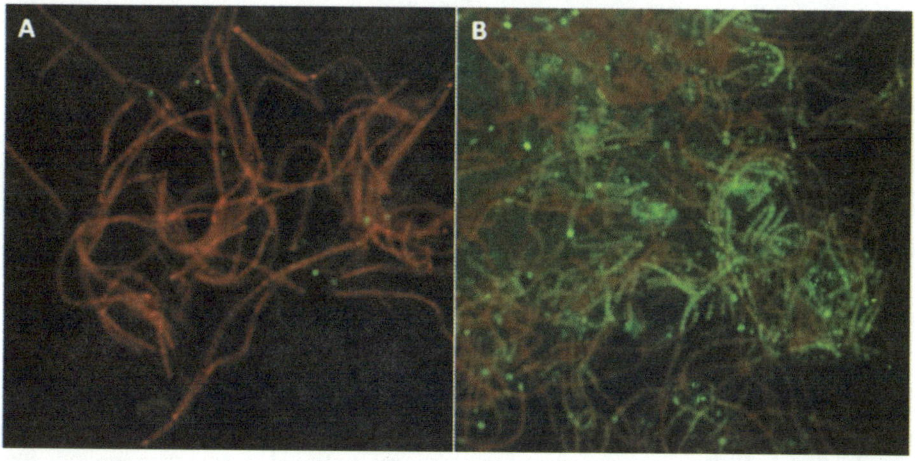

Fig. 3. Production of ROS in the cyanobacterium *Anabaena* sp. after 24h of solar irradiation. Small amount of ROS was observed in cyanobacterial cell under PAR (A) produced under normal metabolic process, whereas elevated level of ROS was observed under UV-B irradiation (B).

Oxidative Damage in Cyanobacteria

In general, oxidative stress occurs as a consequence of imbalance between ROS production and antioxidative defense system (Fig. 4). Molecular oxygen (O_2) is generally unreactive in its ground state, however, during normal metabolic activity as well as a consequence of several environmental factors (such as xenobiotics, temperatures, radiation, etc.) O_2 undergo reactive excited states of free radicals and derivatives (Scandalios 1997). The physically energized form of dioxygen i.e. singlet oxygen ($^1O_2$) are biologically most important form of ROS and about ~22 kcal/mol activation energy is required to raise molecular oxygen (O_2) from its ground state to its first singlet state (Scandalios, 2005).

The molecular O_2 undergo stepwise univalent pathway of reduction resulting in the production of some reactive species such as superoxide radical (O_2^-), hydrogen peroxide (H_2O_2), and the hydroxyl radical ($OH^•$) (Fig. 5). Both O_2^- and $OH^•$ are very reactive and can cause molecular damage, leading to cell death. ROS are shown to be mainly produced by PSI, however, light-driven oxidation of water occurs in PS-II and under certain conditions, PSII contributes to the overall production of ROS in the thylakoid membrane of plants, algae and cyanobacteria (Pospíšil 2009).

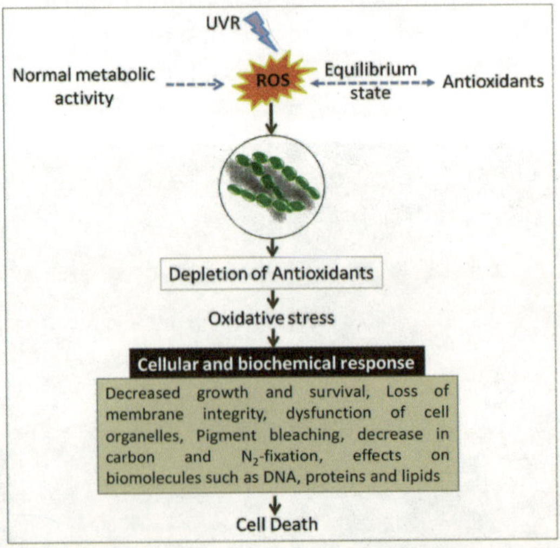

Fig. 4: UV-induced oxidative stress as a result of imbalance between antioxidants (AO) and ROS level. In natural conditions, organisms balance the ROS, produced by the general metabolic activity by means of production of antioxidants maintaining the redox equilibrium. Under severe oxidative stress condition the equilibrium state (i.e., AO=ROS) is disturbed due to overproduction of ROS and/or depletion of antioxidant systems leading to severe cellular as well as biochemical (proteins, lipids, and nucleic acids) alterations and cell death.

The general mechanisms of ROS-induced damage of biomolecules as well as dysfunction of other cellular activities inside the cell have been well documented (Imlay 2003; Rinalducci et al. 2008; Latifi et al. 2009; Blot et al. 2011). Both UV-B and UV-A radiation have direct and indirect (via oxidative stress) effects on cyanobacteria and cause chronic and acute physiological stress, ultimately resulting in inhibition of growth and cell division, bleaching of photosynthetic pigments, decreased nitrogen fixation and energy production, inhibition of photosynthesis as well as alteration in the native structure of proteins and DNA (Sinha et al. 1996; 1997; 2005; Sinha and Häder 2002; Gao et al. 2007; Lesser 2008).

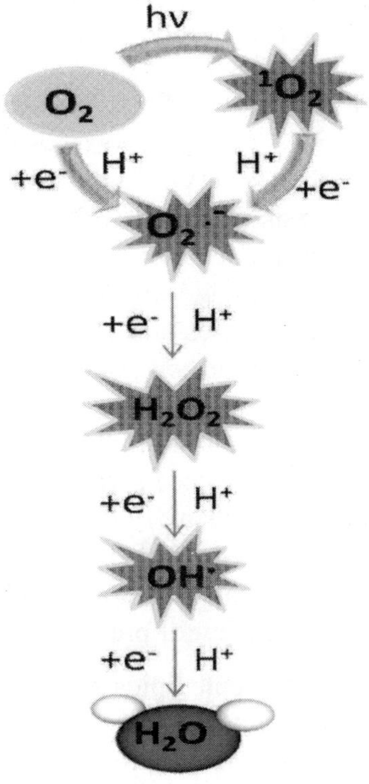

Fig. 5: Sequential univalent pathways of oxygen reduction leading to formation of water and various intermediate ROS.

DNA is one of the most important biomolecules, and its stability is of great importance for proper functioning and existence of all living forms. UVR (mainly UV-B) is one of the potent carcinogens that can alter the normal life processes by inducing a variety of mutagenic and cytotoxic DNA lesions. Several DNA lesions have been identified that are developed either due to direct absorption of UV-B radiation by the native DNA molecule or indirectly by oxidative stress. Cis-syn cyclobutane pyrimidine dimers (CPDs), pyrimidine (6-4) pyrimidone photoproducts (6-4 PPs) and their Dewar isomers are the most important damaging lesions produced by the UVR (Rastogi et al., 2010b). UV-induced oxidative stress may also induce DNA-DNA and DNA-protein cross links, base modifications and translocations (Mercer et al. 2007). It has been reported that DNA damage caused by UV radiation or ROS such as $OH^{\cdot}$ free radical results in ATM mediated phosphorylation of BID protein that induces cell-cycle arrest in S-phase (Kamer et al. 2005; Zinkel et al. 2005). Furthermore, oxidative stress-induced single- and/or double-strand breaks in the native DNA molecule have also been reported in cyanobacteria (He and Häder 2002b; Rastogi et al. 2011b). A fluorimetric analysis of DNA unwinding (FADU) assay was used by He et al. (2002) to quantify the DNA strand breaks induced by ROS under UV-A and UV-B exposure in the cyanobacterium *Anabaena* sp. Moreover, Rastogi et al. (2011b) have also reported UVR-induced DNA strand breaks in the cyanobacterium *Anabaena variabilis* PCC7937. These findings suggested that both UV-A and UV-B radiation results in an evident oxidative stress and enhances the double-strand breaks in the genomic DNA. Nevertheless, UV-A and visible light energy (up to 670–700 nm) are able to generate singlet oxygen ($^1O_2$) that can damage DNA via indirect photosensitizing reactions (Alscher et al. 1997; Hargreaves et al. 2007; Jiang et al. 2009). The $OH^•$ can damage all components of DNA molecules such as purine and pyrimidine bases and also the deoxyribose backbone, inhibiting the normal functions of the cell (Halliwell and Gutteridge 2007). A number of oxidation products of purine bases such as 8-oxo-7,8-dihydroguanyl (8-oxoGua), 8-oxo-Ade, 2,6-diamino-4-hydroxy-5-formamidoguanine (FapyGua), FapyAde, and oxazolone have been reported to form upon exposure of DNA to UV-induced ROS (Doetsch et al. 1995; Hall et al. 1996; Cadet et al. 2005). Overall, oxidative-stress-induced DNA lesions constitute ubiquitous threat to the accurate maintenance of the genome, and their incidence and importance have motivated the development of precise cellular repair functions.

The biomolecule lipids are one of the important sources of storage energy and are important structural constituents of the cellular membranes. They play a vital role in tolerance to several physiological stressors in a variety of organisms including cyanobacteria. Alteration in the membrane lipid content is of major importance in response to environmental stresses. Oxidants and free radicals are deleterious for membrane lipids in many ways (Ritter and Yopp 1993). Hydrogen atoms adjacent to olefinic bonds are highly susceptible to oxidative stress and especially those between unconjugated olefinic bonds. Lipids are rich in these bonds and thus are primary targets for oxidative reactions (Singh et al. 2002). Lipids oxidation is problematic as many oxidative chemical reactions are not controlled and constrained by enzymes and may show exponential reaction rates. Lipid peroxidation occurs especially at sites where polyunsaturated fatty acids occur in high concentrations. UV-B-induced lipid peroxidation was observed by DCF fluorescence in the cyanobacterium *Anabaena* sp. (He et al. 2002). Some of the products of the oxidative attack are highly reactive species that modify proteins and DNA (McIntyre et al. 1999). Oxidative damage to lipids occurs by means of several mechanisms of ROS reacting with fatty acids in the membrane lipid bilayer, leading to membrane leakage and cell death.

UV-B-induced oxidative stress can also affect protein components of the cell (Nishiyama et al. 2011). It has been shown that UV-B can damage D1 protein and also inhibit its synthesis leading to impairment of PSII repair process

(Ivanov et al. 2000; Sicora et al. 2006). Oxidative stress may damage proteins by means of site specific amino acid modifications, fragmentation of the peptide chain, aggregation of cross-linked reaction products, increased susceptibility to proteolysis, loss of enzymatic function and oxidation of specific amino acids. The activity of ribulose-1,5-biphosphatase carboxylase-oxygenase (RUBISCO), involved in the Calvin cycle of CO_2, was shown to be directly inhibited by UV-B in the N_2-fixing cyanobacterium *Anabaena* BT2 (Kumar et al. 2003). Overall, UV-induced oxidative stress may alter a number of biochemical as well as physiological activity of the cell and furthermore, organisms evolve a complex array of mechanisms for adapting to the minor and major variations in the environment.

Defense Strategies Against Oxidative Stress

Survival in the fluctuating environment of increasing/decreasing oxidative stress under UV-B radiation suggested a mitigating process and the induction of self-defense by cyanobacteria against UV-B stress. Oxidative stress induced by moderate UV-B irradiance may stimulate a protective response from irradiated organisms to reverse the negative effect and enable them to thrive and survive. Recent advances in molecular biology have made it possible to investigate the fundamental mechanisms operated by organisms to cope with several environmental stresses. Several strategies have been developed by cyanobacteria to avoid the production of ROS or to scavenge them once produced (He et al. 2002; Scandalios 2005; Kirilovsky 2007; Berera et al. 2009; Latifi et al. 2009; Rastogi et al. 2011b; Rastogi and Sinha 2011; Muramatsu and Hihara 2012; Kirilovsky and Kerfeld 2012). Some of the common defence strategies adopted by cyanobacteria against UV-mediated oxidative stress, such as energy dissipation, photoprotection by UV-protective/screening compounds as well as antioxidant system are discussed below.

Energy-dissipation Mechanisms

Energy dissipation is an ubiquitous and fundamental process in photosynthesis (Demmig-Adams and Adams 2006; Kirilovsky 2007). Under high light condition, photoprotective mechanisms are activated that allow dissipation of excess energy in the form of heat by means of different pathway (Fig. 6). To cope with certain stresses of high irradiations, nutrient (e.g., iron) limitations and other photooxidative stress conditions (such as heat and salt stress), some genes such as *isiA* and *hli* have been shown to express in cyanobacteria (Vinnemeier et al. 1998; Boekema et al. 2001; Havaux et al. 2005). The gene *isiA* encoding the protein IsiA (or CP430) has been reported in cyanobacteria where it forms PSI-IsiA supercomplexes (Andrizhiyevskaya et al. 2002; Berera et al. 2009).

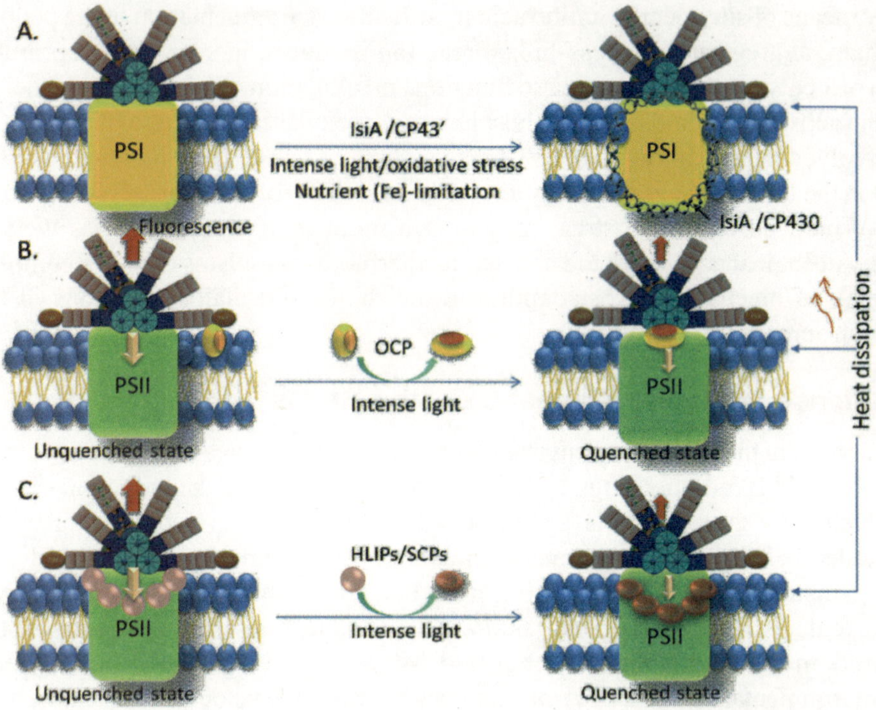

Fig. 6: Diagrammatic presentation of oxidative stress induced photoprotective mechanisms in cyanobacteria by means of IsiA (CP432) (A), OCP (B) and HLIPs (SCPs) proteins (details in text) (For more details readers are suggested to see the references of Kirilovsky, 2007; Berera et al., 2009; Latifi et al., 2009; Gorbunov et al., 2011; Kirilovsky and Kerfeld, 2012).

It has been found that IsiA proteins are not directly bound to a photosystem and function as a light harvester and have great capacity to dissipate excited-state energy leading to photoprotection under stress conditions (Andrizhiyevskaya et al. 2002; Yeremenko et al. 2004; Kouril et al. 2005). Possible roles of IsiA (Iron-starvation-inducible protein) in photoprotection include the dissipation of energy to prevent overexcitation of PSII (Wilson et al. 2007) and dissipation of the energy collected by phycobilisomes (Joshua et al. 2005). It has been shown that most of the cyanobacteria can dissipate energy from phycobilisomes by means of photoactive soluble orange carotenoid protein (OCP) (Wu and Krogmann 1997; Kerfeld 2004; Wilson et al. 2006; 2008; Gorbunov et al. 2011; Kirilovsky and Kerfeld 2012). Under high light, structural changes in the carotenoid have been observed leading to the formation of a red active form. The active red form of carotenoid protein interacts with the cyanobacterial light-harvesting antenna, phycobilisome, leading to a decrease in fluorescence emission of the phycobilisome and energy getting to the reaction centres (Kirilovsky and Kerfeld 2012).

Absorption of light in excess of that required for maximum CO_2 fixation leads to the formation of a high concentration of Chl-triplet states, mainly within the PSII reaction centre. Chl-triplet states are excellent $^1O_2$ sensitizers and are extremely harmful for survival of photosynthetic organisms including cyanobacteria. The water soluble OCP plays a vital role in quenching of Chl-singlet-excited states (known as nonphotochemical quenching: NPQ) formed under blue light or high irradiances and provides photoprotection (Demmig-Adams and Adams 1992; Horton et al. 1996; Polívka and Sundström 2004; Latifi et al. 2009). In the cyanobacterium, *Synechocystis* PCC 6803, OCP was involved in NPQ induction and mediated photoprotective energy dissipation through interaction with the phycobilisomes core (Wilson et al. 2006; Kirilovsky 2007). Conversely, it has been proposed that carotenoids may quench Chl-excited-state energy via a charge-separation mechanism involving the xanthophylls zeaxanthin (Holt et al. 2005). Berera et al. (2006) have reported that carotenoids can quench the excited-state energy of a neighbouring tetrapyrrole by accepting energy from the latter and dissipating it as heat. Recently, Berera et al. (2009) have established the mechanism responsible for the very pronounced Chl fluorescence quenching observed in the cyanobacterium *Synechocystis* PCC 6803 grown under iron-depleted growth conditions and suggested the same functional mechanisms in cyanobacterial IsiA and light-harvesting complex II (LHCII) of green plants. Another protein known as fluorescence recovery protein (FRC) encoded by the *slr1964* gene has been reported in the cyanobacterium *Synechocystis* PCC 6803. FRC plays an important role in dislocating the active red orange carotenoid protein from the phycobilisomes and also inactivating the free active red orange carotenoid into inactive orange form (Kirilovsky and Kerfeld 2012). Overall, the photosynthetic organisms have devised a general and universal mechanism of energy dissipation where by carotenoids act as energy sinks by accepting energy and dissipating it as heat.

The high light-inducible proteins (HLIPs) encoded by *hli* gene, also designated as small CAB-like proteins (SCPs) have also been reported in cyanobacteria to play an important role in photoprotection by energy-dissipation mechanism (Fig. 6). These proteins show similarity to the light-harvesting chlorophyll *a/b*-binding proteins of plants. Yao et al. (2007) showed the presence of small CAB-like proteins associated with PSII in *Synechocystis* PCC 6803. He et al. (2001) were able to express four *hli* genes under high-light stress, low temperature and nutrient starvation in the cyanobacterium *Synechocystis* PCC 6803. Moreover, the quadruple *hli* mutant was incompetent to carry on photosynthetic activity and growth under high light, signifying that HLIPs may possibly play a crucial role in photoprotection (He et al. 2001) by dissipating the excess absorbed energy (Havaux et al. 2003).

Role of MAAs and Scytonemin in Photoprotection

In response to intense solar radiation, cyanobacteria have evolved certain UV absorbing/screening compounds such as mycosporine-like amino acids (MAAs), scytonemin (SCY), and certain other compounds to counteract the toxicity of UVR (Cockell and Knowland 1999; Sinha and Häder 2008; Singh et al. 2010; Rastogi et al. 2010a). MAAs are intracellular, small (<400 Da), colorless and water-soluble secondary metabolites, consisting of cyclohexenone or cyclohexenimine chromophores conjugated with the nitrogen substituent of amino acids or its imino alcohol (Singh et al. 2008). They directly or indirectly absorb the energy of solar radiation and protect organisms exposed to enhanced solar UVR. As a result of UV absorption maximum between 310 and 362 nm, high molar extinction coefficients ($\varepsilon = 28{,}100–50{,}000\ M^{-1}\ cm^{-1}$), capability to dissipate absorbed radiation efficiently as heat without producing ROS, photostability and resistance to several abiotic stressors, MAAs are favoured as efficient photoprotective compounds (Conde et al. 2000; 2007; Gröniger and Häder 2000; Sinha et al. 2000; Whitehead and Hedges 2005). Presently, more than 21 MAAs have been reported from taxonomically diverse organism from terrestrial, freshwater and marine habitats (Sinha et al. 2007; Rastogi et al. 2010a). Some of the common MAAs found in cyanobacteria have been represented in Figure 7. It has been found that MAAs provides protection from UVR not only to their producers but also to primary and secondary consumers through the food chain (Helbling et al. 2002). Induction of MAAs (Fig. 8) under UV-B radiation (Sinha et al. 2001; 2003 Singh et al. 2008; Rastogi et al. 2010d), as well as under the presence of a special photoreceptor required for the induction of MAAs synthesis (Portwich and Garcia-Pichel 2000) are considered as strong indications in favour of UV-B sunscreens function of MAAs. In addition to their role as UV-absorbing compounds, some MAAs also have antioxidant activity with possible applications in the cosmetic and pharmaceutical industries (Torres et al. 2006; Rastogi and Sinha 2009).

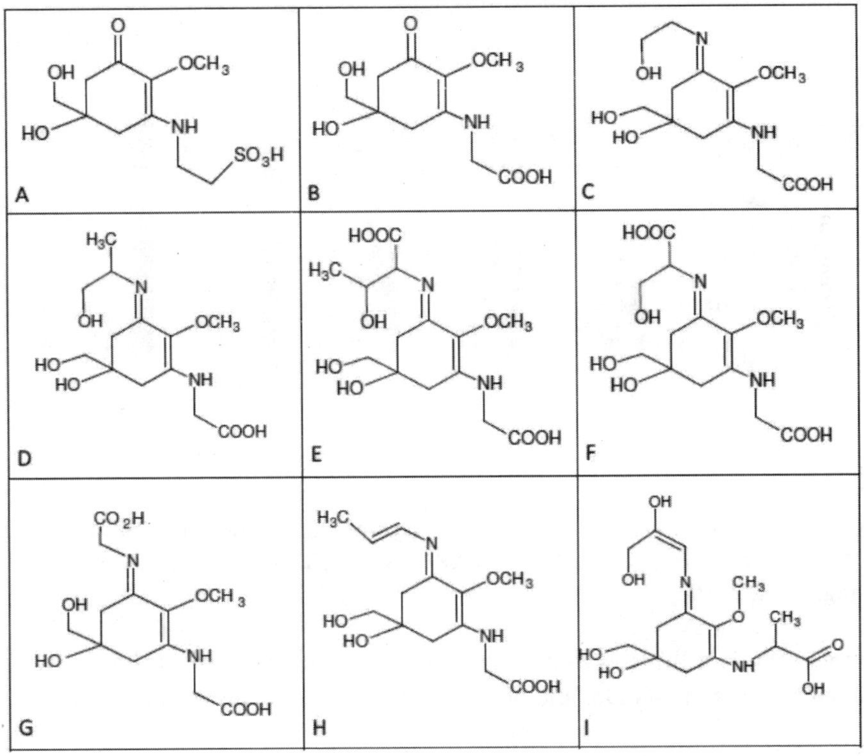

Fig.7. Chemical structure of some common MAAs reported in cyanobacteria. A- Mycosporine-tau(λ_{max}: 309 nm), B- Mycosporine-glycine (λ_{max}: 310 nm), C- Asterina-330 (λ_{max}: 330 nm), D- Palythinol (λ_{max}: 332 nm), E- Porphyra-334 (λ_{max}: 334 nm), F- Shinorine (λ_{max}: 334 nm), G- Mycosporine-2-glycine (λ_{max}: 334 nm), H- Palythene (λ_{max}: 360 nm), I- Euhalothece-362 (λ_{max}: 360 nm).

Recently, de La Coba et al. (2009) reported the antioxidant activity of P-334 and shinorine from *Porphyra* and other red algae. Furthermore, the antioxidant activities of the MAAs, glycine and usujilene have been reported to inhibit lipid peroxidation in aqueous extracts of marine organisms and to scavenge $^1O_2$ generated from certain endogenous photosensitizers (Nakayama et al. 1999; Suh et al. 2003).

In contrast to MAAs, SCY has been restricted to cyanobacterial cells. This pigment was first reported by Nägeli (1849) in some terrestrial cyanobacteria and later termed SCY. It is a yellow-brown lipid soluble pigment located in the extracellular polysaccharide sheath of some cyanobacteria (Sinha and Häder, 2008), and absorbs the UVA irradiation (maximum absorbance at 386 nm) before it reaches the cell (Garcia-Pichel et al. 1992).

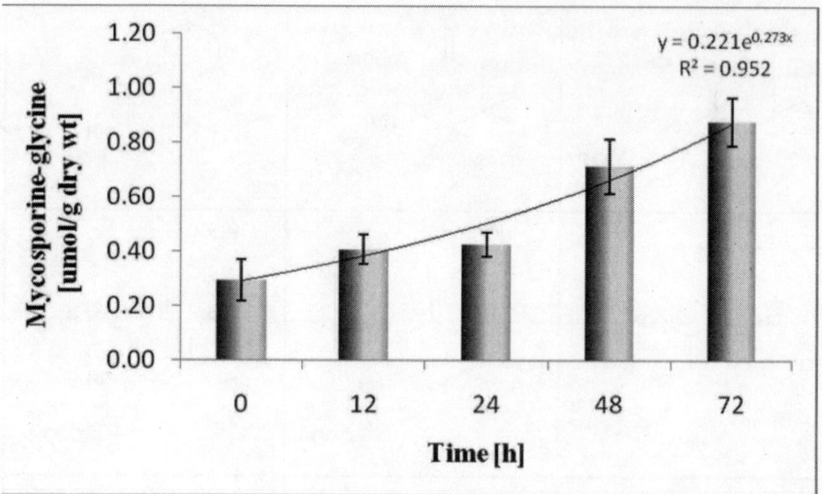

Fig. 8. Induction of MAA, mycosporine glycine concentration in *Scytonema* sp. HKAR-3 after successive intervals of PAR+UV-A+UV-B irradiation (Modified from Rastogi et al., 2010d).

Scytonemin is a dimer composed of indolic and phenolic subunits with a molecular mass of 544 Da; and occurs in both oxidised and reduced form (Fig. 9). The linkage between two subunits in SCY is an olefinic carbon atom that is unique among natural products.

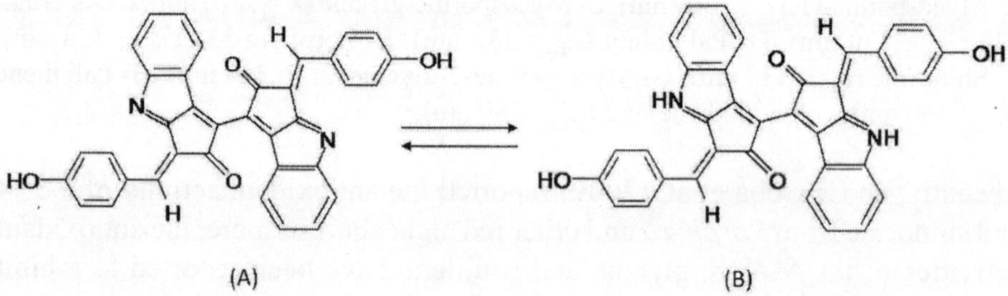

Fig. 9. Chemical structure of scytonemin found in both oxidised (A) and reduced form (B).

Three new derivatives of SCY pigments such as dimethoxyscytonemin, tetramethoxyscytonemin and scytonin (Fig.10) have been derived from the organic extracts of *Scytonema* sp. isolated from Mitaraka Inselberg, French Guyana (Bultel-Poncé et al. 2004). SCY is highly stable and has strong antioxidant activity (Takamatsu et al. 2003); performs its screening activity without any further metabolic investment even after prolonged physiological inactivity; hence, it can be a suitable candidate for cyanobacteria to defend against UVR when other protective mechanisms would be ineffective (Brenowitz and Castenholz 1997; Sinha et al. 1999; Richter et al. 2006; Fleming and Castenholz 2007).

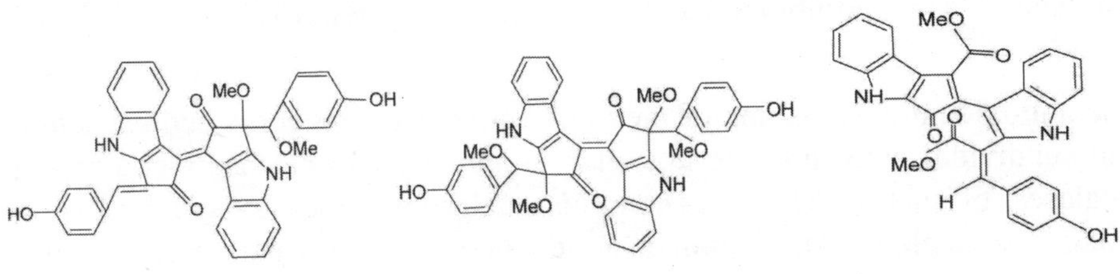

Fig. 10. Structure of scytonemin derivatives. Dimethoxyscytonemin (A), tetramethoxyscytonemin (B), and scytonin (C).

Antioxidant Systems

Photoinhibition induced by ROS is supposed to be higher in cyanobacteria than in other phytoplanktonic organisms (Drabkova et al. 2007). Consequently, cyanobacteria evolved a number of non-enzymatic and enzymatic antioxidant defense systems to minimize the oxidative damage caused by potentially harmful ROS produced by high irradiances and UVR (He and Häder 2002c; Pospíšil 2009; Latifi et al. 2009). The major non-enzymatic antioxidant system includes carotenoids, vitamin C (ascorbic acid: ASA), vitamin E (α-tocopherols: α-TOC) and reduced glutathione, whereas superoxide dismutase (SOD), catalase (CAT), glutathione peroxidase (GPX) as well as the enzymes involved in the ascorbate-glutathione cycle, such as ascorbate peroxidase (APX), monodehydroascorbate reductase (MDHAR), dehydroascorbate reductase (DHAR) and glutathione reductase (GR) are the major enzymatic antioxidants in photosynthetic organisms including cyanobacteria (Sies 1991; He and Häder 2002b; Mallick and Mohn 2000; Wolfe-Simon et al. 2005; Ishikawa and Shigeoka 2008; Bernroitner et al. 2009).

The SOD is one of the most important classes of antioxidant that exists in four different metalloforms such as FeSOD, MnSOD, Cu/ZnSOD and NiSOD, and protects various cellular proteins against oxidative stress (Lesser and Stochaj 1990; Fridovich 1997). It has been observed that Cu/Zn form of SOD is rare among cyanobacteria (Priya et al. 2007). Different forms of SOD have been reported in diverse unicellular cyanobacteria such as *Prochlorococcus* sp. (NiSOD), *Cyanothece* CCY 0110 (MnSOD), *Synechococcus* sp. (FeSOD) as well as filamentous cyanobacteria such as *Anabaena* sp. (FeSOD and MnSOD) and *Trichodesmium* sp. (MnSOD and NiSOD) (Wolfe-Simon et al. 2005). Several studies reported the implication of SODs in protective processes in cyanobacteria (Zhao et al. 2007a; b; Shirkly et al. 2000). SOD scavenges superoxide radicals and converts them to H_2O_2 which is further converted to H_2O and O_2 via a combined catalase-peroxidase system (Tel-Or et al. 1986, Scandalios 2005).

The antioxidant, CAT is one of the most active catalysts produced by nature and subdivided into three groups such as monofunctional haem-containing catalases, bifunctional haem-containing catalase-peroxidases and nonhaem-manganese catalases. Analyses of genome sequence for the catalase-orthologues have revealed their presence in 20 cyanobacterial genomes (Latifi et al. 2009). Tichy and Verma (1999) proposed the protective role of catalase-peroxidase in the cyanobacterium *Synechocystis* sp. PCC 6803 against environmental H_2O_2 production. CAT decomposes H_2O_2 at an extremely rapid rate, corresponding to a catalytic centre activity of about 10^7 min^{-1} (Scandalios 2005). It can degrade H_2O_2 without consuming cellular reducing equivalents and may provide the cell with a very energy efficient mechanism to remove H_2O_2. It has been proposed that CAT may be exclusively suited to regulate the homeostasis of H_2O_2 in the cell (Scandalios 2005).

Another component, glutathione is a small and ubiquitous molecule that plays an important role in several cellular processes besides its role as an antioxidant. Very little is known about the roles of glutathione in photosynthetic organisms including cyanobacteria. Recently, Cameron and Pakrasi (2010) have investigated the role of glutathione in acclimation to environmental and redox perturbations in the cyanobacterium *Synechocystis* sp. PCC 6803. Glutathione protects thiol groups in various enzymes and is also involved in α-TOC and ascorbate regeneration. Glutathione reductase (GR) is considered as a key enzyme in the glutathione-ascorbate cycle, which functions in peroxide scavenging and protection against other oxidative processes in cyanobacteria (Foyer and Halliwell 1976; Serrano et al. 1984; Karni et al. 1984). Exogenous addition of ASA and *N*-acetyl-L-cysteine (NAC) was found to reduce chlorophyll bleaching, damage to the photosynthetic apparatus, lipid peroxidation and DNA strand breaks under UVR-induced oxidative damage in the cyanobacterium *Anabaena* sp. (He and Häder 2002a). Wang et al. (2007, 2008) also reported the involvement of the antioxidant systems such as NAC and ASA in protection of the cyanobacterium *Nostoc* sp. against UV-B radiation. Carotenoids may also augment α-TOC in scavenging peroxy radicals (Burton and Ingold 1984). Tocopherol, a lipid-soluble organic molecule is exclusively synthesized by the oxygenic phototrophs such as plants, green algae, and some cyanobacteria. Recently, Yang et al. (2008) have reported that α-TOC is essential for acquired chill-light tolerance in the cyanobacterium *Synechocystis* sp. PCC 6803. Another group of compounds, polyamines, have also been implicated as free radical scavengers. Polyamines are polycationic molecules found in all living organisms including cyanobacteria. The most commonly existing polyamines inside the cells are the diamine putrescine, the triamine spermidine, the tetramine spermine, where spermidine represents the major polyamine in cyanobacteria (Jantaro et al. 2003; Incharoensakdi et al. 2010) (Fig.11).

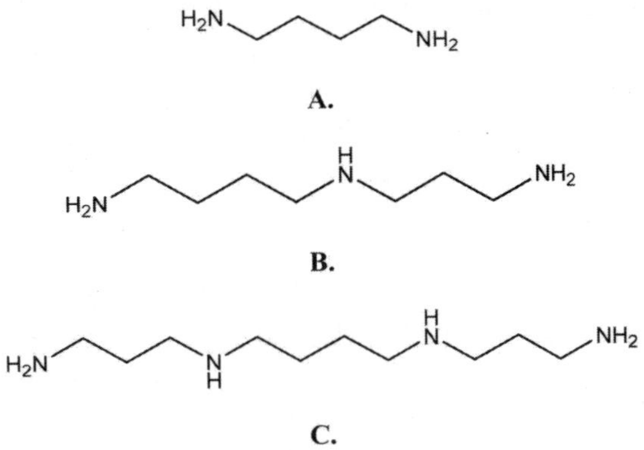

Fig. 11: Structure of some common polyamines in cyanobacteria. Putrescine (A), Spermidine (B), and Spermine (C).

Spermine has been shown to afford the protection of ROS-induced DNA damage (Ha et al. 1998). This was achieved by spermine acting directly as a free radical scavenger. Recently, a study in *Synechocystis* PCC 6803 showed that UVR-induced cell damage was due to a decrease in spermidine content of *Synechocystis* cells (Jantaro et al. 2011). Overall, the antioxidant systems play an indispensible role in mitigating the UV-induced oxidative stress in the photosynthetic organisms including cyanobacteria.

Conclusion

The exciting field of UV-induced oxidative stress in cyanobacteria is only in primitive stage. However, several photosynthetic researches are being carried out on cyanobacteria as model organisms to understand the basic mechanisms of ROS generation and photo-induced oxidative stress in photosynthetic organisms. Absorption of strong light or solar UVR leads to intracellular production of various reactive substances including ROS in photosynthetic organisms (Latifi et al. 2009; Rastogi et al. 2010c). ROS-induced oxidative stress has been found as a root source of several harmful effects on cyanobacteria leading to a number of chronic and acute physiological as well as biochemical alterations. Since, the exact mechanisms of ROS mediated response is still not very clear, future research should be dedicated towards elucidating the biological effects and the specific targets of each ROS species in cyanobacteria and the defined physiological responses that they produce. In response to diverse environmental stresses, cyanobacteria have devised a number of defense mechanisms to combat the direct or indirect damaging effects of UVR (Imlay 2003). OCP as well as HLIPs was found to play a decisive role in photoprotection through dissipating the excess light energy in the form of heat. However, the exact mechanisms of heat dissipation by means of OCP and HLIPs are still to be worked out. Certain UV-absorbing/screening compounds such as MAAs and SCY also provide protection from UVR (Sinha and Häder 2008; Rastogi et al. 2010a). Cyanobacteria evolved a number of enzymatic and non-enzymatic antioxidant defense systems to minimize the oxidative damage caused by ROS under high light and UVR. In the last few years several efforts have already been made in detection of ROS-scavenging enzymes, however more effort should be made to a great extent in order to elucidate the precise role of these antioxidants in vivo to determine potentiality of the particular enzymes to detoxify a range of damaging oxidants.

Acknowledgement

Rajesh P. Rastogi is thankful to the Graduate School and Faculty of Science, Chulalongkorn University for providing financial assistance in the form of Rachadapisaeksompote endowment fund Post-Doctoral fellowship.

References

Alscher RG, Donahue JL and Cramer CL (1997) Reactive oxygen species and antioxidants: relationships in green cells. Physiol Plant 100: 224-233.

Andrizhiyevskaya EG, Schwabe TME, Germano M, D'Haene S, Kruip J, van Grondelle R and Dekker JP (2002) Spectroscopic properties of PSI-*IsiA* supercomplexes from the cyanobacterium *Synechococcus* PCC 7942. Biochim Biophys Acta 1556: 265-272.

Badger MR, von Caemmerer S, Ruuska S and Nakano H (2000) Electron flow to oxygen in higher plants and algae: rates and control of direct photoreduction (Mehler reaction) and rubsico oxygenase. Philos. Trans R Soc. Lond B; Biol Sci 355: 1433-1446.

Berera R, van Stokkum IHM, d'Haene S, Kennis JTM, van Grondelle R, and Jan P. Dekke JP (2009) A mechanism of energy dissipation in cyanobacteria. Biophys J 96: 2261-2267.

Berera R., Herrero C, van Stokkum IHM, Vengris M, Kodis G, Palacios RE, van Amerongen H, van Grondelle R, Gust D, Moore TA, Moore AL and Kennis JTM (2006) A simple artificial light-harvesting dyad as a model for excess energy dissipation in oxygenic photosynthesis. Proc Natl Acad Sci USA 103: 5343-5348.

Bernroitner M, Zamocky M, Furtmüller PG, Peschek GA and Obinger C (2009) Occurrence, phylogeny, structure, and function of catalases and peroxidases in cyanobacteria. J Exp Bot 60: 423-440.

Blot N, Mella-Flores D, Six C, Corguillé GL, Boutte C, Peyrat A, Monnier A, Ratin M, Gourvil P, Campbell DA and Garczarek L (2011) Light history influences the response of the marine cyanobacterium *Synechococcus* sp. WH7803 to oxidative stress. Plant Physiol 156: 1934-1954.

Boekema EJ, Hifney A, Yakushevska AE, Piotrowski M, Keegstra W, Berry S, Michel KP, Pistorius EK and Kruip J (2001) A giant chlorophyll-protein complex induced by iron deficiency in cyanobacteria. Nature 412: 745-748.

Brenowitz S and Castenholz RW (1997) Long-term effects of UV and visible irradiance on natural populations of a scytonemin-containing cyanobacterium (*Calothrix* sp.). FEMS Microbiol Ecol 24: 343-352.

Bultel-Poncé V, Felix-Theodore F, Sarthon C, Ponge J-F and Bodo B (2004) New pigments from the terrestrial cyanobacterium *Scytonema* sp. collected on the Mitaraka Inselberg, French Guyana. J Nat Prod 67: 678-81.

Burton GW and Ingold KU (1984) β-Carotene: an unusual type of lipid antioxidant. Science 224: 569-573.

Cadet J, Sage E and Douki T (2005) Ultraviolet radiation mediated damage to cellular DNA. Mut Res 571: 1-2.

Cameron JC and Pakrasi HB (2010) Essential role of glutathione in acclimation to environmental and redox perturbations in the cyanobacterium *Synechocystis* sp. PCC 6803[1[W][OA]]. Plant Physiol 154: 1672-1685.

Cardozo KHM, Guaratini T, Barros MP, Falcão VR, Tonon AP, Lopes NP, Campos S, Torres MA, Souza AO, Colepicolo P and Pinto E (2007) Metabolites from algae with economical impact. Comp Biochem Physiol C Toxicol Pharmacol 146: 60-78.

Cockell CS and Knowland J (1999) Ultraviolet radiation screening compounds. Biol Rev 74: 311-345.

Conde FR, Churio MS and Previtali CM (2000) The photoprotector mechanism of mycosporine-like amino acids. excited-state properties and photostability of porphyra-334 in aqueous solution. J Photochem Photobiol B Biol 56: 139-144.

Conde FR, Churio MS and Previtali CM (2007) Experimental study of the excited-state properties and photostability of the mycosporine-like amino acid palythine in aqueous solution. Photochem Photobiol Sci 6: 669-674.

De la Coba F, Aguilera J, Figueroa FL, de Galvez MV and Herrera E (2009) Antioxidant activity of mycosporine-like amino acids isolated from three red macroalgae and one marine lichen. J Appl Phycol 21: 161-169.

Demmig-Adams B and Adams WW (2006) Photoprotection in an ecological context: the remarkable complexity of thermal energy dissipation. New Phytol 172: 11-21.

Demmig-Adams B and Adams WW (1992) Photoprotection and other responses of plants to high light stress. Annu Rev Plant Physiol Plant Mol Biol 43: 599-626.

Doetsch PW, Zastawny TH, Martin AM and Dizdaroglu M (1995) Monomeric base damage products from adenine, guanine and thymine induced by exposure of DNA to ultraviolet radiation. Biochemistry 34: 737-742.

Drabkova M, Matthijs HCP, Admiraal W and Marsalek B (2007) Selective effects of H_2O_2 on cyanobacterial photosynthesis. Photosynthetica 45: 363-369.

Fischer WF (2008) Life before the rise of oxygen. Nature 455: 1051-1052.

Fleming ED and Castenholz RW (2007) Effects of periodic desiccation on the synthesis of the UV-screening compound, scytonemin, in cyanobacteria. Environ Microbiol 9: 1448-1455.

Foyer CH and Halliwell B (1976) The presence of glutathione and glutathione reductase in chloroplasts: A proposed role in ascorbic acid metabolism. Planta 133: 21-25.

Fridovich I (1997) Superoxide anion radiacal (O_2^-), superoxide dismutases and related matters. J Biol Chem 272: 18515-18517.

Gao K, Yu H and Brown MT (2007) Solar PAR and UV radiation affects the physiology and morphology of the cyanobacterium *Anabaena* sp. PCC 7120. J Photochem Photobiol B Biol 89: 117-124.

Garcia-Pichel F, Sherry ND and Castenholz RW (1992) Evidence for an ultraviolet sunscreen role of the extracellular pigment scytonemin in the terrestrial cyanobacterium *Chlorogloeopsis* sp. Photochem Photobiol 56: 17-23.

Gorbunov MY, Kuzminov FI, Fadeev VV, Kim JD and Falkowski PG (2011) A kinetic model of non-photochemical quenching in cyanobacteria. Biochim Biophys Acta 1807: 1591-1599.

Gorman AA and Rodgers MA (1992) Current perspectives of singlet oxygen detection in biological environments. J Photochem Photobiol B Biol 14: 159-176.

Gröniger A and Häder D-P (2000) Stability of mycosporine-like amino acids. Recent Res Devel Photochem Photobiol 4: 247-252.

Ha HC, Sirisoma NS, Kuppusamy P, Zweir JL, Woster PM and Casero Jr RA (1998) The natural polyamine spermine functions directly as a free radical scavenger. Proc Natl Acad Sci USA 95: 11140-11145.

Häder D-P, Helbling EW, Williamson CE and Worrest RC (2011) Effects of UV radiation on aquatic ecosystems and interactions with climate change. Photochem Photobiol Sci 10: 242-260.

Hall DB, Holmlin RE and Barton JK (1996) Oxidative DNA damage through long-range electron transfer. Nature 382: 731-735.

Halliwell B and Gutteridge J (2007) Free Radicals in Biology and Medicine, Oxford University Press, Oxford, UK, 4th edition.

Hargreaves A, Taiwo FA, Duggan O, Kirk SH, Ahmad SI (2007) Near-ultraviolet photolysis of b-phenylpyruvic acid generates free radicals and results in DNA damage. J Photochem Photobiol B Biol 89: 110-116.

Havaux M, Guedeney G, He Q and Grossman AR (2003) Elimination of high-light-inducible polypeptides related to eukaryotic chlorophyll *a/b*-binding proteins results in aberrant photoacclimation in *Synechocystis* PCC 6803. Biochim Biophys Acta 1557: 21-33.

Havaux M, Guedeney G, Hagemann M, Yeremenko N, Matthijs HCP and Jeanjean R (2005) The chlorophyll-binding protein IsiA is inducible by high light and protects the cyanobacterium *Synechocystis* PCC6803 from photooxidative stress. FEBS Lett 579: 2289-2293.

He Q, Dolganov N, Bjorkman O and Grossman AR (2001) The High light-inducible polypeptides in *Synechocystis* PCC 6803. Expression and function in high light. J Biol Chem 276: 306-314.

He Y-Y and Häder D-P (2002a) UV-B-induced formation of reactive oxygen species and oxidative damage of the cyanobacterium *Anabaena* sp.: protective effects of ascorbic acid and N-acetyl-L-cysteine. J Photochem Photobiol B Biol 66: 115-124.

He Y-Y and Häder D-P (2002b) Reactive oxygen species and UV-B: effect on cyanobacteria. Photochem Photobiol Sci 1: 729-736.

He Y-Y and Häder D-P (2002c) Involvement of reactive oxygen species in the UV-B damage to the cyanobacterium *Anabaena* sp. J Photochem Photobiol B Biol 66: 73-80.

He Y-Y, Klisch M and Häder D-P (2002) Adaptation of cyanobacteria to UV-B stress correlated with oxidative stress and oxidative damage. Photochem Photobiol 76: 188-196.

Helbling EW, Menchi CF and Villafañe VE (2002) Bioaccumulation and role of UV-absorbing compounds in two marine crustacean species from Patagonia, Argentina. Photochem Photobiol Sci 1: 820-825.

Helman Y, Tchernov D, Reinhold L, Shibata M, Ogawa T, Schwarz R Ohad I and Kaplan A (2003) Genes encoding A-type flavoproteins are essential for photoreduction of O_2 in cyanobacteria. Curr Biol 13: 230-235.

Holt NE, Zigmantas D, Valkunas L, Li XP, Niyogi KK and Fleming GR (2005) Carotenoid cation formation and the regulation of photosynthetic light harvesting. Science 307: 433-436.

Horton P, Ruban AV and Walters RG (1996) Regulation of light harvesting in green plants. Annu Rev Plant Physiol Plant Mol Biol 47: 655-684.

Imlay JA (2003) Pathways of oxidative damage. Annu Rev Microbiol 57: 395-418.

Incharoensakdi A, Jantaro S, Raksajit W and Maenpaa P (2010) Polyamines in cyanobacteria: biosynthesis, transport and abiotic stress response. In: *Current Research, Technology and Education Topics in Applied Microbiology and Microbial Biotechnology* (Ed. Mendez-Vilas A), Microbiology Book Series No. 2, Vol. 1, Formatex, Spain. Pp. 23-32.

Ishikawa T and Shigeoka S (2008) Recent advances in ascorbate biosynthesis and the physiological significance of ascorbate peroxidase in photosynthesizing organisms. Biosci Biotechnol Biochem 72: 1143-1154.

Ivanov AG, Miskiewicz E, Clarke AK, Greenberg BM and Huner NP (2000) Protection of photosystem II against UV-A and UV-B radiation in the cyanobacterium *Plectonema boryanum*: the role of growth temperature and growth irradiance. Photochem Photobiol 72: 772-779.

Jantaro S, Maenpaa P, Mulo P and Incharoensakdi A (2003) Content and biosynthesis of polyamines in salt and osmotically stressed cells of *Synechocystis* sp. PCC 6803. FEMS Microbiol Lett 228: 129-135.

Jantaro S, Pothipongsa A, Khanthasuwan S and Incharoensakdi A (2011) Short-term UV-B and UV-C radiations preferentially decrease spermidine contents and arginine decarboxylase transcript levels of *Synechocystis* sp. PCC 6803. Curr Microbiol 62: 420-426.

Jiang Y, Rabbi M, Kim M, Ke C, Lee W, Clark RL, Mieczkowski PA and Marszalek PE (2009) UVA generates pyrimidine dimers in DNA directly. Biophys J 96: 1151-1158.

Joshua S, Bailey S, Mann NH and Mullineaux CW (2005) Involvement of phycobilisome diffusion in energy quenching in cyanobacteria. Plant Physiol 138: 1577–1585.

Kamer I, Sarig R, Zaltsman Y, Niv H, Oberkovitz G, Regev L, Haimovich G, Lerenthal Y, Marcellus RC and Gross A (2005) Proapoptotic BID is an ATM effector in the DNA-damage response. Cell 122: 593-603.

Karni L, Moss SJ and Tel-Or E (1984) Glutathione reductase activity in heterocysts and vegetative cells of *Nostoc muscorum*. Arch Microbiol 140: 215-217.

Kerfeld CA (2004) Water-soluble carotenoid proteins of cyanobacteria. Arch Biochem Biophys 430: 2-9.

Khetkorn W, Lindblad P and Incharoensakdi A (2010) Enhanced biohydrogen production by the N_2-fixing cyanobacterium *Anabaena siamensis* strain TISTR 8012. Inter J Hydro Ener 35: 12767-12776.

Kirilovsky D (2007) Photoprotection in cyanobacteria: the orange carotenoid protein (OCP)-related non-photochemical-quenching mechanism. Photosynth Res 93: 7-16.

Kirilovsky D and Kerfeld CA (2012) The orange carotenoid protein in photoprotection of photosystem II in cyanobacteria. Biochim. Biophysic Acta 1817: 158-166.

Kouril R, Arteni AA, Lax J, Yeremenko N, D'Haene S, Rögner M, Matthijs HCP, Dekker JP and Boekema EJ (2005) Structure and functional role of supercomplexes of IsiA and photosystem I in cyanobacterial photosynthesis. FEBS Lett 579: 3253-3257.Affiliations

Kramlich JC and Linak WP (1994) Nitrous oxide behavior in the atmosphere, and in combustion and industrial systems. Prog Energy Combust Sci 20: 149-202.

Kumar A, Tyagi MB, Jha PN, Srinivas G and Singh A (2003) Inactivation of cyanobacterial nitrogenase after exposure to ultraviolet-B radiation. Curr Microbiol 46: 380-384.

Latifi A, Ruiz M and Zhang C-C (2009) Oxidative stress in cyanobacteria. FEMS Microbiol Rev 33: 258-278.

Lesser MP (2008) Effects of ultraviolet radiation on productivity and nitrogen fixation in the cyanobacterium, *Anabaena* sp. (Newton's strain). Hydrobiologia 598: 1-9.

Lesser MP and Stochaj WR (1990) Photoadaptation and protection against active forms of oxygen in the symbiotic prokaryote *Prochloron* sp. and its ascidian host. Appl Environ Microbiol 56: 1530-1535.

Mallick N and Mohn FH (2000) Reactive oxygen species: response of algal cells. J Plant Physiol 157: 183-193.

McIntyre TM, Zimmerman GA and Prescott SM (1999) Biologically active oxidized phospholipids. J Biol Chem 274: 25189-25192.

Mehler AH (1951) Studies on reactions of illuminated chloroplasts. I. Mechanisms of the reduction of oxygen and other hill reagents. Arch Biochem Biophys 33: 65-77.

Mercer J, Mahmoudi M and Bennett M (2007) DNA damage, p53, apoptosis and vascular disease. Mut Res 621: 75-86.

Muramatsu M and Hihara Y (2012) Acclimation to high-light conditions in cyanobacteria: from gene expression to physiological responses. J Plant Res 125: 11-39.

Nägeli C (1849) Gattungen einzelliger Algen, physiologisch und systematisch bearbeitet Neue Denkschrift. Allg Schweiz Nat Ges 10: 1-138.

Nakayama R, Tamura Y, Kikuzaki H and Nakatani N (1999) Antioxidant effect of the constituents of susabinori (*Porphyra yezoensis*). J Ame Oil Chem Soc 76: 649-653.

Nishiyama Y, Allakhverdiev SI, Murata N (2006) A new paradigm for the action of reactive oxygen species in the photoinhibition of photosystem II. Biochim Biophys Acta - Bioenergetics 1757: 742-749.

Nishiyama Y, Allakhverdiev SI and Murata N (2011) Protein synthesis is the primary target of reactive oxygen species in the photoinhibition of photosystem II. Physiol Plantarum 142: 35-46.

Patterson CO and Myers J (1973) Photosynthesis production of hydrogen peroxide by *Anacystis nidulans*. Plant Physiol 51: 104-109.

Polívka T and Sundström V (2004) Ultrafast dynamics of carotenoid excited states. From solution to natural and artificial systems. Chem Rev 104: 2021-2071.

Portwich A and Garcia-Pichel F (2000) A novel prokaryotic UVB photoreceptor in the cyanobacterium *Chlorogloeopsis* PCC 6912. Photochem Photobiol 71: 493-498.

Pospíšil P (2009) Production of reactive oxygen species by photosystem II. Biochim Biophys Acta 1787: 1151-1160.

Priya B, Premanandh J, Dhanalakshmi RT, Seethalakshmi T, Uma L, Prabaharan D and Subramanian G (2007) Comparative analysis of cyanobacterial superoxide dismutases to discriminate canonical forms. BMC Genomics 8: 435.

Rastogi RP and Sinha RP (2009) Biotechnological and industrial significance of cyanobacterial secondary metabolites. Biotechnol Adv 27: 521-539.

Rastogi RP, Richa, Sinha RP, Singh SP and Häder D-P (2010a) Photoprotective compounds from marine organisms. J Ind Microbiol Biotechnol 37: 537-558.

Rastogi RP, Richa, Kumar A, Tyagi MB and Sinha RP (2010b) MolecularMechanisms of ultraviolet radiation-induced DNA damage and repair. J Nucleic Acids 2010, Article ID 592980, 32, doi:10.4061/2010/592980.

Rastogi RP, Singh SP, Häder D-P, and Sinha RP (2010c) Detection of reactive oxygen species (ROS) by the oxidantsensing probe 22 , 72 -dichlorodihydrofluorescein diacetate in the cyanobacterium *Anabaena variabilis* PCC 7937. Biochem Biophys Res Commun 397: 603-607.

Rastogi RP, Richa, Singh SP, Häder D-P and Sinha RP (2010d) Mycosporine-like amino acids profile and their activity under PAR and UVR in a hot-spring cyanobacterium *Scytonema* sp. HKAR-3. Aus J Bot 58: 286-293.

Rastogi RP and Sinha RP (2011) Solar ultraviolet radiation-induced DNA damage and protection/repair strategies in cyanobacteria. Inter. J. Pharma and Bio Sci. 2: B271-B288.

Rastogi RP, Richa, and Sinha RP (2011a) Biotechnological exploitations of cyanobacteria. In: *Plant Genome* (Eds. Roy BK, Chaudhary BR and Sinha RP), Narosa Publishing House, New Delhi. pp. 43-57.

Rastogi RP, Singh SP, Häder D-P and Sinha RP (2011b) Ultraviolet-B-induced DNA damage and photorepair in the cyanobacterium *Anabaena variabilis* PCC 7937. Environ Exp Bot 74: 280-288.

Richter PR, Sinha RP and Häder DP (2006). Scytonemin-rich epilithic cyanobacteria survive acetone treatment Res. Devel. Photochem Photobiol 6: 107-119.

Rinalducci S, Pedersen JZ and Zolla L (2008) Generation of reactive oxygen species upon strong visible light irradiation of isolated phycobilisomes from *Synechocystis* PCC 6803. Biochim Biophys Acta 1777: 417–424.

Ritter D and Yopp JH (1993) Plasma membrane lipid composition of the halophilic cyanobacterium *Aphanothece halophytica.* Arch Microbiol 159: 435-439.

Scandalios JG (1997) Oxidative stress and the molecular biology of antioxidant defenses. Cold Spring Harbor Laboratory Press, Plainview, NY, USA.

Scandalios JG (2005) Oxidative stress: molecular perception and transduction of signals triggering antioxidant gene defences. Brazilian J Med Biol Res 38: 995-1014.

Serrano A, Rivas J and Losada M (1984) Purification and properties of glutathione reductase from the cyanobacterium *Anabaena* sp. strain 7119. J Bacteriol 158: 317-324.

Shirkly B, Kovarcik DP, Wright DJ, Wilmoth G, Prickett TF, Helm RF, Gregory EM and Potts M (2000) Active Fe-containing superoxide dismutase and abundant *sodF* mRNA in *Nostoc commune* (cyanobacteria) after years of desiccation. J Bacteriol 182: 189-197.

Sicora CI, Appleton SE, Brown CM, Chung J, Chandler J, Cockshutt AM, Vass I and Campbell DA (2006) Cyanobacterial psbA families in *Anabaena* and *Synechocystis* encode trace, constitutive and UVB-induced D1 isoforms. Biochim Biophys 1757: 47-56.

Sies H (1991) Oxidative Stress: Oxidants and Antioxidants. London, UK: Academic Press.

Singh SC, Sinha RP and Häder D-P (2002) Role of lipids and fatty acids in stress tolerance in cyanobacteria. Acta Protozool 41: 297-308.

Singh SP, Kumari S, Rastogi RP, Singh KL and Sinha RP (2008) Mycosporine-like amino acids (MAAs): chemical structure, biosynthesis and significance as UV-absorbing/screening compounds. Ind J Exp Biol 46: 7-17.

Singh SP, Häder D-P and Sinha RP (2010) Cyanobacteria and ultraviolet radiation (UVR) stress: mitigation strategies. Ageing Res Rev 9: 79-90.

Sinha RP and Häder D-P (2002) Life under solar UV radiation in aquatic organisms. Adv Space Res 30: 1547-1556.

Sinha RP and Häder D-P (2008) UV-protectants in cyanobacteria. Plant Sci 174: 278–289.

Sinha RP, Ambasht NK, Sinha JP, Klisch M and Häder D-P (2003) UVB-induced synthesis of mycosporine-like amino acids in three strains of *Nodularia* (cyanobacteria). J Photochem Photobiol B Biol 71: 51-58.

Sinha RP, Klisch M, Vaishampayan A and Häder D-P (1999). Biochemical and spectroscopic characterization of the cyanobacterium *Lyngbya* sp. inhabiting Mango (*Mangifera indica*) trees: Presence of an ultraviolet-absorbing pigment, Scytonemin. Acta Protozool 38: 291-298.

Sinha RP, Klisch M, Gröniger A and Häder D-P (2000) Mycosporine-like amino acids in the marine red alga *Gracilaria cornea*-effects of UV and heat. Environ Exp Bot 43: 33-43.

Sinha RP, Klisch M, Helbling EW and Häder D-P (2001) Induction of mycosporine-like amino acids (MAAs) in cyanobacteria by solar ultraviolet-B radiation. J Photochem Photobiol B Biol 60: 129-135.

Sinha RP, Kumar A, Tyagi MB and Häder D-P (2005) Ultraviolet-B-induced destruction of phycobiliproteins in cyanobacteria. Physiol Mol Biol Plants 11: 313–319.

Sinha RP, Kumari S and Rastogi RP (2008) Impacts of ultraviolet-B radiation on cyanobacteria: photoprotection and repair. J Sci Res 52: 125-142.

Sinha RP, Singh N, Kumar A, Kumar HD, Häder M and Häder D-P (1996) Effects of UV irradiation on certain physiological and biochemical processes in cyanobacteria. J Photochem Photobiol B Biol 32: 107-113.

Sinha RP, Singh N, Kumar A, Kumar HD and Häder D-P (1997) Impacts of ultraviolet-B irradiation on nitrogen-fixing cyanobacteria of rice paddy fields. J Plant Physiol 150: 188-193.

Sinha RP, Singh SP and Häder D-P (2007) Database on mycosporines and mycosporine-like amino acids (MAAs) in fungi, cyanobacteria, macroalgae, phytoplankton and animals. J Photochem Photobiol B Biol 89: 29-35.

Stanier RY and Cohen-Bazire G (1977) Phototrophic prokaryotes: the cyanobacteria. Annu Rev Microbiol 31: 225-274.

Suh HJ, Lee HW and Jung J (2003) Mycosporine glycine protects biological systems against photodynamic damage by quenching singlet oxygen with a high efficiency. Photochem Photobiol 78: 109-113.

Takamatsu S, Hodges TW, Rajbhandari I, Gerwick WH, Hamann MT and Nagle DG (2003) Marine natural products as novel antioxidant prototypes. J Nat Prod 66: 605-608.

Tel-Or E, Huflejt ME and Packer L (1986) Hydroperoxide metabolism in cyanobacteria. Arch Biochem Biophys 246: 396-402.

Tichy M and Vermaas W (1999) In Vivo role of catalase-peroxidase in *Synechocystis* sp. strain PCC 6803. J Bacteriol 181: 1875-1882.

Torres A, Enk CD, Hochberg M and Srebnik M (2006) Porphyra-334, a potential natural source for UVA protective sunscreens. Photochem Photobiol Sci 5: 432-435.

Vicente JB, Gomes CM, Wasserfallen A and Teixeira M (2002) Module fusion in an A-type flavoprotein from the cyanobacterium *Synechocystis* condenses a multiplecomponent pathway in a single polypeptide chain. Biochem Biophys Res Commun 294: 82-87.

Vincent WF and Neale PJ (2000) Mechanisms of UV damage to aquatic organisms. In: *The Effects of UV Radiation on Marine Ecosystems* (Eds. de Mora SJ, Demers S and Vernet M), Cambridge Univ. Press, Cambridge. pp. 149-176.

Vinnemeier J., Kunert A and Hagemann M (1998) Transcriptional analysis of the isiAB operon in salt-stressed cells of the cyanobacterium *Synechocystis* sp. PCC 6803. FEMS Microbiol Lett 169: 323-330.

Wang G, Hu C, Li D, Zhang D, Li X, Chen K and Liu Y (2007) The response of antioxidant systems in *Nostoc sphaeroides* against UV-B radiation and the protective effects of exogenous antioxidants. Adv Space Res 39: 1034-1042.

Wang G, Chen K, Chen L, Hu C, Zhang D and Liu Y (2008) The involvement of the antioxidant system in protection of desert cyanobacterium *Nostoc* sp. against UV-B radiation and the effects of exogenous antioxidants. Ecotoxicol Environ Safety 69: 150-157.

Whitehead K and Hedges JI (2005) Photodegradation and photosensitization of mycosporine-like amino acids. J Photochem Photobiol 80: 115-121.

Wilson A, Ajlani G, Verbavatz JM, Vass I, Kerfeld C and Kirilovsky D (2006) A soluble carotenoid protein involved in phycobilisome-related energy dissipation in cyanobacteria. Plant Cell 18: 992-1007.

Wilson A., Boulay C., Wilde A., Kerfeld C.A and Kirilovsky D (2007) Light-induced energy dissipation in iron-starved cyanobacteria: roles of OCP and IsiA proteins. Plant Cell 19: 656-672.

Wilson A., Punginelli C., Gall A., Bonetti C., Alexandre M., Routaboul J-M, Kerfeld CA, van Grondelle R, Robert B, Kennis JTM and Kirilovsky D (2008) A photoactive carotenoid protein acting as light intensity sensor. Proc Natl Acad Sci USA 105: 12075-12080.

Wolfe-Simon F, Grzebyk D, Schofield O and Falkowski PG (2005) The role and evolution of superoxide dismutases in algae. J Phycol 41: 453-465.

Wu YP and Krogmann DW (1997) The orange carotenoid protein of *Synechocystis* PCC 6803. Biochim Biophys Acta 1322: 1-7.

Yang Y, Yin C, Li W, and Xu X (2008) á-Tocopherol Is Essential for Acquired Chill-Light Tolerance in the Cyanobacterium *Synechocystis* sp. Strain PCC 6803. J Bacteriol 190: 1554-1560.

Yao D., Kieselbach T, Komenda J, Promnares K, Prieto MA, Tichy M, Vermaas W and Funk C (2007) Localization of the small CAB-like proteins in photosystem II. J Biol Chem 282: 267-276.

Yeremenko N, Kouril R, Ihalainen JA, D'Haene S, van Oosterwijk N, Andrizhiyevskaya EG, Keegstra W, Dekker HL, Hagemann M, Boekema EJ, Matthijs HC and Dekker JP (2004) Supramolecular organization and dual function of the IsiA chlorophyll-binding protein in cyanobacteria. Biochemistry 43: 10308-10313.

Zhao W, Guo Q and Zhao J (2007a) A membrane-associated Mn superoxide dismutase protects the photosynthetic apparatus and nitrogenase from oxidative damage in the cyanobacterium *Anabaena* sp. PCC 7120. Plant Cell Physiol 48: 563-572.

Zhao W, Ye Z and Zhao J (2007b) RbrA, a cyanobacterial rubrerythrin, functions as a FNR-dependent peroxidase in heterocysts in protection of nitrogenase from damage by hydrogen peroxide in *Anabaena* sp. PCC 7120. Mol Microbiol 66: 1219-1230.

Zinkel SS, Hurov KE, Ong C, Abtahi FM, Gross A and Korsmeyer SJ (2005) A role for proapoptotic BID in the DNA damage response. Cell 122: 579-591.

5

UV-absorbing Compounds in Cyanobacteria

Richa, Minu Kesheri, Vinod K. Kannaujiya and Rajeshwar P. Sinha

Abstract

The substantial loss in the shielding ozone layer has resulted in an increase in the solar ultraviolet radiation (UVR) on the Earth's surface, which is detrimental for all living organisms ranging from prokaryotes to eukaryotes, including humans. This has augmented the scientific concern and interest in searching for the photoprotective compounds from diverse organisms. Cyanobacteria, the primitive O_2-evolving photosynthetic prokaryotes are solely dependent upon the solar radiation for their energy requirement and thereby have to always face the UV-B stress. UVR affects cyanobacteria either directly or indirectly by inducing oxidative stress. In response, they have evolved the capacity to synthesize, accumulate and metabolize a variety of UV-absorbing compounds such as, mycosporine-like amino acids (MAAs), scytonemin, carotenoids, biopterin glucoside, as a strategy to reduce the direct and indirect damaging effects of UVR. MAAs forms the most common class of UV-absorbing/screening compound in cyanobacteria. Scytonemin, a predominant UV-A photoprotective pigment has been exclusively reported to be synthesized in cyanobacteria. Carotenoids are important components of the photosynthetic apparatus that serve both light-harvesting and photoprotective functions. The biosynthesis of these UV-absorbing compounds is affected by several environmental factors. This chapter focuses on various UV-absorbing compounds present in cyanobacteria, their biosynthetic pathway, photoprotective role and commercial applications.

Keywords: Ultraviolet radiation (UVR); Mycosporine-like amino acid (MAAs); Scytonemin; Carotenoids; Programmed cell death (PCD).

Introduction

The anthropogenically released atmospheric pollutants such as chlorofluorocarbons (CFCs), chlorocarbons (CCs) and organobromides (OBs) are continuously depleting the stratospheric ozone layer that shields the Earth from the deleterious shorter-wavelength solar radiation (Crutzen 1992; Lubin and Jensen 1995). As a consequence, there is an increase in ultraviolet-B (UV-B) radiation (280-315 nm) reaching onto the Earth's suface, which has aroused the scientific concern in the past few decades. Natural production of considerable amounts of reactive nitrogen species (RNS) such as nitric oxide (NO^-), peroxynitrite ($ONOO^-$) and nitrous oxide (N_2O), from unpolluted terrestrial and aquatic ecosystems or from anthropogenic sources also contribute to the depletion of the ozone layer (Kramlich and Linak 1994). The processes of ozone depletion has been reported at mid latitudes but are more pronounced in the Antarctic region where ozone levels have been reported to decline by more than 70% during late winter and early spring in the polar vortex (Smith et al. 1992). Highly energetic ultraviolet-C (100-280 nm) radiation does not reach the Earth because it is absorbed by atmospheric ozone and O_2. The incoming solar radiation consists of mainly highly energetic UV-B which is lethal for most sun-exposed organisms ranging from prokaryotes to eukaryotes, including humans (Sinha and Häder 2002; Häder et al. 2007; Sinha et al. 2008a; b; Hansson and Hylander 2009) as it is readily absorbed by the biomolecules such as nucleic acids and proteins (Häder et al. 2007) and indirectly *via* the production of reactive oxygen species (ROS). In contrast, UV-A irradiation has indirect effects via energy transfer from UV-A-stimulated chromophores to the DNA target or *via* photosensitized (chlorophylls and phycobilins can act as photosensitizers) production of ROS.

Cyanobacteria, the largest group of Gram-negative, oxygenic photoautotrophic prokaryotes have cosmopolitan distribution ranging from Arctic to Antarctic regions and are important biomass producers in both aquatic and terrestrial ecosystems (Stanier and Cohen- Bazire 1977; Häder et al. 2007). Members of cyanobacteria are valuable sources of various natural products of medicinal and industrial importance (Rastogi and Sinha 2009). In addition, these ecologically important organisms are dominant flora of the wetland soils, especially in rice-paddy field where they contribute to the fertility as a natural biofertilizers (Vaishampayan et al. 2001) by virtue of their ability to fix atmospheric nitrogen in the presence of the enzyme nitrogenase. Harvesting of solar energy to perform photosynthesis and nitrogen fixation exposes these cyanobacteria to lethal doses of UV-B and UV-A (315 - 400 nm) radiations in their natural lit habitats. The highly energetic ultraviolet radiation (UVR) directly or indirectly affects the aquatic cyanobacteria as they can penetrate deeper into the water column up to 20 m and to a few centimetres in lakes and rivers. UVR has been reported to adversely affect several key processes of cyanobacteria such as, survival, growth, pigmentation, photosynthetic oxygen production, motility, nitrogen metabolism, phycobiliprotein composition and $14CO_2$ uptake (Sinha et al. 1995a; b; Sinha et al. 1996; Sinha et al. 2005; Xue et al. 2005; Häder et al. 2007). To counteract the deleterious effects of UVR mainly UV-B, these photosynthetic organisms have evolved certain defensive mechanisms such as avoidance, scavenging of reactive oxygen species (ROS) by antioxidants, synthesis of UV-absorbing/screening compounds and repair and resynthesis and programmed cell death (PCD) (Fig. 1). Cyanobacteria were present on the Earth during Precambrian era when there was no oxygen in the atmosphere (Fischer et al. 2008) and thus the presence of UV-screening compounds such as MAAs and scytonemin might have played an important role in protecting these organisms from the lethal UVR (Cockell and Knowland 1999). This chapter deals with the occurrence and biosynthesis of various UV-absorbing compounds from cyanobacteria that nullify the effect of UVR.

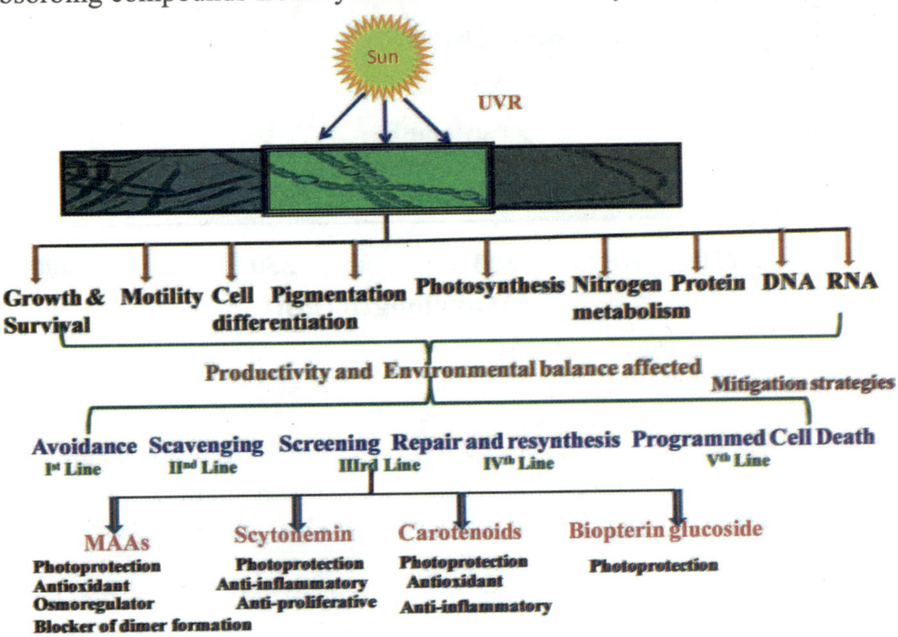

Fig. 1: Detrimental effects of UVR on cyanobacteria and possible mitigation strategies employed by them.

(A) Mycosporine-like amino acids (MAAs)

Mycosporine-like amino acids (MAAs) are an important group of UV-absorbing, secondary metabolites containing cyclohexenone or cyclohexenimine chromophore conjugated with the nitrogen substituent of an amino acid or its imino alcohol (Singh et al. 2008a). They are small (<400 Da), colorless, water-soluble compound with glycine moiety at the third carbon atom of the ring system. In addition, some MAAs also contain sulfate esters or glycosidic linkages through the imine substituent. However, in some corals glycine has been replaced by methyl amine (mycosporine-methyl-amine-serine, mycosporine-methylamine-threonine (Teai et al. 1997) or to an amine group (palythine-serine, and palythine-threonine) (Teai et al. 1997; Carignan et al. 2009). They absorb maximally in the UV-range (310-362) (Fig. 2) and variations in the attached side groups and nitrogen substituents are mainly responsible for the different spectral properties of the compound.

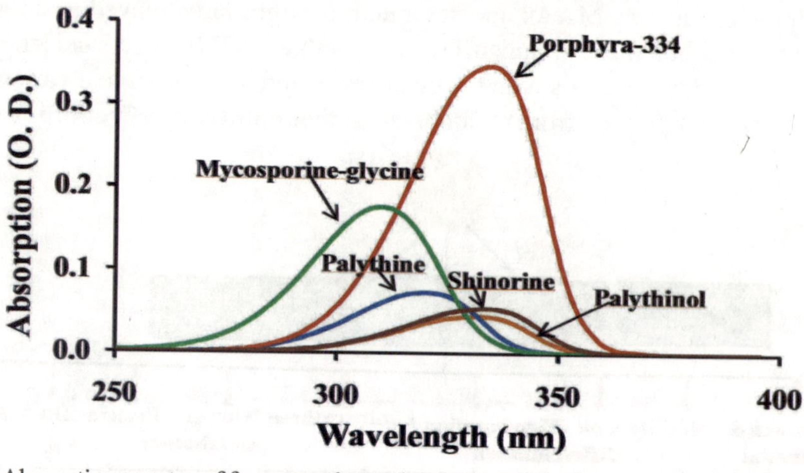

Fig. 2: Absorption spectra of few cyanobacterial mycosporine-like amino acids (MAAs).

High molar extinction coefficients ($\varepsilon = 28{,}100 - 50{,}000\ M^{-1}\ cm^{-1}$), photostability in both distilled and sea water in the presence of photosensitizers and high resistance against physico-chemical stressors like temperature, strong UVR, various solvents as well as pH make them successful photoprotectants in various habitats and organisms. MAAs protects the cells by absorbing highly energetic UVR and then dissipating excess energy in the form of heat to their environment (Conde et al. 2004). These compounds prevent 3 out of 10 photon from hitting cytoplasmic targets in cyanobacteria (Garcia- Pichel et al. 1993) and this supports the fact that the presence of MAAs protects vital functions in phytoplankton from deleterious short wavelength radiation (Klish et al. 2001; Singh and Sinha 2011). MAAs was originally identified in fungi as having a role in UV induced sporulation (Leach 1965) and hypothetically designated P-310. The first P-310 substance was isolated from *Stereum hirsutum,* and Favre et al. (1976) elucidated its structure as 2-methoxy-3-bis (hydroxymethyl) methylamino-5-hydroxy-5hydroxymethyl-2-cyclohexene-1 (i.e. mycosporine-serinol). An evolutionary divide appears between the parent class of mycosporines expressed by fungi exclusively as oxo-carbonyl chromophores (absorbing UV-B) and MAAs in aquatic organisms and terrestrial cyanobacterial containing mainly imino-carbonyl constituents (absorbing UV-B and principally the shorter wavelengths of UV-A). The synthesis of MAAs is restricted to bacteria, cyanobacteria, phytoplankton and macroalgae and does not occur in animals (metazoa), as they lack its major biosynthetic pathway (shikimate pathway), and take up MAAs through their diet or symbionts. However, the presence of genes encoding enzymes for the shikimate pathway in the genome of an marine animal, the starlet sea anemone *Nematostella vectensis* has challenged this conventional view (Starcevic et al. 2008). Thus, MAAs provide protection from UVR to their producers, as well as to primary and secondary consumers through the food chain (Helbling et al. 2002).

Occurrence of MAAs in Cyanobacteria

MAAs have been reported in diverse organisms, including heterotrophic bacteria (Arai et al. 1992), cyanobacteria and micro/macroalgae. Many animals such as arthropods, rotifers, molluscs, fish, cnidarians, tunicates, eubacteriobionts, poriferans, nemertineans, echinodermates, platyhelminthes, polychaetes, bryozoans and protozoans have also been reported to protect themselves from UVR by MAAs (Sinha et al. 2007). Shibata et al. (1969) for the first time reported the presence of MAAs in a cyanobacterium (likely *Trichodesmium*) from the Great Barrier Reef. This compound was named as S-320 and was assumed to have a role in photoprotection. Later on this UV-absorbing compound was identified in a number of macroalgae (Yoshida and Sivalingam 1969; Sivalingam et al. 1974), corals (Maragos 1972; Dunlap and Chalker 1986) and phytoplanktons (Carreto et al. 1989; Vernet et al. 1989). Till date, 22 MAAs have been reported from terrestrial, aquatic and freshwater microorganisms with known absorbance maxima and retention time during high performance liquid chromatography (HPLC). Some of the common MAAs found in cyanobacteria have been represented in Table 1. In addition, to this several partially characterized and unknown MAAs have also been detected with the subsequent development in chromatographic techniques (Carreto et al. 2001; 2005; Volkmann et al. 2003; Carignan et al. 2009).

Certain *Anabaena* sp. has been reported to synthesize the MAA shinorine (Sinha et al. 1998; Sinha et al. 2007; Sinha and Häder 2008). Among the four cyanobacterial species (e.g., *A. variabilis* PCC 7937, *Anabaena* sp. PCC 7120, *Synechocystis* sp. PCC 6803 and *Synechococcus* sp. PCC 6301), screened for MAAs by Singh et al. (2010a), only *A. variabilis* PCC 7937 was able to synthesize MAA. Whereas, no MAAs has been reported by this group of workers in *Synechocystis* sp. PCC 6803. However, a novel MAA tentatively identified as dehydroxyl-usujirene was isolated from the cyanobacteria *Synechocystis* sp. (Zhang et al. 2007). Shinorine and Porphyra-334 are the most common class of MAAs that are present in both freshwater and marine cyanobacteria (Karsten and Garcia-Pichel 1996; Sinha et al. 2001; Sinha et al. 2003; Volkmann et al. 2006; Sinha et al. 2007; Sinha and Häder 2008; Singh et al. 2010b). Singh et al. (2010b) reported the occurrence of mycosporine-glycine, porphyra-334 and shinorine in *Anabaena doliolum*, a common rice-field cyanobacterium. Three species of *Nodularia* (*Nodularia spumigena*, *Nodularia baltica* and *Nodularia harveyana*) from the Baltic Sea has been reported to synthesize MAAs porphyra-334 and shinorine upon UV-B irradiation (Sinha et al. 2003). Garcia-Pichel et al. (1998) found a common complement of MAAs in 13 strains of unicellular halophilic cyanobacteria. A symbiotic cyanobacteria, *Prochloron* sp. forming association with ascidians contained shinorine and traces of mycosporine-glycine and palythine (Dionisio-Sese et al. 1997). Screening of *Microcoleus* strains revealed the presence of Shinorine and three unidentified compounds having absorption maxima at 332, 344 and 346 nm (Karsten and Garcia-Pichel 1996). Recently, Volkmann et al. (2006) have reported the occurence of a novel MAA containing the amino acid alanine (2-(e)-2,3-dihydroxipro-1-enylimino-mycosporine-alanine) in an unicellular cyanobacteria *Euhalothece* sp. inhabiting in hypersaline pond. Volkmann et al. (2003; 2006) reported that mycosporine-alanine, mycosporine-glutaminol, mycosporine-glutaminol-glucoside provides protection to the terrestrial cyanobacteria against harmful UVR. Sommaruga and Garcia-Pichel (1999) have reported certain MAAs such as asterina-330, shinorine, palythinol and mycosporine-glycine in epilithic cyanobacteria from a freshwater lake.

Microcystis aeruginosa, the freshwater bloom forming cyanobacteria also contains high concentrations of MAAs including shinorine and several other unknown MAAs (Liu et al. 2004). The diazotrophic ocean bloom forming cyanobacteria *Trichodesmium* spp. were also reported to contain high amounts of asterina-330 and shinorine. In addition, to this mycosporine-glycine, porphyra-334 and palythene were also found to be present. However, two of the most abundant compounds were unidentified (Subramanian et al. 1999). *Synechococcus* sp. and *Prochloroccocus* sp. constituting the major component of phytoplanktons at the surface of the Brazil Current synthesizes diverse class of MAAs such as, shinorine, palythine, porphyra-334, mycosporine-glycine, palythenic acid and shinorine-methyl ester (Carreto et al. 2005). The filamentous bloom-forming cyanobacterium *Aphanizomenon flos-aquae* is marketed today as a food supplement and has been used as a source for the preparation of porphyra-334 (Torres et al. 2006).

Table 1: Occurrence of some common MAAs in cyanobacteria

MAAs	Structure	Absorbance (λmax) (nm)
Mycosporine-taurine	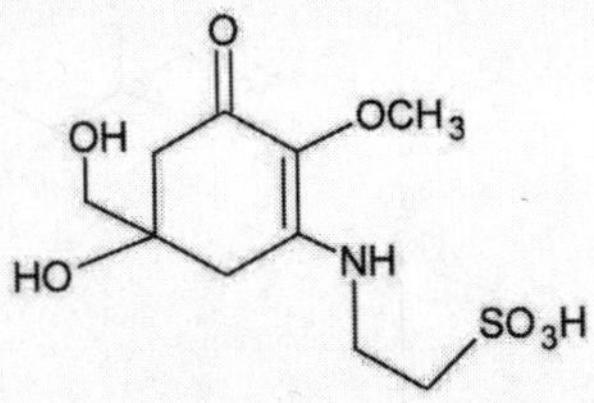	309
Mycosporine-glycine	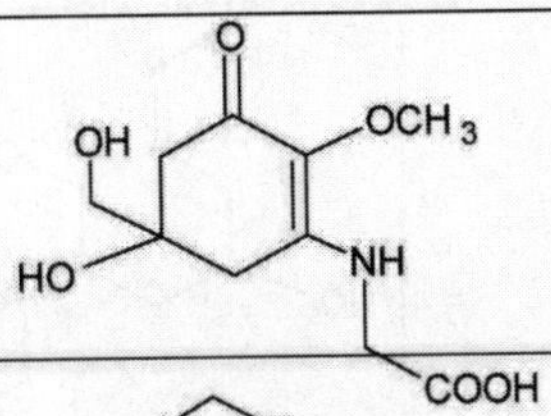	310
Asterina-330	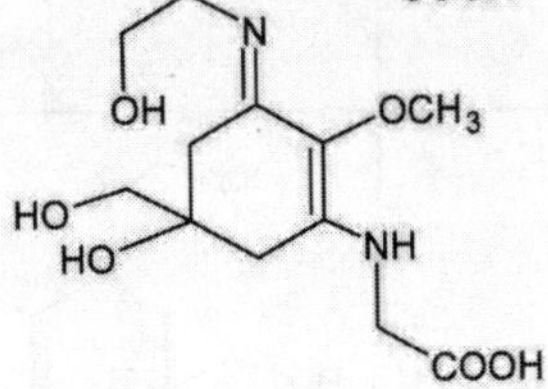	330
Palythinol	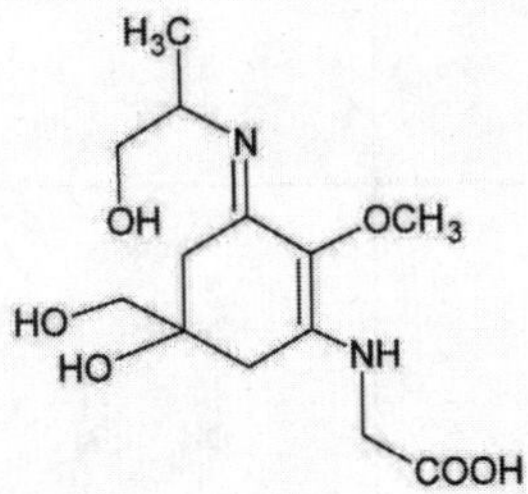	332
Porphyra-334	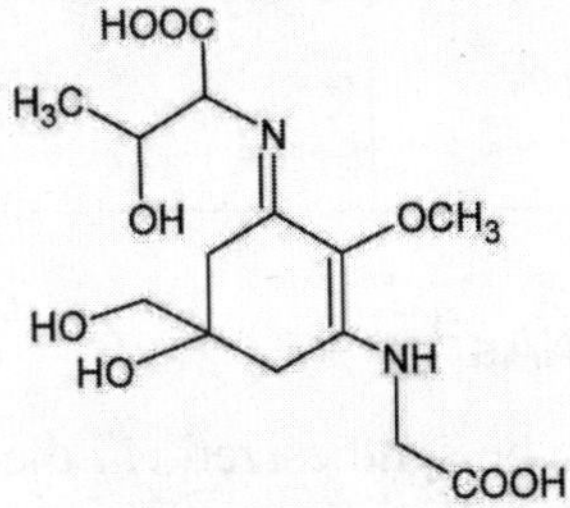	334

Contd.

Shinorine	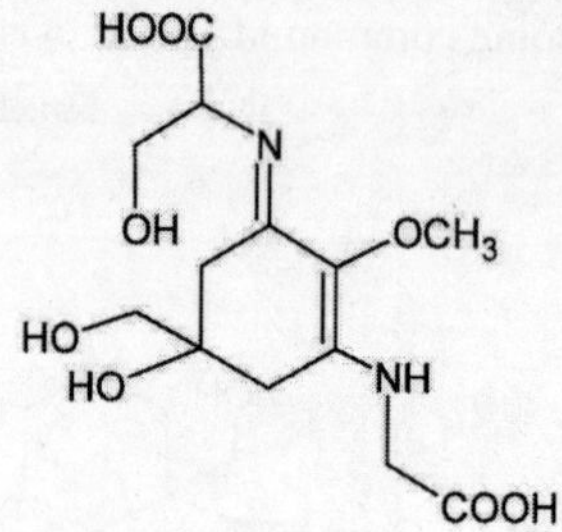	334
Mycosporine-2-glycine	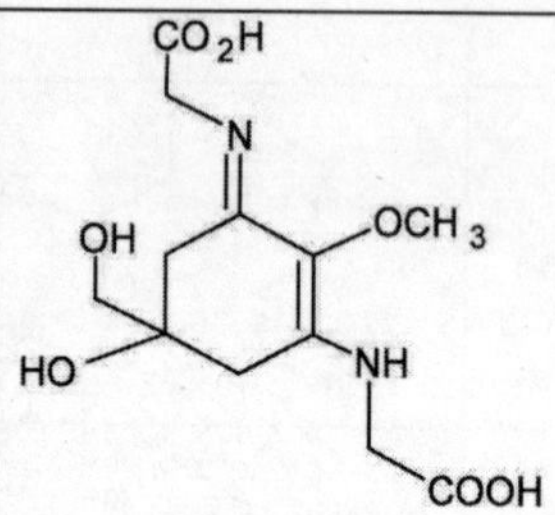	334
Palythene	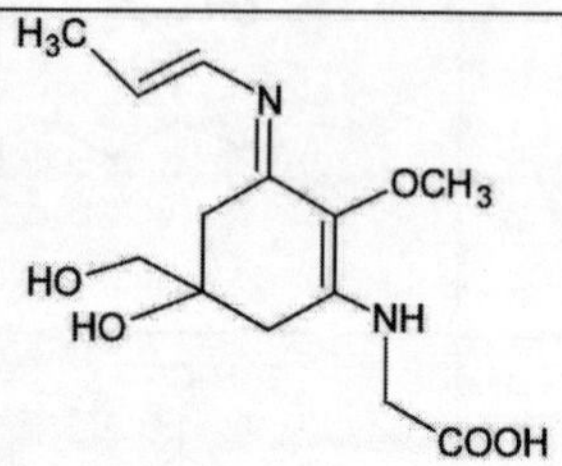	360
Euhalothece-362	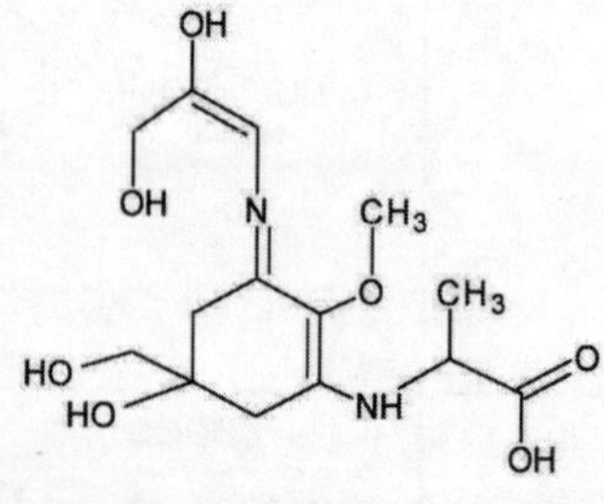	362

Biosynthesis of MAAs

MAAs are assumed to be synthesized *via* the first part of shikimate pathway (Fig. 3), but there is still controversy regarding its precise mechanism. 3-dehydroquinate (DHQ), the shikimate pathway intermediate serves as a precursor for the synthesis of fungal mycosporines (Favre-Bonvin et al. 1987) and MAAs *via* gadusols (Bandaranayake 1998; Shick and Dunlap 2002; Singh et al. 2008a). The blockage of MAAs synthesis in *Stylophora pistillata* by N- phosphonomethylglycine (glyphosate), a specific inhibitor of the shikimate pathway, provides the direct evidence for their synthesis in marine organisms *via* this route (Shick et al. 1999). Gadusol (3,5,6-trihydroxy- 5-hydroxymethyl-2-methoxycyclohex-2-en-1-one) and deoxygadusol (3,5-dihydroxy-5-hydroxymethyl-2-methoxycyclohex-2-en-1-one) are structurally and biosynthetically related to the MAAs that have been found in roes of cod and other marine fish (Chioccara et al. 1980; Plack et al. 1981) as well as in eggs, ovaries and larvae of various marine invertebrates (Grant et al. 1980; Bandaranayake and Rocher 1999). The genomic mining of four fully sequenced cyanobacteria by Singh et al. (2010a) revealed that YP_324358 (predicted DHQ synthase) and YP_324357 (O-methyltransferase), which were present only in *A. variabilis* PCC 7937 and were missing in other studied cyanobacteria, including *Synechocystis* sp. PCC 6803. Phylogenetic analysis revealed that these two genes are transferred from a cyanobacteria donor to dinoflagellates and finally to metazoan (Starcevic et al. 2008; Singh et al. 2010a) by lateral gene transfer, suggesting their possible contribution in MAAs synthesis. Singh et al. (2010a) also suggested that the protein structure for YP_324358 is different from the chemically characterized DHQ synthase of *Aspergillus nidulans* (Wu et al. 2007), contrary to the YP_324879, which was predicted to be similar to the DHQ synthase of *A. nidulans*. Based on these findings it can be proposed that the YP_324358 and YP_324357 gene products are involved in the biosynthesis of deoxygadusol, the common core of all MAAs.

The YP_324879 gene product is exclusively involved in the shikimate pathway catalyzing the formation of dehydroquinate, while the YP_324358 gene product in conjunction with the YP_324357 gene product (O-methyltransferase) catalyzes the formation of deoxygadusol, which is the core of all MAAs. Recently, Portwich and Garcia-Pichel (2003) have suggested that the central ring of the MAAs present in the cyanobacterium *Chlorogloeopsis* sp. strain 6912 has its origin in the shikimate pathway, suggesting that the biosynthesis of the MAAs would be very similar or possibly identical in eukaryotic and prokaryotic organisms. However, the recent results of Balskus and Walsh (2010) challenged the established assumption that MAA biosynthesis involved a Shikimate pathway intermediate, as these authors failed to observe any 6-deoxygadusol production when the putative substrate 3-dehydroquinate was incubated with the dehydroquinate synthase (DHQS) and O-methyltransferase (O-MT) homologues (NpR 5600 and NpR 5599 respectively) from *Nostoc punctiforme* (ATCC 29133) and the typical cofactors S-adenosylmethionine (SAM), nicotinamide adenine dinucleotide (NAD+) and Co^{2}+. Mycosporine-glycine is considered as a prime MAA synthesized in the shikimate pathway that further goes through chemical and/or biochemical conversions to produce secondary MAAs (Davidson et al. 1994; Carreto et al. 2005; Callone et al. 2006). Portwich and Garcia-Pichel (2003) have observed the specific incorporation of ^{14}C-glycine and ^{14}C-serine into the consequent side chains of mycosporine glycine and shinorine using radiolabelled amino acids, indicating that these free amino acids are their direct precursors.

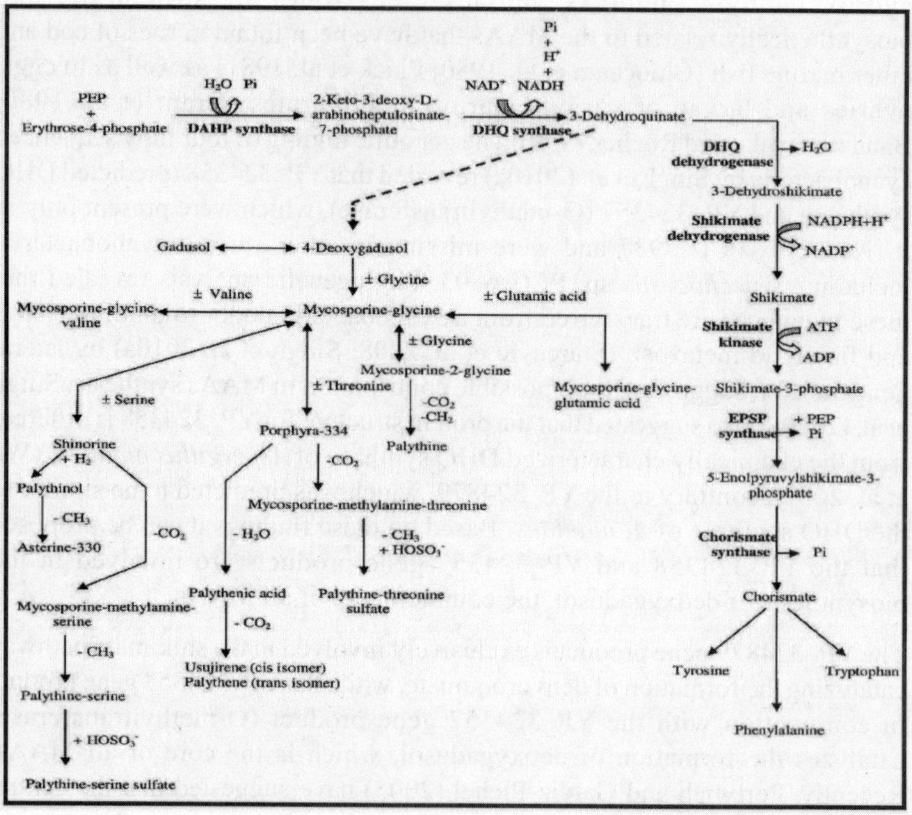

Fig. 3: Shikimate pathway for MAAs biosynthesis and their possible interconversion. Broken line represents the putative biosynthetic connection between Dehydroquinate, gadusols and MAAs. PEP: Phosphoenolpyruvate; DAHP: 3-deoxy-D-arabinoheptulosinate-7-phosphate; DHQ: dehydroquinate; EPSP: 5-enolpyruvylshikimate-3-phosphate (adapted from Singh et al. 2008a).

The high diversity of MAAs present in marine organisms is assumed to be derived from the transformation of mycosporine-glycine, porphyra-334, shinorine and other MAA bisubstituted by amino acids (Carreto et al. 2001; 2005). Carreto et al. (1990) also reported the role of varying levels of UVR in the interconversions of different MAAs. The conversion of primary MAAs, mycosporine-glycine into two MAAs, porphyra-334 and shinorine, under the influence by PAR and UVR in the rice-field cyanobacterium *Anabaena doliolum* has been worked out by Singh et al. (2008b) (Fig. 4).

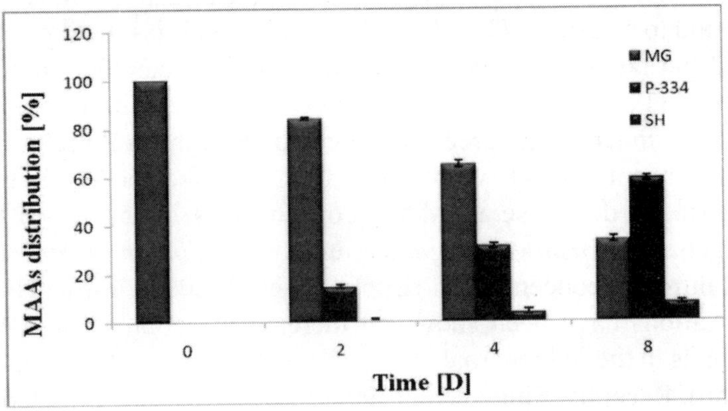

Fig. 4: Biotransformation of MAAs in *A. doliolum* under the influence of PAR, PAR+UV-A and PAR+UV-A+UV-B (Modified from Singh et al. 2008b).

Effect of Abiotic factors on synthesis of MAAs

Several environmental factors such as different wavelength bands of UVR, desiccation, nutrients, salt concentration, osmotic stress, light as well as dark periods have been found to have pronounced effect on MAAs synthesis in cyanobacteria. In cyanobacteria, MAAs can be induced by PAR, UVA and UVB radiation (Singh et al. 2008b). However, UV-B has the most pronounced effect in comparison with the other wavelength ranges (Castenholz and Garcia-Pichel 2000; Singh et al. 2008b). The same fact has been earlier reported by Sinha et al. (2001; 2002). UVB wavelengths (280–315 nm) were found to be more effective in inducing MAAs in the dinoflagellate *G. dorsum* (Carreto et al. 1990). There is an direct correlation between MAAs contents and natural or experimental levels of UV exposure (Garcia-Pichel and Castenholz 1993; Karsten et al. 1998),evidenced by the higher MAAs concentrations in tropical corals species from shallow waters than in isolates of the same species living at greater depths (Dunlap et al. 1986). It has been found that light is an essential factor for MAAs biosynthesis. Sinha et al. (2001) has observed the circadian behaviour of MAAs under alternating light and dark periods. Under natural solar radiation, increasing concentrations of shinorine, a bisubstituted MAA, were found only during the light periods, whereas more or less constant values of shinorine concentrations were found during and at the end of the dark period in *Nostoc commune* (Sinha et al. 2001). This suggests that synthesis of MAAs is an energy dependent process and depends on solar energy for its maintenance in natural habitats. However, the role of PAR and photosynthesis in MAAs production is still a topic of a debate, since biosynthesis of MAAs has been reported in the absence of PAR in the cyanobacterium *Chlorogloeopsis* PCC 6912 (Portwich and Garcia- Pichel 2000). Accumulation of MAAs has also been found to be affected by the different light qualities. Korbee et al. (2005a) reported the favourable role of blue light in the accumulation of the MAAs porphyra- 334, palythine and asterina-330, while shinorine was found to accumulate under white, green, yellow and red light in Porphyra leucosticta. Nutrient concentrations have also been found to affect the MAAs biosynthesis. A significant decrease in MAA content was observed in the marine dinoflagellates *Akashiwo sanguinea* and *Gymnodinium* cf. *instriatum* under limited nitrogen concentration (Litchman et al. 2002). Enriched ammonium concentrations have been shown to increase the contents of UV-absorbing compounds in the red macroalga *Porphyra* sp. (Pienado et al. 2004; Korbee et al. 2005b). Recently, Singh et al. (2008c) also reported the induction of MAA synthesis by salt and ammonium in a concentration-dependent manner without UV stress in the cyanobacterium *Anabaena variabilis* PCC 7937. The MAAs composition is also highly influenced by nitrogen status. Mycosporine-glycine, a least N-rich MAA was present in higher concentrations in N-starved cells of *A. sanguinea* (Litchman et al. 2002), *A. tamarense* and *Karenia brevis* (Frame 2004). Recently, Singh et al. (2010c) showed for the first time that synthesis and bioconversion of a primary mycosporine-like amino acid (MAA) into a secondary MAA is regulated by sulfur deficiency in the cyanobacterium *A. variabilis* PCC 7937. However, till date there are no reports of P-limitation on synthesis and accumulation of MAAs. In addition, osmotic stress, salt stress and temperature have also been found to affect the synthesis of MAAs in cyanobacteria (Portwich and Garcia-Pichel 2000; Singh et al. 2010b). Osmotic stress, alone or in combination with UVB radiation, can induce the synthesis of MAAs in some cyanobacteria (Portwich and Garcia-Pichel 1999; Singh et al. 2008c). Salt stress, alone, without PAR or UV radiation has been found to induce MAAs in *Chlorogloeopsis* sp. PCC 6912, which can tolerate upto 70% sea water salinity. Desiccation has been reported to have less effect on accumulation of UV-absorbing compounds in intertidal cyanobacteria (Fleming and Castenholz 2007). Whereas, higher concentration of MAAs was observed in intertidal red alga *Porphyra haitanensis* following desiccation stress. The local climatic conditions and depth zonation also affects the MAAs concentration in a particular organism.

Functions

MAAs has high absorption in the UV region with possible applications in cosmetics and tolieteries as UV protectors and activators of cell proliferation (Conde et al. 2000; Shick and Dunlap 2002; Whitehead and Hedges 2005; Torres et al. 2006). Rastogi and Sinha (2009) have reported the anti-inflammatory and antioxidant properties of MAAs. MAAs shows a high antioxidant activity scavenging superoxide anions and inhibiting lipid peroxidation (Coba et al. 2007a; b) resulting from UV-induced production of reactive oxygen species (ROS) (Suh et al. 2003; Oren and Gunde-Cimerman 2007). The role of P-334 and shinorine in maintaining the antioxidant defense system of the skin as well as Hsp70 expression has been worked out by Coba et al. (2009). Oyamada et al. (2008) have reported that the MAAs such as shinorine (SH), porphyra-334 (P-334) and mycosporine-glycine have potentials to protect the fibroblast cells from UVR-induced cell death. A combination of P-334 and shinorine from *Porphyra umbilicalis*, has been reported to suppress the UV-induced aging in human skin (Daniel et al. 2004). The antioxidant activities of the MAAs glycine and usujilene have been reported to inhibit lipid peroxidation in aqueous extracts of marine organisms and to scavenge 1_{O2} generated from certain endogenous photosensitizers (Nakayama et al. 1999; Suh et al. 2003). MAAs can block the production of certain cytotoxic DNA lesions such as, 6–4photoproduct and cyclobutane pyrimidine dimer (CPD) formation (Misonou et al. 2003). Tetrahydropyridine derivatives, a synthetic analogue of MAAs have been commercialized as suncare products (Bird et al. 1987; Chalmers et al. 1990; Dunlap et al. 1998). Moreover, a product called Helioguard® 365 that contains mycosporine-like amino acids from the red alga *Porphyra umbilicalis* has been commercialized (Cardozo et al. 2007). In addition, MAAs have also been used in the manufacture of several non-biological materials such as photostabilizing additives in plastics, paints and varnishes (Bandaranayake 1998). The role of MAA (porphyra-334) as a potent photoprotectant and antioxidant has been demonstrated by Richa (2014).

(B) Scytonemin

Scytonemin constitutes the second important class of UV-absorbing compounds in cyanobacteria. It is a yellow-brown lipid soluble pigment having a molecular mass of 544 Da, present in the extracellular polysaccharide sheath of some cyanobacteria (Fig. 5) (Desikachary 1959; Geitler 1932; Garcia-Pichel and Castenholz 1991; Garcia-Pichel and Belnap 1996). This photoprotective compound was first reported by Nägeli (1849) in some terrestrial cyanobacteria. The presence of scytonemin has been reported in a terrestrial cyanobacteria *Tolypothrix byssoidea,* showing maxium absorption at 260 and 384 nm (Adhikary and Sahu, 1998). Scytonemin is a dimeric compound composed of indolic and phenolic subunits linked by an olefinic carbon atom, a unique natural product. Scytonemin, generally occurs in green oxidized form, but depending on the redox and acid-base conditions during the extraction process two other forms such as the oxidized (yellow; e.g., fuscochlorin) and the reduced (red; e.g., fuscorhodin) has been reported by Garcia-Pichel and Castenholz (1991). Recently, scytonemin has been extracted from from *Lyngbya* sp. inhabiting rocks at the shore near Sao Fransisco do Sul, Brasil (Richter et al. 2006). Recently, three new pigments, dimethoxyscytonemin, tetramethoxyscytonemin and scytonin from the organic extracts of *Scytonema* sp. has been reported to be derived from the scytoneman skeleton of scytonemin (Bultel-Poncé et al. 2004). Their structures were confirmed on the basis of ^{1}H and ^{13}C NMR and MS experiments. Scytonemin has an *in vivo* absorption maximum at 370 nm in the UV-A region, but shows considerable absorption in the UV-B region also (Richa and Sinha 2011). Purified scytonemin has a highest absorption at 386 nm, although it also absorbs significantly at 252, 278 and 300 nm, and so it probably helps cyanobacteria to survive lethal UV (UV-A and UV-B) radiation. Microspectrophotometric measurements of the transmittance of pigmented sheath and the quenching of ultraviolet excitation of phycocyanin fluorescence suggest that the pigment is effective in shielding the cells from incoming UV radiation and is able to reduce the entry of UV-A radiation into the cells to around 90% (Garcia-Pichel et al. 1992; Brenowitz and Castenholz 1997). Scytonemin alone is sufficient in providing significant protection against UV-C damage (Dillon and Castenholz 1999). It has been reported by Castenholz (1997) that scytonemin constitute more than 5% of the total cellular weight in cyanobacteria. It is highly stable against various stressors such as temperature, strong UV radiation etc. and able to persist very long in terrestrial cyanobacterial crusts or dried mats (Brenowitz and Castenholz 1997; Quesada et al. 1999) and performs its screening activity without any further metabolic investment even under prolonged physiological inactivity (e.g. desiccation) (Singh et al. 2010b).

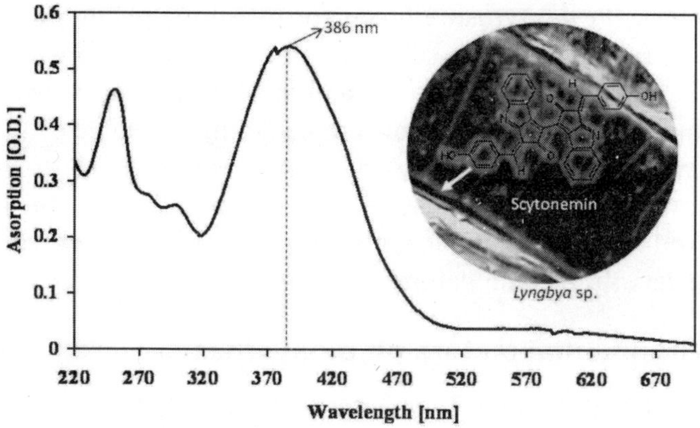

Fig. 5: Presence of scytonemin in the extracellular sheath of *Lyngbya* sp. (In the inset) and its absorption spectrum showing peaks maximum at 386 nm.

Biosynthesis

Scytonemin is thought to be synthesized from metabolites of aromatic amino acid biosynthesis, mainly tyrosine and tryptophan (Strickland et al. 1970) and can be induced under high photon fluence rate (Garcia-Pichel and Castenholz, 1991). Soule et al. (2007) created a scytonemin-deficient *Nostoc punctiforme* ATCC 29133 mutant using random transposon mutagenesis and reported a specific region of the genome revealing a cluster of 18 gene (NpR1276 to NpR1259) to be involved in scytonemin synthesis. Soule et al. (2009a) reported the upregulation of the transcriptional unit consisting of these 18 genes by UV radiation. Six putative genes from the scytonemin gene cluster are expected to be involved in the assembly of scytonemin from central metabolites (Soule et al. 2009 b). The comparison of this gene clusters revealed that the redundant copies of genes were responsible for aromatic amino acid biosynthesis. Based on findings from evolutionary phylogenetic analysis and cosmopolitan distribution of scytonemin gene cluster among several cyanobacterial lineages, it was confirmed that an ancient evolutionary origin was the best way to explain the distribution of this gene cluster (Sorrels et al. 2009). The probable biosynthetic route for the scytonemin biosynthesis was framed by Balskus and Walsh (2008) and they identified the acyloin reaction as a key step in constructing the carbon framework. (Fig. 6). Soule et al. (2007) identified two enzymes encoded by ORF NpR1275 and NpR1276 gene cluster involved in the initial step of scytonemin biosynthesis (Singh et al. 2010b), were functionally characterized by Balskus and Walsh (2008). However, the gene products of NpR1263 and NpR1269 ORFs are still to be functionally characterized. NpR1275, similar to leucine dehydrogenase, was shown to catalyze the oxidative deamination of tryptophan to form indole-3-pyruvic acid

(IPA). IPA, along with the tyrosine precursor p-hydroxyphenylpyruvic acid, acts as the substrate for highly selective thiamine diphosphate (ThDP)-dependent NpR1276 catalyzed acyloin reaction (Balskus and Walsh 2008). A comparative study of four other cyanobacterial strains by Soule et al. (2009b) revealed that two supplementary conserved clusters (NpF5232 to NpF5236) and a putative two-component regulatory system (NpF1277 and NpF1278) are possibly involved in scytonemin biosynthesis and regulation, respectively, on the basis of conservation and location.

Effect of environmental factors on biosynthesis

UV-A and PAR at high photon fluence rates (Garcia-Pichel and Castenholz 1991) sturdily induces scytonemin biosynthesis. Among the various examined light qualities, UV-A treatment has been found to be very efficient in inducing the synthesis of scytonemin, whereas blue, green or red light at the same fluence rates do not cause any significant increase in scytonemin (Garcia-Pichel and Castenholz 1991). The effect of bright light in promoting the production of scytonemin could be prevented by treating the cells with formaldehyde, chloramphenicol and keeping the cultures at 4°C during the exposure to high light. The inhibition of scytonemin synthesis in the presence of chloramphenicol

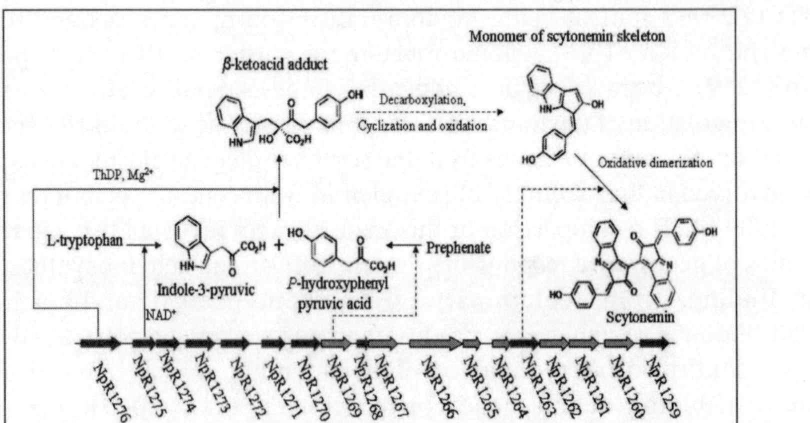

Fig. 6: Biosynthetic pathway for scytonemin and corresponding gene products involved in each step. Continuous arrows represent gene products which are functionally characterized while gene products indicated by broken arrows are still to be functionally characterized for their involvement in corresponding steps. Gray arrows represent ORFs involved in the biosynthesis of aromatic amino acids while brown arrows represent ORFs of unknown function. The model is based on the informations from Soule et al. (2007) and Balskus and Walsh (2008). (Adapted from Singh et al. 2010b).

indicates that protein synthesis pathway is involved in its production. The effect of other correlated stress factors including heat, osmotic and oxidative stress on the synthesis of scytonemin has been worked out by Dillon et al. (2002) in the cyanobacterium *Chroococcidiopsis* sp. and showed that both increases in temperature and oxidative stress in combination with UV-A have a synergistic effect on high production of scytonemin.

In contrast, osmotic stress causes a decrease in scytonemin synthesis both in the presence and absence of UV-A irradiation. Nutrients concentration also affects the scytonemin biosynthesis. Fleming and Castenholz (2008) have shown the effects of nitrogen sources on scytonemin synthesis in the cyanobacterium *Nostoc punctiforme* PCC 73102. The cultures were found to produce concentrated form of scytonemin when utilized N_2, followed by nitrate and then ammonium. A variable and even negative correlation between UV-flux and scytonemin content in *Scytonema* sp., but a positive correlation in *Rivularia* sp. was observed (Pentecost 1993). Brenowitz and Castenholz (1997) examined that high amount of scytonemin is essential for natural photosynthesis under high UV-radiation in a monospecific population of *Calothrix* sp. Fleming and Castenholz (2007) has demonstrated the effect of desiccation on the scytonemin content. He found that the cells of *Nostoc punctiforme* hydrated for two days, in between desiccation periods had higher scytonemin synthesis compared to the cells hydrated for one day. However, in *Chroococcidiopsis* periodic desiccation inhibits scytonemin synthesis.

Functions

Scytonemin is a potent UV-screening compound with maximum absorbance in the range of UV-A. The synthesis of scytonemin by the cyanobacteria is considered to be the most ancient measure against UV protection. The photoprotective role of scytonemin has been reported in a number of cyanobacteria (Garcia-Pichel and Castenholz 1991; Sinha et al. 1999) and cyanobacterial lichens growing under elevated light intensity (Büdel et al. 1997). Garcia-Pichel et al. (1992) has reported the UV-sunscreen role of scytonemin in terrestrial cyanobacterium *Chlorogloeopsis* sp. The scytoneman skeleton of scytonemin that is closely linked to nostodione A, is among the unique natural product (Proteau et al. 1993). Scytonemin has dual kinase inhibitory activity that may be pharmacologically important in treating acute and possibly chronic disorders of inflammation and proliferation. Certain striking structural features such as lack of chirality, multiple dissection points and easily modifiable phenolic groups contribute to its anti-proliferative property. Scytonemin showed a concentration-dependent inhibition of *polo*-like kinase 1 (PLK1) activity, with an IC50 of 2 mM and plays a significant role in regulating the G2/M transition in the cell cycle (Stevenson et al. 2002). In contrast, scytonemin was found to reduce the GST-*polo*-like kinase 1 activity in a time-independent fashion. Scytonemin also inhibits other cell cycle kinases with similar potency to that of PLK1. In human T-cell leukemia Jurkat cells, scytonemin repressed cell proliferation (IC50 = 7.8 lM) and induced apoptosis in 24% of the cells (at 3lM). The cyclic peptide scytonemin A from *Scytonema* sp. has been shown to be a calcium agonist (Helms et al. 1988).

(C) Carotenoids

Carotenoids are multifunctional yellow, orange or red hydrophobic substances that play an important role in light-harvesting and photoprotection by serving as direct quenchers of singlet oxygen and other toxic reactive oxygen species (ROS) and dissipating excess of energy in the form of heat in the photosynthetic apparatus (Young et al. 1991; Edge et al. 1997). They are highly conjugated 40-carbon structures formed by the polymerization of five-carbon isoprene units. The polyene chain in carotenoids contains up to 15 conjugated double bonds, a feature that is responsible for their characteristic absorption spectra and specific photochemical properties (Goodwin 1980). One or both ends of the carbon skeleton may go through cyclization to form ring β-ionone ring end groups, which additionally may be substituted by oxo, hydroxy or epoxy groups at dissimilar positions to form the different xanthophylls, the oxygenated carotenoids (Solomons and Bulux 1994). Although ε-rings occur in the *Prochlorococcus* sp., in which α-carotene is the principal carotene (Strickforth et al. 2003). Certain cyanobacteria produces the monocyclic xanthophyll myxoxanthophyll or the acyclic xanthophylls oscillaxanthin, in which the ψ ends of the molecules are hydroxylated at the C-1 and C-2 positions and an additional double bond is added at the C-3,4 position (Liaane-Jensen and Andrews 1972). About 600 different carotenoids have been reported from diverse organisms including cyanobacteria, bacteria, fungi, phytoplankton, macroalgae, plants and animals (Polivka and Sundström 2004; Coesel et al. 2008) except animals, as they lack the capacity to synthesize these compounds endogenously (Moeller et al. 2005). Thus, vertebrates depend on dietary carotenoid sources for making their retenoids, such as retinal, retinol and retenoic acid. The predominant carotenoid, astaxanthin in marine animals is formed by the modification of plant carotenoid obtained through the diet.

Distribution of carotenoids

β-carotene is the major non-oxygenated carotenoid associated with both the photosystems in most of the cyanobacteria (Jordan et. al 2001; Barber and Iwata 2005), and the cyanobacterial cytochrome b_6f complex also contains one carotenoid molecule (Boronowsky et al. 2001, Wenk et al. 2005). In cyanobacteria, carotenoids are also associated with proteins devoid of chlorophyll. These water-soluble carotenoid binding proteins were first described by Holt and Krogmann (1981) and later found in several different cyanobacterial species (Kerfeld, 2004a, b). Carotenoids are also associated with cyanobacterial Photosystem II (PS II) (Newman and Sherman, 1978). There are two β- carotene molecules in the reaction center core of PS II (Newell et al. 1993) which protect chlorophyll P680 from photodamage (De Las Rivas et al. 1993). Cyanobacteria does not synthesize ε-ring and therefore contain â, but not α- or δ -, carotenes and their oxygenated forms. Glycosylated carotenoids are most commonly present in cyanobacteria (Hirschberg and Chamovitz 1994). The diketocarotenoid, canthaxanthin, is also widespread in cyanobacteria. The xanthophylls of cyanobacteria are mainly located in the thylakoid, cytoplasmic, and outer membranes. *Synechococcus* sp. strain PCC 6301 contains primarily β-carotene and zeaxanthin in its thylakoid membranes (Murata et al. 1981), whereas, zeaxanthin and nostoxanthin predominates in its cell envelope and cytoplasmic membranes in similar proportions (Murata et al. 1981; Omata and Murata 1983). Jurgens and Weckesser (1985) reported the presence of myxoxanthophyll and certain other carotenoids in the outer membrane of *Synechocystis* sp. strain PCC 6714. A novel, bicyclic aromatic carotenoid, synechoxanthin has been reported in *Synechococcus* sp. strain PCC 7002 (Graham et al. 2008). A substantial amount of β-carotene and the xanthophylls cryptoxanthin and isocriptoxanthin were detected in PS I pigment-protein complexes of the thermophilic cyanobacterium *Synechococcus elongatus* (Coufal et al. 1989). An average of 1.3 molecules of β-carotene has been reported in the isolated complexes of PSI from *Phormidium luridum* (Thornber et al. 1976). Olie and Potts (1986) reported the predominance of β-carotene and echinenone in the fresh samples of *Nostoc commune*, while large amount of canthaxanthin, β- and γ- carotenes were present in the desiccated samples.

Biosynthesis

The biosynthetic pathway of caroteniods is almost similar in plants and cyanobacteria. Carotenoids are synthesized from the precusors of isoprenoid (Fig. 7). Isopentenyl pyrophosphate (IPP) (C_5 molecule) formed from mevalonate, is the building block for all long-chain isoprenoids. IPP isomerizes to dimethylallyl pyrophosphate (DMAPP), with three additional IPP molecules to yield a condensation product geranylgeranyl pyrophosphate (GGPP) (C_{20} molecule). The first step in the carotenoid biosynthesis is the head-to-head condensation of two molecules of GGPP following removal of pyrophosphate to yield a colorless C_{40} 15-*cis*-phytoene molecule. The reaction is catalyzed by phytoene synthase, a soluble enzyme encoded by *crtB* gene in both cyanobacteria and plants (Chamovitz et al. 1992). All the subsequent steps of the biosynthetic pathway occur in the membranes. Four dehydrogenation reactions convert phytotene to lycopene, with each desaturation increasing the number of conjugated bonds by two. It has been established that in cyanobacteria, algae and plants the desaturations, from 15-*cis*-phytoene to ζ-carotene are catalyzed by a single membrane-bound enzyme, phytoene desaturase (Beyer et al. 1989). In cyanobacteria algae and plants ζ-carotene is converted to lycopene *via* neurosporene. However, little is known about the enzymatic mechanism. Two cyclization reactions convert lycopene to β-carotene. Evidence suugests that in *Synechococcus*sp. strain PCC7942 (Cunninghamet al. 1993), as well as in plants (Camara and Dogbo 1986), these two cyclizations are catalyzed by a single membrane-bound enzyme, lycopene cyclase. Cyanobacteria carry out only the β-cyclization therefore does not

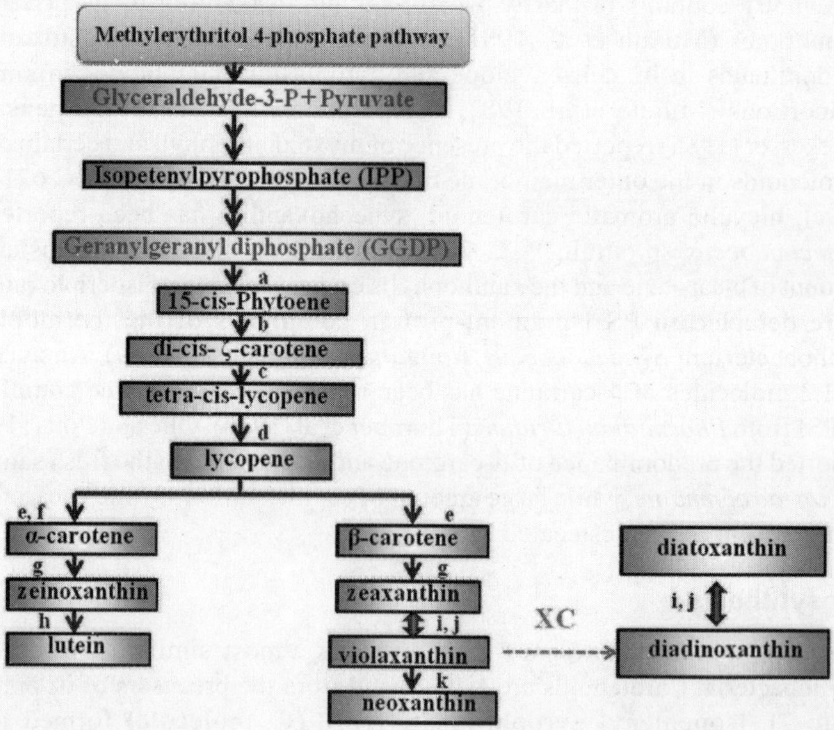

Fig. 7: Biosynthetic pathway of certain carotenoids. The *alphabets* indicates the enzymes involved in the pathway. *a*: Phytoene synthase, *b*: phytoene desaturase, *c*: ζ –carotene desaturase, *d*: carotenoid isomerase enzyme, *e*: β-carotene cyclise, *f*: ε-cyclase, *g*: β-carotene hydroxylase, *h*: ε-carotene hydroxylase, *i*: violaxanthin deepoxidase, *j*: zeaxanthin epoxidase, κ: neoxanthin synthase. XC: Xanthophyll cycle (Modified from Penna and Pogson 2006; Rastogi et al. 2010).

contain ε-,δ- and α-carotene and their oxygenated derivatives. A dinucleotide-binding domain was found to be associated in the lycopene cyclase polypeptide of *Synechococcus* sp. strain PCC 7942, implicating NAD(P) or FAD as coenzymes with lycopene cyclase (Hirschberg and Chamovitz 1994). The ends of the β-ring may be further substituted by various oxygen-containing side groups, such as hydroxy-, methoxy-, oxo-, epoxy-, aldehyde or carboxylic acid moieties, forming the various xanthophyll species. The light-induced de-epoxidation of violaxanthin and the reverse reaction that takes place in the dark, are known as the 'xanthophyll cycle' (Demmig-Adams and Adams 1992). In contrast to algae and plants, cyanobacteria lacks the xanthophyll cycle. However, they do contain ample quantities of zeaxanthin and other xanthophylls that can support photoprotection of chlorophyll. Several studies have been recently published regarding the biosynthetic pathway of carotenoids in diverse organisms (Cardozo et al. 2007; Cunningham et al. 2007; Coesel et al. 2008).

Effect of environmental factors on synthesis

Carotenoid biosynthesis is a regulated process, being controlled by environmental and developmental factors. Changes in carotenoid content and composition, which occur in response to light intensity and light quality, are one of the adaptation mechanisms to high-light that exist in acclimatized cyanobacteria. The changes in the carotenoid pattern in response to UV-B radiation has been reported in *Nostoc commune* and myxoxanthophyll and echinenone were suggested to act as outer membrane-bound UV-B photoprotectors (Ehling-Schulz et al. 1997). Exposure of *Oscillatoria* sp. strain BG 091600 to UV-A at an irradiance of 10 Wm^{-2} resulted in a slight increase in the carotenoid content (Wachi et al. 1995). Götz et al. (1999) reported that the increase in the amount of carotenoids by genetic manipulation has positive effect on the photosynthetic activity of the cyanobacterium *Synechococcus* sp. against UV-B irradiation. PAR and UVR dependent induction of carotenoids (Smith et al. 1992) in phytoplanktons has been reported to play an important role in the reduction of photoinhibition of photosynthesis (Brown et al. 1999). Steiger et al. (1999) reported the light-dependent formation of carotenoids in the cyanobacterium *Synechocystis* PCC 6803. They investigated that myxoxanthophyll is preferentially synthesized in high light exhibiting the strongest protection and echinenone is the most stable carotenoid under photo-oxidative conditions in the test organism. The enhancement in the carotenoid pool in response to high irradiance conditions has also been worked out by Schagerl and Müller (2006) in four freshwater cyanobacterial strains. They found that *Anabaena torulosa* contained higher amounts of every carotenoid, while *Nostoc* sp. mainly increased zeaxanthin, and myxoxanthophyll. *Anabaenopsis elenkinii* produced exceptionally high amounts of myxoxanthophyll and β-carotene under higher irradiances whereas, *Anabaena cylindrica* generally showed less variation of carotenoids under different irradiances. An increase in the relative abundance of myxoxanthophyll and zeaxanthin, and a decrease in the relative abundance of echinenone and were observed in *Oscillatoria agardhii* following exposure to high intensity of white light (Millie et al. 1990). Carotenoid composition of *Synechococcus* sp. strain WH 7803 has also been reported to be affected by light quality. The ratio of β-carotne to chlorophyll *a* was found to be constant under daylight, white or blue-green light (Bidigare et al. 1989). However, zeaxanthin was two-fold higher in cells grown in blue-green light than in those grown under white light or daylight illumination. Certain other factors such as, salinity and temperature have also been found to affect the biosynthesis of carotenoids in some cyanobacteria. An increase in carotenoid content, with a concomitant depletion in phycocyanin has been observed in *Pseudanabaena* sp. strain M2 and *Oscillatoria splendida* following nitrogen deficiency during growth (Canto de Loura et al. 1987). Increase in salinity has been reported to alter the carotenoid content in the freshwater cyanobacterium *Synechococcus* sp. strain PCC 6311 (Khomutov et al. 1990). The increase in the amount of carotenoids and proportion of unsaturated fatty acids in the membrane lipids resulted in lowered growth temperature of a thermophilic cyanobacterium *Synechococcus* sp. (Miller et al. 1987).

Functions

Carotenoid pigments have been reported to serve two major functions: as accessory pigments for light harvesting and in prevention of photooxidative damage in the photosynthetic organisms. They absorb the photon and transfer the energy to chlorophyll, thus supporting the light harvesting in the range of 450-570 nm (Cogdell and Gardiner 1993). They are also important components of the photosynthetic reaction centers. The occurrence and the photoprotective role of orange carotenoid protein (OCP) in cyanobacteria has been worked out by Boulay et al. (2008). The soluble carotenoid-binding protein (OCP) encoded by the *slr*1963 gene in *Synechocystis* sp. PCC 6803 has been reported to play a vital role in non-photochemical quenching (NPQ), resulting in a decrease of PSII-related fluorescence emission (Fig. 8).

The amount of energy transferred from the phycobilisomes to the photosystems is reduced and the OCP acts as the photoreceptor and as the mediator of this antenna-related process (Kirilovsky 2007). Carotenoids have also been reported to protect the gonads and eggs of four sea urchin species against the harmful effects of UV-B (Lamare and Hoffman 2004). A photoprotective function of the ketocarotenoid canthaxanthin has been shown in genetically modified

Synechococcus (Albrecht et al. 2001). The xanthophyll zeaxanthin plays an important role in protecting green plant against photoinhibition (Demmig-Adams 1990; Niyogi, 1999) and it has been speculated that it could play a similar role in cyanobacteria (Demmig- Adams et al. 1990). β- carotene has been shown to play a vital role in the assembly of an active PS II in green algae (Humbeck et al. 1989). Carotenoids also have substantial influence on the phototactic responses in the cyanobacterium *Anabaena variabilis,* contributing to the removal of singlet oxygen radicals ($^1O_2$) (Nultsch and Schuchart 1985). Carotenoids also provide protection against photo-oxidation processes in the photosynthetic apparatus that are caused by the excited triplet state of chlorophyll, as reported in *Synechococcus* sp. (Schlodder and Brettel 1988). They directly provide protection against UVR-induced photooxidation in the skin and inflammation, ulcers due to *Helicobacter pylori* infection and age-related diseases (Sies and Stahl 2004; Cardozo et al. 2007). It has been shown that β-carotene can modulate UV-A-induced gene expression in human keratinocytes (Wertz et al. 2005). Carotenoids are strong antioxidants (Cockell and Knowland 1999) with the potential of reducing the incidence of several UV-mediated diseases. In addition, carotenoids being non-toxic, have various commercial applications. They are used as natural food coloring agents, food additives, drugs, enhancers of the color of egg yolks, in improving the health and fertility of cattle, and in the cosmetic industries (Jha and Zi-rong 2004; Stahl and Sies 2007).

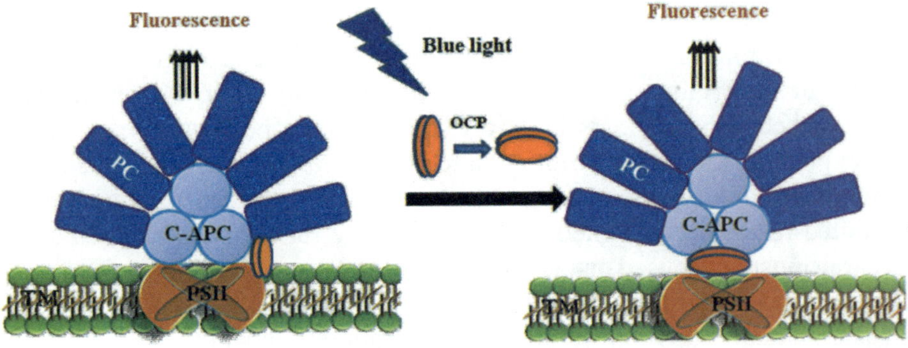

Fig. 8: Showing the role of orange carotenoid protein (OCP) in non-photochemical quenching (NPQ). The absorption of blue light by the carotenoid of the OCP induces conformational changes in the carotenoid and in the protein allowing the energy transfer from the phycobilisome to the OCP in which the energy is thermally dissipated. Under these conditions less energy arrives to the reaction center II and less light comes back as fluorescence (Based on Kirilovsky 2007).

Other UV-absorbing compounds

A UV-A absorbing pigment, biopterin glucoside (BG; a compound chemically related to the pteridine pigments found in butterfly wings), with absorption maxima at 256 and 362 nm, has been purified from a marine planktonic cyanobacterium Oscillatoria sp. collected from the coastal areas of Japan (Matsunaga et al. 1993). The same flouresent pigment (biopterin-α-glucoside) has also been isolated from the cyanobacterium *Spirulina platensis,* where they prevented the decolorization of the photosynthetic pigments and carotenoids by UVR (Nohuchi et al. 1999). Certain cyanobacteria such as, *Anacystis nidulans*, *Anabaena variabilis*, *Nostoc muscorum* and *Nostoc maseaum* were reported to produce pteridines in relatively large concentrations to counteract the damaging effects of UVR.

Conclusion

Highly energetic solar UV-B reaching onto the Earth's surface adversely affects a number of key physiological and biochemical processes such as, growth, survival, pigmentation, photosynthesis, nitrogen fixation, proteins and DNA in cyanobacteria. Certain cytotoxic and mutagenic pyrimidine dimers have also been demonstrated in cyanobacteria following UV-B exposure. To counteract the damaging effects of UVR, cyanobacteria have evolved the capacity to synthesize, metabolize and accumulate a variety of UV-absorbing compounds, including MAAs, scytonemin, carotenoids, and several other compounds. MAAs are a ubiquitous group of sunscreen compounds present in a large number of cyanobacteria that can potentially reduce the detrimental effects of UVR. In addition, scytonemin, carotenoids and biopterin glucoside also play an important role in diminishing the direct and indirect damaging effects of environmental UVR. The amount and composition of these photoprotective compounds in an organism is often positively regulated by UV-exposure and other abiotic factors. Moreover, these UV-absorbing compounds have several other therapeutic properties that can be commercially exploited for the development of biotechnology for human welfare. Further research must be focused on the biosynthesis and photoprotective role of these and several other photoprotective compounds so that a number of pharmaceuticals and biotechnological applications can be found.

Acknowledgements

Richa is thankful to Department of Science and Technology, Govt. of India, New Delhi for providing the financial support in form of a fellowship under the project (No. SR/WOS-A/LS-140/2011). The work was also partially supported by the financial assistance provided to Vinod K. Kannaujiya and

Minu Kesheri by CSIR and UGC (Govt. of India, New Delhi), respectively, in the form of SRF.

References

Albrecht M, Steiger S, Sandmann G (2001) Expression of a ketolase gene mediates the synthesis of canthaxanthin in *Synechococcus* leading to tolerance against photoinhibition, pigment degradation and UV-B sensitivity of photosynthesis. Photochem Photobiol 73: 551-555.

Arai T, Nishijima M, Adachi K, Sano H (1992) Isolation and structure of a UV absorbing substance from the marine bacterium *Micrococcus* sp. AK-334. Marine Biotechnology Institute, Tokyo, pp 88-94.

Balskus EP, Walsh CT (2008) Investigating the initial steps in the biosynthesis of cyanobacterial sunscreen scytonemin. J Am Chem Soc 130: 15260-115261.

Balskus EP, Walsh CT (2010) The genetic and molecular basis for sunscreen biosynthesis in cyanobacteria. Science 329: 1653-1656.

Bandaranayake WM (1998) Mycosporines: are they nature's sunscreens? Nat Prod Rep 15: 159-172.

Bandaranayake WM, des Rocher A (1999) Role of secondary metabolites and pigments in the epidermal tissues, ripe ovaries, viscera, gut contents and diet of the sea cucumber *Holothuria atra*. Mar Biol 133: 163-169.

Barber J, Iwata S (2005) Refined X-ray structure if photosystem II and its implications, p. 469–489. *In* T. Wydrzynski, and K. Satoh (ed.*), Advances in photosynthesis and respiration*, vol. 22. Photosystem II: the light driven water:plastiquinone oxidoreductase. Springer, Dordrecht, The Netherlands.

Beyer P, Mayer M, Kleinig H (1989) Molecular oxygen and the state of geometric isomerism of intermediates are essential in the carotene desaturation and cyclization reactions in daffodil chromoplasts. Eur J Biochem 184: 141-150.

Bidigare RR, Schofield O, Prezelin BB (1989) Influence of zeaxanthin on quantum yield of photosynthesis of *Synechococcus* clone WH7803 (DC2). Mar Ecol Prog Ser 56: 177-188.

Bird G, Fitzmaurice N, Dunlap WC, Chalker BE, Bandaranayake WM (1987) Sunscreen compositions and compounds for use therein. International patent application PCT/AU87/00330, publication no. WO 88/02251. Australian patent 595075. ICI Australia Operations Pty Ltd and Australian Institute of Marine Science, Townsville.

Boronowsky U, Wenk SO, Schneider D, Jäger C, Rögner M. (2001) Isolation of membrane protein subunits in their native state: evidence for selective binding of chlorophyll and carotenoid to the *b*6 subunit of the cytochrome b_6f complex. Biochim Biophys Acta 1506: 55-66.

Boulay C, Abasova L, Six C, Vass I, Kirilovsky D (2008) Occurrence and function of the orange carotenoid protein in photoprotective mechanisms in various cyanobacteria. Biochim et Biophy Acta 1777: 1344-1354.

Brenowitz S, Castenholz RW (1997) Long-term effects of UV and visible irradiance on natural populations of a scytonemin containing cyanobacterium (*Calothrix* sp.). FEMS Microbiol Ecol 24: 343-352.

Brown BE, Ambarsari I, Warner ME, Fitt WK, Dunne RP, Gibb SW, Cummings DG (1999) Diurnal changes in photochemical efficiency and xanthophyll concentrations in shallow water reef corals: evidence for photoinhibition and photoprotection. Coral Reefs 18: 99-105.

Büdel B, Karsten U, Garcia-Pichel F (1997) Ultraviolet-absorbing scytonemin and mycosporine-like amino acid derivatives in exposed, rock-inhabiting cyanobacterial lichens. Oecologia 112: 165-172.

Bultel-Poncé V, Felix-Theodore F, Sarthon C, Ponge J-F, Bodo B (2004) New pigments from the terrestrial cyanobacterium *Scytonema* sp. collected on the Mitaraka Inselberg, French Guyana. J Nat Prod 67: 678-681.

Callone AI, Carignan M, Montoya NG, Carreto JI (2006) Biotransformation of mycosporine like amino acids (MAAs) in the toxic dinoflagellate *Alexandrium tamarense.* J Photochem Photobiol B Biol 84: 204-212.

Camara B, Dogbo O (1986) Demonstration and Solubilization of lycopene cyclase from *Capsicum* chromoplast membranes. Plant Physiol 80: 172-184.

Canto de Loura I, Dubacq JP, Thomas JC (1987) The effects of nitrogen deficiency on pigments and lipids of cyanobacteria. Plant Physiol 83: 838-843.

Cardozo KHM, Guaratini T, Barros MP, Falcão VR, Tonon AP, Lopes NP, Campos S, Torres MA, Souza AO, Colepicolo P, Pinto E (2007) Metabolites from algae with economical impact. Comp Biochem Physiol C 146: 60-78.

Carignan MO, Cardozo KHM, Oliveira-Silva D, Colepicolo P, Carreto JI (2009) Palythine-treonine, a major novel mycosporine-like amino acid (MAA) isolated from the hermatypic coral *Pocillopora capitata*. J Photochem Photobiol B 94: 191-200.

Carreto JI, Carignan MO, Montoya NG (2001) Comparative studies on mycosporine-like amino acids, paralytic shellfish toxins and pigment profiles of the toxic dinoflagellates *Alexandrium tamarense, A. catenella* and *A. minutum*. Mar Ecol Progr Ser 223: 49-60.

Carreto JI, Carignan MO, Montoya NG (2005) A high resolution reverse-phase liquid chromatography method for the analysis of mycosporine-like amino acids (MAAs) in marine organisms. Mar Biol 146: 237.

Carreto JI, de Marco S, Lutz VA (1989) UV-absorbing pigments in the dinoflagellates *Alexandrium excavatum* and *Prorocentrum micans*: Effects of light intensity. In *Red Tides: Biology, Environmental Science and Toxicology*; (Eds. Okaichi, T., Anderson, D.M., Nemoto, T.) Elsevier, New York, NY, USA. pp. 37-40.

Carreto JI, Lutz VA, Dc Marco SG, Carignan MO (1990) Fluence and wavelength dependence of mycosporine-like amino acid synthesis in the dinoflagellate *Alexandrium excavatum*. In: *Toxic marine phytoplankton* (Eds. Graneli E, Edler L, Sundström B, Anderson DM), Elsevier, New York. pp. 275-279.

Castenholz RW (1997) Multiple strategies for UV tolerance in cyanobacteria. The Spectrum 10: 10-16.

Castenholz RW, Garcia-Pichel F (2000) Cyanobacterial responses to UV-radiation. In: *The Ecology of Cyanobacteria: Their Diversity in Time and Space* (Eds. Whitton BA, Potts M), Kluwer: Dordrecht, The Netherlands. pp. 591-611.

Chalmers PJ, Fitzmaurice N, Rigg DJ, Thang SH, Bird G (1990) UVabsorbing compounds and compositions. International Patent Application PCT/AU90/00078, publication no. WO 90/09995. Australian patent 653495. ICI Australia Operations Pty Ltd and Australian Institute of Marine Science, Townsville.

Chamovitz D, Misawa N, Sandmann G, Hirschberg J (1992) Molecular cloning and expression in *Escherichia coli* of a cyanobacterial gene coding for phytoene synthase, a carotenoid biosynthesis enzyme. FEBS Lett 296: 305-310.

Chioccara F, Della Gala A, De Rosa M, Novellino E, Prota G (1980) Mycosporine aminoacids and related compounds from the eggs of fishes. Bulletin de la Societe´ Chimique de Belgique 89: 1101-1106.

Coba FDL, Aguilera J, Figueroa FL (2007a) Use of mycosporine-type amino acid shinorine as an antioxidant. Intl Patent WO2007/026038A2.

Coba FDL, Aguilera J, Figueroa FL (2007b) Use of mycosporine-type amino acid Porphyra-334 as an antioxidant. Intl Patent WO2007/026035 A2.

Coba FDL, Aguilera J, Figueroa FL, de Gálvez MV, Herrera E (2009) Antioxidant activity of mycosporine-like amino acids isolated from three red macroalgae and one marine lichen. J Appl Phycol 21:161-169.

Cockell CS, Knowland J (1999) Ultraviolet radiation screening compounds. Biol Rev 74: 311-345.

Coesel S, Obornßk M, Varela J, Falciatore A, Bowler C (2008) Evolutionary origins and functions of the carotenoid biosynthetic pathway in marine diatoms. PLoS ONE 3: e2896.

Cogdell RJ, Gardiner AT (1993) Functions of carotenoids in photosynthesis. Meth Enzymol 214: 185-193.

Conde FR, Churio MS, Previtali CM (2004) The deactivation pathways of the excited-states of the mycosporine-like amino acids shinorine and porphyra-334 in aqueous solution. Photochem Photobiol Sci 3: 960-967.

Crutzen PJ (1992) Ultraviolet on the increase. Nature 356: 104-105.

Cunningham FX Jr, Chamovitz D, Misawa N, Gantt E, Hirschberg J (1993) Cloning and functional expression in *Escherichia coli* of a cyanobacterial gene for lycopene cyclase, the enzyme that catalyzes the biosynthesis of β- carotene. FEBS Lett 328:130-138.

Cunningham FX, Lee H, Gantt E (2007) Carotenoid biosynthesis in the primitive red alga *Cyanidioschyzon merolae*. Eukaryot Cell 6: 533-545.

Daniel S, Cornelia S, Fred Z (2004) UV-A sunscreen from red algae for protection against premature skin aging. Cosmet Toilet Manuf Worldw 139-143.

Davidson AT, Brarnich D, Marchant HJ, McMinn A (1994) Effects of UV-B irradiation on growth and survival of Antarctic marine diatoms. Mar Biol 119: 507-515.

De Las Rivas J, Telfer A, Barber J (1993) 2-coupled carotene molecules protect P680 from photodamage in isolated Photosystem II reaction centres. Biochim Biophys Acta 1142: 155-164.

Demmig-Adams B (1990) Carotenoids and photoprotection in plants: a role for the xanthophyll zeaxanthin. Biochim Biophys Acta 1020: 1-24.

Demmig-Adams B, Adams WW III (1992) Photoprotection and other responses of plants to high light stress. Ann. Rev. Plant Physiol Plant Mol Biol 43: 599-626.

Demmig-Adams B, AdamsWW III, Czygan F-C, Schreiber U, Lange OL (1990) Differences in the capacity for radiationless energy dissipation in the photochemical apparatus of the green and blue-green algal lichens associated with differences in carotenoid composition. Planta 180: 582-589.

Desikachary TV (1959) Cyanophyta. New Delhi, India: Indian Council of Agriculture Research, 1-686.

Dillon JG, Castenholz RW (1999) Scytonemin, a cyanobacterial sheath pigment, protects against UVC radiation: implications for early photosynthetic life. J Phycol 35: 673-681.

Dillon JG, Tatsumi CM, Tandingan PG, Castenholz RW (2002) Effect of environmental factors on the synthesis of scytonemin, a UV-screening pigment, in a cyanobacterium (*Chroococcidiopsis* sp.). Arch Microbiol 177: 322-331.

Dionisio-Sese ML, Ishikura M, Maruyama T, Miyachi S (1997) UV-absorbing substances in the tunic of a colonial ascidian protect its symbiont, *Prochloron* sp., from damage by UV-B radiation. Mar Biol 128: 455-461.

Dunlap WC, Chalker BE (1986) Identification and quantification of near-UV absorbing compound (S-320) in a hermatypic scleractinian. Coral Reefs 5: 1-5.

Dunlap WC, Chalker BE, Bandaranayake WM, Wu Won JJ (1998) Nature's sunscreen from the Great Barrier Reef, Australia. J Cosmet Sci 20: 41-51.

Dunlap WC, Chalker BE, Oliver JK (1986) Bathymetric adaptations of reef-building corals at Davies Reef, Great Barrier Reef. Australia, III. UV-B absorbing compounds J Exp Mar Biol Ecol 104: 239-248.

Edge R, McGarvey DJ, Truscott TG (1997) The carotenoids as anti-oxidants-a review. J Photochem Photobiol B: Biol 41: 189-200.

Ehling-Schulz M, Bilger W, Scherer S (1997) UV-B induced synthesis of photoprotective pigments and extracellular polysaccharides in the terrestrial cyanobacterium *Nostoc commune*. J Bacteriol 179:1940-1945.

Favre-Bonvin J, Arpin N, Brevard C (1976) Structure de la mycosporine (P310). Can J Chem 54: 1105-1113.

Favre-Bonvin J, Bernillon J, Salin N, Arpin N (1987) Biosynthesis of mycosporines: mycosporine glutaminol in *Tricothecium roseum.* Phytochemistry 26: 2509-2514.

Fleming ED, Castenholz RW (2007) Effects of periodic desiccation on the synthesis of the UV-screening compound, scytonemin, in cyanobacteria. Environ Microbiol 9: 1448-1455.

Fleming ED, Castenholz RW (2008) Effects of nitrogen source on the synthesis of the UV screening compound, scytonemin, in the cyanobacterium *Nostoc punctiforme* PCC 73102. FEMS Microbiol Ecol 63: 301-308.

Frame ER (2004) Mycosporine-like amino acids (MAAs) in bloom forming phytoplankton: The influence of nitrogen, ultraviolet radiation and species composition. Ph.D. Thesis, University of California, San Diego, CA, USA.

Garcia-Pichel F, Belnap J (1996) Microenvironments and microscale productivity of cyanobacterial desert crusts. J Phycol 32: 774-782.

Garcia-Pichel F, Castenholz RW (1991) Characterization and biological implications of scytonemin, a cyanobacterial sheath pigment. J Phycol 27: 395-409.

Garcia-Pichel F, Castenholz RW (1993) Occurrence of UV-absorbing mycosporine-like amino acids among cyanobacterial isolates and an estimate of their screening capacity. Appl Environ Microbiol 59: 163-169.

Garcia-Pichel F, Nu¨bel U, Muyzer G (1998) The phylogeny of unicellular, extremely halotolerant cyanobacteria. Arch Microbiol 169: 469-482.

Garcia-Pichel F, Sherry ND, Castenholz RW (1992) Evidence for a UV sunscreen role of the extracellular pigment scytonemin in the terrestrial cyanobacterium *Chlorogloeopsis* sp. Photochem Photobiol 56: 17-23.

Garcia-Pichel F, Wingard CE, Castenholz RW (1993) Evidence regarding the UV sunscreen role of a mycosporine-like compound in the cyanobacterium *Gloeocapsa* sp. Appl Environ Microbiol 59: 170-176.

Geitler L (1932) Cyanophyceae (Blaulagen). In: Rabenhorst L (ed.) Kryptogamen-Flora von Deutschland, 14th edn, Osterreich und der Schweiz Leipzig: Akademische Verlags Gesellschaft, pp. 1-1196.

Goodwin TW (1980) Carotenoids. In Encyclopedia of Plant Physiology (Eds Bell E A and Charlwood BV), Springer-Verlag, Berlin, vol. 8, pp. 257-281.

Götz T, Windhovel U, Böger P, Sandmann G (1999) Protection of photosynthesis against UV-B radiation by carotenoids in transformants of the cyanobacterium *Synechococcus* PCC 7942. Plant Physiol 120: 599-604.

Graham JE, Lecomte JT, Bryant DA (2008) Synechoxanthin: an aromatic C40 xanthophyll is a major carotenoid in the cyanobacterium *Synechococcus* sp. PCC 7002. J Nat Prod 71: 1647-1650.

Grant PT, Plack PA, Thomson RH (1980) Gadusol, a metabolite from fish eggs. Tetrahedron Lett 21: 4043-4044.

Häder D-P, Kumar HD, Smith RC, Worrest RC (2007) Effects of solar UV radiation on aquatic ecosystems and interactions with climate change. Photochem Photobiol Sci 6: 267-285.

Hansson L-A, Hylander S (2009) Effects of ultraviolet radiation on pigmentation, photoenzymatic repair, behavior, and community ecology of zooplankton. Photochem Photobiol Sci 8: 1266-1275.

Helbling EW, Menchi CF, Villafañe VE (2002) Bioaccumulation and role of UV-absorbing compounds in two marine crustacean species from Patagonia, Argentina. Photochem Photobiol Sci 1: 820-825.

Helms GL, Moore RE, Niemczura WP, Patterson GML, Tomer KB, Gross ML (1988) Scytonemin A, a novel calcium antagonist from a blue-green alga. J Org Chem 53: 1298-1307.

Hirschberg J, Chamovitz D (1994) Carotenoids in Cyanobacteria. In: *The Molecular Biology of Cyanobacteria* (Ed Bryant DA), Kluwer Academic Publishers, The Netherlands. pp. 559-579.

Holt TK, Krogmann DW (1981) A carotenoid protein from cyanobacteria. Biochim Biophys Acta 637: 408-414.

Humbeck K, Römer S and Senger H (1989) Evidence for the essential role of carotenoids in the assembly of an active Photosystem II. Planta 179: 242-250.

Jha RK, Zi-rong X (2004) Biomedical compounds from marine organisms. Mar Drugs 2: 123-146.

Jordan P, Fromme P, Witt HT, Klukas O, Sanger W, Krauss N (2001) Three-dimensional structure of cyanobacterial photosystem I at 2.5 Å resolution. Nature 411: 909-911.

Jurgens UJ, Weckesser J (1985) Carotenoid-containing outer membrane of *Synechocystis* sp. strain PCC6714. J Bacteriol 164: 384-389.

Karsten U, García Pichel F (1996) Carotenoids and mycosporine-like amino acids compounds in members of the genus *Microcoleus* (Cyanobacteria): A chemosystematic study. System Appl Microbiol 19: 285-294.

Karsten U, Garcia-Pichel F (1996) Carotenoids and mycosporine- like amino acid compounds in members of the genus *Microcoleus* (cyanobacteria): a chemosystematic study. Syst Appl Microbiol 19: 285-294.

Karsten U, Sawall T, Wiencke C (1998) Survey of the distribution of UV-absorbing substances in tropical macroalgae. Phycol Res 46: 271-279.

Kerfeld CA (2004a) Structure and function of the water-soluble carotenoid-binding proteins in cyanobacteria. Photosynth Res 81: 215-225.

Kerfeld CA (2004b) Water-soluble carotenoid proteins of cyanobacteria. Arch Biochem Biophys 430: 2-9.

Khomutov G, Fry IV, Huflejt ME, Packer L (1990) Membrane lipid composition, fluidity and surface charge changes in response to growth of the fresh water cyanobacterium *Synechococcus* 6311 under high salinity. Arch Biochem Biophys 277: 263-267.

Kirilovsky D (2007) Photoprotection in cyanobacteria: the orange carotenoid protein (OCP)-related non-photochemical-quenching mechanism. Photosynth Res 93: 7-16.

Klisch M, Sinha RP, Richter PE, Häder D-P (2001) Mycosporine-like amino acids (MAAs) protect against UV-B-induced damage in *Gyrodinium dorsum Kofoid*. J Plant Physiol 158: 1449-1454.

Korbee N, Figueroa FL, Aguilera FJ (2005a) Effect of light quality on the accumulation of photosynthetic pigments, proteins and mycosporine-like amino acids in the red alga *Porphyra leucosticta* (Bangiales, Rhodophyta). J Photochem Photobiol B Biol 80: 71-78.

Korbee N, Huovinen P, Figueroa FL, Aguilera J, Karsten U (2005b) Availability of ammonium influences photosynthesis and the accumulation of mycosporine-like amino acids in two *Porphyra* species (Bangiales, Rhodophyta). Mar Biol 146: 645-654.

Kramlich JC and Linak WP (1994) Nitrous oxide behaviour in the atmosphere, and in combustion and industrial systems. Prog Energy Combust Sci 20: 149.

Lamare MD, Hoffman J (2004) Natural variation of carotenoids in the eggs and gonads of the echinoid genus, Strongylocentrotus: implications for their role in ultraviolet radiation photoprotection. J Exp Mar Biol Ecol 312: 215-233.

Liaaen-Jensen S, Andrews AG (1972) Microbial carotenoids. Annu Rev Microbiol 26: 225-248.

Litchman E, Neale PJ, Banaszak AT (2002) Increased sensitivity to ultraviolet radiation in nitrogen-limited dinoflagellates: photoprotection and repair. Limnol Oceanogr 47: 86-94.

Liu Z, Häder DP, Sommaruga R (2004) Occurrence of mycosporine-like amino acids (MAAs) in the bloom-forming cyanobacterium *Microcystis aeruginosa.* J Plankton Res 26: 963-966.

Lubin D and Jensen EH (1995) Effects of clouds and stratospheric ozone depletion on ultraviolet radiation trends. Nature 377: 710-713.

Maragos JE (1972) A study of the ecology of Hawaiian reef corals. Ph.D. Thesis, University of Hawaii, Honolulu, HI, USA.

Matsunaga T, Burgess JG, Yamada N, Komatsu K, Yoshida S, Wachi Y (1993) An ultraviolet (UV-A) absorbing biopterin glucoside from the marine planktonic cyanobacterium *Oscillatoria* sp. Appl Microbiol Biotechnol 39: 250-253.

Miller M, Pedersen JZ, Cox RP (1987) Effect of growth temperature on membrane properties in a thermophilic cyanobacterium (*Synechococcus* sp.). In: *Progress in Photosynthesis Research* (Ed Biggins J), Martinus-Nijhoff, Dordrecht. Vol II, pp. 789-792.

Millie DF, Ingram DA, Dionigi CP (1990) Pigment and photosynthetic response of *Oscillatoria agardhii* (*Cyanophyta*) to photon flux density and spectral quality. J Phycol 26: 660-666.

Misonou T, Saitoh J, Oshiba S, Tokitomo Y, Maegawa M, Inoue Y, Hori H, Sakurai T (2003) UV-absorbing substance in the red alga *Porphyra yezoensis* (Bangiales, Rhodophyta) block thymine dimer production. Mar Biotechnol 5: 194-200.

Moeller RE, Gilroy S, Williamson CE, Grad G (2005) Dietary acquisition of photoprotective compounds (mycosporine-like amino acids, carotenoids) and acclimation to ultraviolet radiation in a freshwater copepod. Limnol Oceanogr 50: 427-439.

Murata N, Sato N, Kuwabara Y (1981) Separation and characterization of thylakoid and cell envelop of the blue-green alga (cyanobacterium) *Anacystis nidulans*. Plant Cell Phys 22: 855-866.

Nägeli C (1849) Gattungen einzelliger Algen, physiologisch und systematisch bearbeitet Neue Denkschrift. Allg. Schweiz. Nat Ges 10: 1-138.

Nakayama R, Tamura Y, Kikuzaki H, Nakatani N (1999) Antioxidant effect of the constituents of susabinori (*Porphyra yezoensis*). JAOCS 76: 649-653.

Newell RW, van Amerongen H, Barber J, van Grondelle R (1993) Spectroscopic characterisation of the reaction center of Photosystem II using polarised light: Evidence for excitons in PS II reaction centers. Biochim Biophys Acta 1057: 232-238.

Newman PJ, Sherman LA (1978) Isolation and characterization of Photosystem I and II membrane particles from the bluegreen alga *Synechococcus cedrorum.* Biochim Biophys Acta 503: 343-361.

Niyogi KK (1999) Photoprotection revisited: genetic and molecular approaches. Annu Rev Plant Physiol Plant Mol Biol 50: 333-359.

Noguchi Y, Ishii A, Matsushima A, Haishi D, Yasumuro Ken-ichi, Moriguchi T, Wada T, Kodera Y, Hiroto M, Nishimura H, Sekine M, Inada Y (1999) Isolation of Biopterin-α-glucoside from *Spirulina (Arthrospira) platensis* and its physiologic function. Mar Biotechnol 1: 207-210.

Nultsch W, Schuchart H (1985) A model of the phototactic reaction chain of cyanobacterium *Anabaena variabilis.* Arch Microbiol 142: 180-184.

Olie J, Potts M (1986) Purification and biochemical analysis of the cytoplasmic membrane from the desiccation-tolerant cyanobacterium *Nostoc commune* UTEX 584. Appl Environ Microbiol 52: 706-710.

Omata T, Murata N (1983) Isolation and characterization of the cytoplasmic membranes from the blue-green alga (cyanobacterium) *Anacystis nidulans*. Plant Cell Phys 24: 1101-1112.

Oren A, Gunde-Cimerman N (2007) Mycosporines and mycosporine-like amino acids: UV protectants or multipurpose secondary metabolites? FEMS Microbiol Lett 269:1-10.

Oyamada C, Kaneniwa M, Ebitani K, Murata M, Ishihara K (2008) Mycosporine-like amino acids extracted from Scallop (*Patinopecten yessoensis*) ovaries: UV protection and growth stimulation activities on human cells. Mar Biotech 10:141-150.

Peinado NK, Dßaz RTA, Figueroa FL, Helbling EW (2004) Ammonium and UV radiation stimulate the accumulation of mycosporine-like amino acids in *Porphyra columbina* (Rhodophyta) from Patagonia, Argentina. J Phycol 40: 248-259.

Penna DD, Pogson BJ (2006) Vitamin Synthesis in Plants: Tocopherols and Carotenoids. Annu Rev Plant Biol 57: 711-738.

Pentecost A (1993) Field relationships between scytonemin density, growth and irradiance in cyanobacteria occurring in low illumination regimes. Microb Ecol 26: 101-110.

Plack PA, Fraser NW, Grant PT, Middleton C, Mitchell AI, Thomson RH (1981) Gadusol, an enolic derivative of cyclohexane-1, 3-dione present in the roes of cod and other marine fish. Biochemical J 199: 741-747.

Polivka T, Sundström V (2004) Ultrafast dynamics of carotenoid excited states-from solution to natural and artificial systems. Chem Rev 104: 2021-2071.

Portwich A, Garcia-Pichel F (1999) Ultraviolet and osmotic stress induce and regulate the synthesis of mycosporines in the cyanobacterium *Chlorogloeopsis* PCC 6912. Arch Microbiol 172: 187-192.

Portwich A, Garcia-Pichel F (2000) A novel prokaryotic UVB photoreceptor in the cyanobacterium Chlorogloeopsis PCC 6912. Photochem Photobiol 71: 493-498.

Portwich A, Garcia-Pichel F (2003) Biosynthetic pathway of mycosporines (mycosporine-like amino acids) in the cyanobacterium *Chlorogloeopsis* sp. strain PCC 6912. Phycologia 42: 384-392.

Proteau PJ, Gerwick WH, Garcia-Pichel F, Castenholz RW (1993) The structure of scytonemin, an ultraviolet sunscreen pigment from the sheath of cyanobacteria. Experimentia 49: 825-829.

Quesada A, Vincent WF, Lean-David RS (1999) Community and pigment structure of arctic cyanobacterial assemblages: the occurrence and distribution of UV-absorbing compounds. FEMS Microbiol Ecol 28: 315-323.

Rastogi RP, Richa, Sinha RP, Singh SP, Häder D-P (2010) Photoprotective compounds from marine organisms. J Ind Microbiol Biotechnol 37: 537-558.

Rastogi RP, Sinha RP (2009) Biotechnological and industrial significance of cyanobacterial secondary metabolites. Biotechnol Adv 27: 521-539.

Richa, Sinha RP (2011) UV-mediated stress and its mitigation in cyanobacteria. Intern. J. Plant Animal Env Sci 1: 155-166.

Richa (2014) Physiological and Biochemical studies on mycosporine-like amino acids (MAAs) in cyanobacteria. Ph. D. Thesis, Banaras Hindu University, Varanasi.

Richter PR, Sinha RP, Häder D-P (2006) Scytonemin-rich epilithic cyanobacteria survive acetone treatment. Curr Trends Microbiol 2:13-19.

Schagerl M, Müller B (2006) Acclimation of chlorophyll *a* and carotenoid levels to different irradiances in four freshwater cyanobacteria. J Plant Physiol 163: 709-716.

Schlodder E, Brettel K (1988) Primary charge separation in closed Photosystem II with a lifetime of 11 ns. Flash-absorption spectroscopy with oxygen-evolving Photosystem II complexes from *Synechococcus*. Biochim Biophys Acta 933: 22-34.

Shibata K (1969) Pigments and a UV-absorbing substance in coral and a blue-green alga living on the Great barrier reef. Plant Cell Physiol 10: 325-335.

Shick JM, Dunlap WC (2002) Mycosporine-like amino acids and related gadusols: biosynthesis, accumulation, and UV protective function in aquatic organisms. Annu Rev Physiol 64: 223-262.

Shick JM, Romaine-Lioud S, Ferrier-Pagès C, Gattuso JP (1999) Ultraviolet-B radiation stimulated shikimate pathway dependent accumulation of mycosporine-like amino acids in the coral *Stylophora pistillata* despite decrease in its population of symbiotic dinoflagellates. Limnol Oceanogr 44: 1667-1682.

Sies H, Stahl W (2004) Nutritional protection against skin damage from sunlight. Annu Rev Nutr 24: 173-200.

Singh KL and Sinha RP (2011) UV-absorbing compounds in algae. In: *Advances in Life Sciences* (Eds. Sinha, R.P., Sharma, N.K. and Rai, A.K.), I.K. International Publishing House Pvt. Ltd., New Delhi, India. pp. 213-239.

Singh SP, Kumari S, Rastogi RP, Singh KL, Sinha RP (2008a) Mycosporine-like amino acids (MAAs): chemical structure, biosynthesis and significance as UV-absorbing/screening compounds. Ind J Exp Biol 46:7-17.

Singh SP, Sinha RP, Klisch M, Häder D-P (2008b) Mycosporine-like amino acids (MAAs) profile of a rice-field cyanobacterium *Anabaena doliolum* as influenced by PAR and UVR. Planta 229: 225-233.

Singh SP, Klisch M, Sinha RP, Häder DP (2008c) Effects of abiotic stressors on synthesis of the mycosporine-like amino acid shinorine in the cyanobacterium *Anabaena variabilis* PCC 7937. Photochem Photobiol 84: 1500-1505.

Singh SP, Klisch M, Sinha RP, Häder DP (2010a) Genome mining of mycosporine-like amino acid (MAA) synthesizing and non-synthesizing cyanobacteria: A bioinformatics study. Genomics 95: 120-128.

Singh SP, Häder D-P, Sinha RP (2010 b) Cyanobacteria and ultraviolet radiation (UVR) stress: mitigation strategies. Ageing Res Rev 9: 79-90.

Singh SP, Klisch M, Sinha RP, Häder DP (2010c) Sulfur deficiency changes mycosporine-like amino acid (MAA) composition of *Anabaena variabilis* PCC 7937: A possible role of sulfur in MAA bioconversion. Photochem Photobiol 86: 862-870.

Sinha RP, Ambasht NK, Sinha JP, Klisch M, Häder D-P (2003) UVB-induced synthesis of mycosporine-like amino acids in three strains of *Nodularia* (cyanobacteria). J Photochem Photobiol B Biol 71: 51-58.

Sinha RP, Häder D-P (2002) UV-induced DNA damage and repair: a review. Photochem Photobiol Sci 1: 225-236.

Sinha RP, Häder DP (2008) UV-protectants in cyanobacteria. Plant Sci 174: 278-289.

Sinha RP, Klisch M, Gröniger A, Häder DP (1998) Ultraviolet-absorbing/screening substances in cyanobacteria, phytoplankton and macroalgae. J Photochem Photobiol B 47: 83-94.

Sinha RP, Klisch M, Helbling EW, Häder D-P (2001) Induction of mycosporine-like amino acids (MAAs) in cyanobacteria by solar ultraviolet-B radiation. J Photochem Photobiol B: Biol 60: 129-135.

Sinha RP, Klisch M, Vaishampayan A, Häder DP (1999). Biochemical and spectroscopic characterization of the cyanobacterium *Lyngbya* sp. inhabiting Mango (*Mangifera indica*) trees: Presence of an ultraviolet-absorbing pigment, Scytonemin. Acta Protozool 38: 291-298.

Sinha RP, Kumar A, Tyagi MB, Häder DP (2005) Ultraviolet-B-induced destruction of phycobiliproteins in cyanobacteria. Physiol Mol Biol Plants 11: 313-319.

Sinha RP, Kumari S, Rastogi RP (2008a) Impacts of ultraviolet-B radiation on cyanobacteria: photoprotection and repair. J Sci Res 52: 125-142.

Sinha RP, Rastogi RP, Ambasht NK, Häder D-P (2008b) Life of wetland cyanobacteria under enhancing solar UV-B radiation. Proc Natl Acad Sci India B 78: 53-65.

Sinha RP, Lebert M, Kumar A, Kumar HD, Häder DP (1995a) Disintegration of phycobilisomes in a rice field cyanobacterium *Nostoc* sp. following UV irradiation. Biochem Mol Biol Int 37: 697- 706.

Sinha RP, Lebert M, Kumar A, Kumar HD, Häder D-P (1995b) Spectroscopic and biochemical analyses of UV effects on phycobiliproteins of *Anabaena* sp. and *Nostoc commune*. Bot Acta 180: 87-92.

Sinha RP, Singh N, Kumar A, Kumar HD, Häder M, Häder DP (1996) Effects of UV-irradiation on certain physiological and biochemical processes in cyanobacteria. J Photochem Photobiol B Biol 32: 107-113.

Sinha RP, Singh SP, Häder D-P (2007) Database on mycosporines and mycosporine-like amino acids (MAAs) in fungi, cyanobacteria, macroalgae, phytoplankton and animals. J Photochem Photobiol B Biol 89: 29-35.

Sinha RP, Sinha JP, Gröniger A, Häder D-P (2002) Polychromatic action spectrum for the induction of a mycosporine-like amino acid in a rice-field cyanobacterium, *Anabaena* sp. J Photochem Photobiol B Biol 66: 47-53.

Sivalingam PM, Ikawa T, Yokohama Y, Nisizawa K (1974) Distribution of a 334 UV-absorbing-substance in algae, with special regard of its possible physiological roles. Bot Mar 17: 23-29.

Smith RC, Pre′zelin BB, Baker KS, Bidigare RR, Boucher NP, Coley T, Karentz D, MacIntyre S, Matlick HA, Menzies D, Ondrusek M, Wan Z, Waters KJ(1992) Ozone depletion: ultraviolet radiation and phytoplankton biology in Antarctic waters. Science 255: 952-959.

Smith RC, Prézelin BB, Baker KS, Bidigare RR, Boucher NP, Coley T (1992) Ozone depletion: ultraviolet radiation and phytoplankton biology in Antarctic waters. Science 255: 952-959.

Solomons NW, Bulux J (1994) Plant sources of pro-vitamin A and human nutriture. Nutr Rev 51: 199-204.

Sommaruga R, Garcia-Pichel F (1999) UV-absorbing mycosporine-like compounds in planktonic and benthic organisms from a high-mountain lake. Arch Hydrobiol 144: 255-269.

Sorrels CM, Proteau PJ, Gerwick WH (2009) Organization, evolution, and expression analysis of the biosynthetic gene cluster for scytonemin, a cyanobacterial UV-absorbing pigment. Appl Environ Microbiol 75: 4861-4869.

Soule T, Garcia-Pichel F, Stout V (2009a) Gene expression patterns associated with the biosynthesis of the sunscreen scytonemin in *Nostoc punctiforme* ATCC 29133 in response to UVA radiation. J Bacteriol 191: 4639-4646.

Soule T, Palmer K, Gao Q, Potrafka RM, Stout V, Garcia-Pichel F (2009b) A comparative genomics approach to understanding the biosynthesis of the sunscreen scytonemin in cyanobacteria. BMC Genomics 10: 336-345.

Soule T, Stout V, Swingley WD, Meeks JC, Garcia-Pichel F (2007) Molecular genetics and genomic analysis of scytonemin biosynthesis in *Nostoc punctiforme* ATCC 29133. J Bacteriol 189: 4465-4472.

Stahl W, Sies H (2007) Carotenoids and flavonoids contribute to nutritional protection against skin damage from sunlight. Mol Biotechnol 37: 26-30.

Stanier RY, Cohen-Bazire G (1977) Phototrophic prokaryotes: the cyanobacteria. Annu Rev Microbiol 31: 225-274.

Starcevic A, Akthar S, Dunlap WC, Shick JM, Hranueli D, Cullum J, Long PF (2008) Enzymes of the shikimic acid pathway encoded in the genome of a basal metazoan, *Nematostella vectensis*, have microbial origins. PNAS 105: 2533-2537.

Starcevic A, Akthar S, Dunlap WC, Shick JM, Hranueli D, Cullum J, Long PF (2008) Enzymes of the shikimic acid pathway encoded in the genome of a basal metazoan, *Nematostella vectensis*, have microbial origins. Proc Natl Acad Sci USA 105: 2533-2537.

Steiger S, Schäfer L, Sandmann G (1999) High-light-dependent upregulation of carotenoids and their antioxidative properties in the cyanobacterium *Synechocystis* PCC 6803. J Photochem Photobiol B Biol 52: 14-18.

Stevenson CS, Capper EA, Roshak AK, Marquez B, Eichman C, Jackson JR, Mattern M, Gerwick WH, Jacobs RS, Marshall LA (2002) The identification and characterization of the marine natural product scytonemin as a novel antiproliferative pharmacophore. JPET 303: 858-866.

Stickforth P, Steiger S, Hess WR, Sandmann G (2003) A novel type of lycopene ε-cyclase in the marine cyanobacterium *Prochlorococcus marinus* MED4. Arch. Microbiol. 179: 409-415.

Strickland EH, Horwitz J, Billups C (1970) Near-ultraviolet absorption bands of tryptophan, studies using indole and 3-methylindole as models. Biochemistry 9: 4914-4921.

Subramanian A, Carpenter EJ, Karentz D, Falkowski PG (1999) Bio-optical properties of the marine diazotrophic cyanobacteria *Trichodesmium* spp. I. Absorption and photosynthetic action spectra. Limnol Oceanogr 44: 608-617.

Suh H-J, Lee H-W, Jung J (2003) Mycosporine glycine protects biological systems against photodynamic damage by quenching singlet oxygen with a high efficiency. Photochem Photobiol78: 109-113.

Teai T, Raharivelomanana P, Bianchini JP, Faura R, Martín PMV, Cambon A (1997) Structure de deux nouvelles iminomycosporines isolées de *Pocillopora eydouxy*. Tetrahedron Lett 38: 5799-5800.

Thornber JP, Alberte RS, Hunter FA, Shiozawa JA, Kan K-S (1976) The organization of chlorophyll in the plant photosynthetic unit. Brookhaven Symp Biology 28: 132-148.

Torres A, Enk CD, Hochberg M, Srebnik M (2006) Porphyra- 334, a potential natural source for UVA protective sunscreens. Photochem Photobiol Sci 5: 432-435.

Vaishampayan A, Sinha RP, Häder D-P, Dey T, Gupta AK, Bhan U, Rao AL (2001) Cyanobacterial biofertilizers in rice agriculture. Bot Rev 67: 453-516.

Vernet M, Neori A, Haxo FT (1989) Spectral properties and photosynthetic action in red-tide populations of *Prorocentrum micans* and *Gonyaulax polyedra*. Mar Biol 103: 365-371.

Volkmann M, Gorbushina AA (2006) A broadly applicable method for extraction and characterization of mycosporines and mycosporine-like amino acids of terrestrial, marine and freshwater origin. Microbiol Lett 255: 286-295.

Volkmann M, Gorbushina AA, Kedar L, Oren A (2006) Structure of euhalothece-362, a novel red-shifted mycosporine-like amino acid, from a halophilic cyanobacterium (*Euhalothece* sp). Microbiol Lett 258: 50-54.

Volkmann M, Whitehead K, Rütters H, Rullkötter J, Gorbushina AA (2003) Mycosporine-glutamicol-glucoside: A natural UV-absorbing secondary metabolite of rock-inhabiting microcolonial fungi. Rapid Commun Mass Spectrom 17: 897-902.

Wachi Y, Burgess JG, Iwamoto K, Yamada N, Nakamura N, Matsunaga T (1995) Effect of ultraviolet-A (UV-A) light on growth, photosynthetic activity and production of biopterin glucoside by the marine UV-A resistant cyanobacterium *Oscillatoria* sp. Biochimica et Biophysics Acta 1244: 165-168.

Wenk SO, Schnneider D, Boronowsky U, Jager C, Klughammer C, de Weerd FL, Roon H, Vermaas WF, Dekker JP, Rogner M (2005) Functional implications of pigments bound to a cyanobacterial *b6f* complex. FEBS J 272: 582-592.

Wertz K, Hunziker-Buchwald P, Seifert N, Riss G, Neeb M, Steiner G, Goralczyk R (2005) Beta-carotene interferes with ultraviolet light A-induced gene expression by multiple pathways. J Invest Dermatol 124: 428-434.

Whitehead K, Hedges JI (2005) Photodegradation and photosensitization of mycosporine-like amino acids. J Photochem Photobiol 80:115-121.

Wu S, Skolnick J, Zhang Y (2007) Ab initio modeling of small proteins by iterative TASSER simulations, BMC Biol 5: 17.

Xue L, Zhang Y, Zhang T, An L, Wang X (2005) Effect of enhanced Ultraviolet-B radiation on algae and cyanobacteria. Crit Rev Microbiol 31: 79-89.

Yoshida T, Sivalingam PM (1969) UV-absorbing substance in red alga, *Porphyra yezoensis* Ueda. J Tokyo Univ Fish 2: 427-434.

Young AJ (1991) The photoprotective role of carotenoids in higher plants. Physiol Plantarum 83: 702-708.

Zhang L, Li L, Wu Q (2007) Protective effects of mycosporine-like amino acids of *Synechocystis* sp. PCC 6803 and their partial characterization. J Photochem Photobiol B 86: 240-245.

6

Genomics and Structural Analysis of Cyanobacterial Phycobiliproteins

Vinod K. Kannaujiya, Richa and Rajeshwar P. Sinha

Abstract

Synonymous and non-synonymous substitution result in isomerization and deformity in protein structure. Both play more or less important role in evolution. Interestingly, cyanobacterial phycobiliproteins (PBPs) are composed of proteins which are coded by codons by mRNA translation. These proteins are categorized into three groups e.g., C-PC, C-PE and C-APC having α and β subunits. The skeleton and arrangement of amino acids in subunits can provide structural and functional integrity of PBPs as well as GC_3 associated substitution of synonymous and non-synonymous codons. The amino acid composition varies highly under GC_3 nucleotides in Wobble position of codons. We have analyzed the PBPs GC percentage of 11 fully sequenced cyanobacterial genomes. GC contents ranges from 47.7 to 61.8 % in C-PC, C-PE and C-APC in each α and β subunits. At codons level, α subunits of C-PC, C-PE and C-APC expressed GCC (Alanine), Tyrosine (TAC), CTG (Leucine), ATG (Methionine), AAC Asparagine), GAC (Aspartic acid) while in β subunits GGC (Glycine) and CGC (Arginine) are more expressed as compared to other amino acids. These results suggest genomic constraint linked with GC constituents select the codon for particular amino acids. We expect that codon level study particularly in cyanobacteria will provide a novel approach to explore the evolution of proteins at codon levels and will also solve the redundancy of codons in genes and genomics.

Keywords: Genome; Cyanobacteria; Phycobiliproteins; Codon usage; Amino acids; GC_3.

Introduction

Cyanobacteria that originated about 3.5 billion years ago (Sinha and Häder 1996; Guan et al. 2007). are phylogenetically a primitive group of Gram-negative prokaryotes which are capable of oxygenic photosynthesis. They are found in both symbiotic and non-symbiotic association with several organisms (Sinha and Häder 1996). Photosynthetic organisms utilize various type of light harvesting complex that are capable of absorbing the broad spectrum of photonic energy for the enhancement of photosynthesis (Sinha et al. 2005). The photosynthetic prokaryotes such as cyanobacteria and eukaryotic red algae have light harvesting processes which are governed by chromophore containing macromolecular proteins called phycobilisomes (PBSs). Photosynthesis is initiated by the transfer of energy from PBSs to Chl *a* (Adir et al. 2006). Cyanobacterial PBSs are composed of phycobiliproteins (PBPs) and linker proteins. PBPs include C-phycocyanin (C-PC), C-allophycocyanin (C-APC) and C-phycoerythrin (C-PE) (Fig. 1). These are vibrantly colored as blue (C-PC), red (C-PE) and bluish green (C-APC) and constitute about 24 % of dry mass of the cell (Glazer 1985; Grossman et al. 1993; Cai et al. 2001). The PBPs are identified by the wavelength maxima such as C-PC (610-620 nm), C-PE (540-570 nm) and C-APC (650-655 nm). The transfer of radiation energy from PBPs to the thylakoid membrane follows the sequence PE→PC→APC→ reaction centre (Glazer 1989).

PBPs are variable in terms of intensity and their structure in cyanobacteria, red algae and cryptomonads (Grossman et al. 1995). The basic structural element of these PBPs is composed of two different polypeptide chains but probably evolves from common ancestor and is divided into two chains (Thomas and Passaquet 1999). Each PBPs are composed of hetero-dimer subunits (chains) α and β, which are similar in concentration but varies in amino acids constituents and linear tetra-pyrrole chromophore content at different cystein position. The α subunits (12-20 kDa) are generally smaller then â

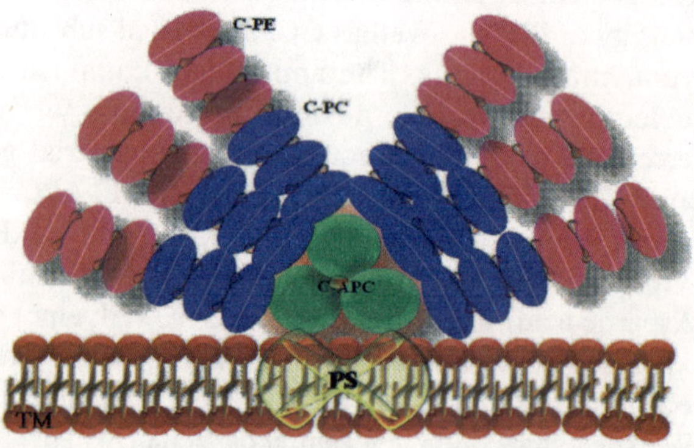

Fig. 1: Structure of phycobiliproteins complex anchored to thylakoid membrane. They are associated with core proteins of allophycocyanin (C-APC) which surround several outwardly oriented rods made up of stacked discs of phycocyanin (C-PC) and phycoerythrin (C-PE).

subunits (15-22 kDa) and each contains one to four chromophores(s) attached with cystein residue at different position of both the subunits (Sinha et al. 1995a; b). The stabilization of protein structure depends on the amino acid composition. Interestingly the G+C nucleotide composition in different genome varies from species to species such as in bacteria it ranges from 22.5% to 72.0% (Kanehisa and Goto 2000).

Prokaryotic nucleotide composition is affected by the natural selection and mutational bias which results in skewness of the genome towards GC or AT (Naya et al. 2001). Variable codon usage results in variation in the amino acids that is also controlled by the drift (Hershberg and Petrov 2008). But the selection is too small ($\sim 10^{-6}$) for desired changes in amino acids or proteins frame work, so it requires a large population size for effective change (dos Reis and Wernisch 2009). The GC variation also depends on the rate of recombination at a particular site (Spencer et al. 2006; Duret and Arndt 2008).

Sueoka (1961) provided first evidence of correlation between the nucleotide and amino acid content of bacterial cells. Porter (1995) identified correlations between the nucleotide composition of DNA and the amino acid content of the expressed proteins. Both coding and non-coding bias originated due to nucleotide variation in particular genome (Musto et al. 1997). Coding bias resulted in the change in amino acid constituents and destabilized the structure whereas non-coding bias does not show the variation in amino acids. Interestingly, the change in few bases can not show the change in amino acids (Lobry 1997). The reason behind the non variable changes in amino acid bases is due to the silence nature of synonymous codons variation at third position (Wada 1992). The non-synonymous codon changes at the third position leads to marked variation in the amino acid composition and protein structural integrity (Singer and Hickey 2000; de Miranda et al. 2000; Knight et al. 2001).

Bharanidharan et al. (2004) have studied 115 bacterial nucleotide frequency in relation to amino acids composition and have demonstrated that Arg, Val, Asp, Glu, Ser, and Cys amino acids varies from genome to genome but Thr, Phe, Lys, and Asn are similar from the predicted value. The amino acid GARP is coded by GC-rich codons [(GGA, GGC, GGT, GGG), (GCT, GCA, GCC, GCG), (CGC, CGA, CGT, CGG, AGA, AGG) and (CCT, CCA, CCG, CCC)]. All these codons are not expressed simultaneously and depend on the preference of tRNA (de Miranda et al., 2000). Another segment is GC-poor amino acids (FYMINK) that codes [(TTT, TTC), (TAT, TAC), (ATG), (ATT, ATC, ATA), (AAT, AAC) and (AAA, AAG)] respectively. These variations in the amino acid are due to selection pressure and mutation bias. Viral codon usage has also shown profound variation in *Picornaviridae*, *Pestivirus*, plant viruses, and vertebrate DNA viruses (Zhang et al 2011). Harrison and Charlesworth

(2011) have reported the GC bias codon usage in eukaryotic *Saccharomyces cerevisiae* and established relationship between recombination frequencies. All of these studies, despite the fact that they were based on a wide variety of different nucleotide composition and amino acids, provide evidence for a significant correlation between gene and proteins. Cyanobacteria have also been found to show the codons variability in the genome as described by Li and Watanabe (2002) in the studied 15 cyanobacterial strains and have reported that the total G+C constituent variation ranges from 35% to 56.4%.

Codons variability in particular subunits of PBPs plays a critical role in capturing the light energy of different wavelength for photosynthesis. The variation in the composition and structural integrity of PBPs in all cyanobacterial species suggests that there is a need to investigate from genome to gene and amino acid to codon level to explore the unknown mystery.

Synonymous and Non-synonymous evolution of proteins

The DNA sequence does not use synonymous codons with equal frequency from species to species. The random use of synonymous codons has been recognized in unicellular and multicellular organisms (Grantham et al. 1981). Apart from that tRNA abundance varies in different genomes during different stages of growth (Dong 1996). The degree of codon bias linked to the acceptance of tRNA and particular gene expression affects the translational efficiency (Blumer 1991).

Highly expressed genes use specific synonymous codons which have abundant tRNA while weakly expressed gene use unspecified synonymous codons (Ikemura et al 1985). Synonymous codons and their coding mRNA sequences carry potential structural information concerning the encoded protein due to the degeneracy of the genetic code (Guisez et al. 1993). Experimentally it has been shown that mRNA consists of specific codons which are translated more ideally than artificially modified codons (Robinson et al. 1984, Sorensen et al. 1989, Xia 1998). Apart from synonymous substitution of codons, there are non-synonymous substitutions which also play an important role in protein structure. Non-synonymous codons results in change in the amino acid in respect to codons variability. Gabriel et al. (2004) attributed that positive selection at non-synonymous and weak selection at synonymous codon usage sites in *D. melanogaster* and *D. yakuba* which comprises of 630 gene pairs. According to several reports both synonymous and non-synonymous substitution have important role in long as well as short-term evolution (Fig. 2). When directly compared to protein structures such as in PBPs, their mRNA sequences carry structural information concerning the encoded protein due to the degeneracy of the genetic code which can be at different nucleotide position. The codon expression frequencies for particular protein helix differ in *E. coli* (Thanaraj et al. 1996a; b). However, Brunak et al. (1996) have reported that position of rare codons under the structural location of protein cannot be confirmed by correlation. Adzhubei et al. (1996) reported that the codon usage may influence the secondary structure of proteins.

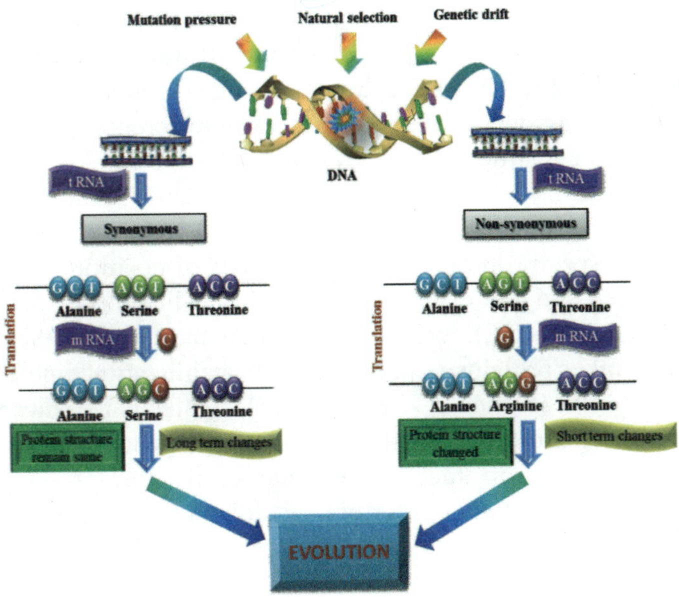

Fig. 2: Synonymous *vs* non-synonymous substitutions in contrast to mutation, genetic drift and natural selection in protein evolution (PDB ID: 1BNA for DNA).

Molecular architecture of PBPs

All PBPs are heterodimeric in nature, formed by the association between α and β subunits. Both α and β subunits are commonly found in trimers $(\alpha\beta)_3$ and hexamers $(\alpha\beta)_6$ form in the fundamental assembly units of PBPs. Hexamers and monomers are inter-convertable, with the equilibrium between the forms depending upon pH (MacColl et al. 1971). Each PBPs have own brilliantly colors and light absorption properties due to the presence of thio- ether which are covalently linked cysteine residues that binds to the chromophore. The C-PC α are formed due to the attachment of tetrapyrrole chromophore to α 84 position through a Cys-84 position *via* thio-ether linkage whereas C-PC β have two chromophores one binded to α84 and other to β155 at the cystein position (Glazer 1985). The *cpcE* and *cpcF* genes transcribe to produce CpcE and CpcF proteins and form a heterodimeric lyase structure that attaches to C-PC Cys-84 (Zhou 1992; Fairchild and Glazer 1994). It has been studied that in

Scynechococcus sp. PCC 7002, which contains seven different sites in cells, may require 16 polypeptides for the process of bilin attachment (Tandeau de Marsac et al. 1990; Zhou et al. 1992)

Chromophore attached PBPs have been divided into phycocyanobilin, phycoerythrobilin, phycourobilin and phycobiliviolin in different cyanobacterial species. Colorless linker proteins accounts for about 15 % and it divided into Group A (70-120 kDa) and Group B (25-70 kDa) (Sinha et al. 2005) in various cyanobacteria. The structural association of α and β subunits are rods and core which are strongly associated with at least seven different linker proteins and molecular chaperons. Structurally aggregated C-PC have linker proteins under the central hole of 25-50 Å diameters (Liu et al. 2005; Adir 2005). Linker proteins have three functions: (i) organization and stabilization, (ii) mode of attachment, (iii) spectral properties. On the basis of position, linker proteins have been divided into four classes: core-membrane linkers (L_{CM}), core linkers (L_C), rod–core linkers (L_{RC}), rod linkers (L_R) (Yu and Glazer 1982). CpcG1 is the major L_{RC} linker proteins required for PBS stabilization whereas, CpcG2 act as antenna to PSI (Kondo et al 2009; 2007). Molecular studies reveal that PBPs polypeptides are often clustered on the large genome fragment where genes for PC and APC are localized (Lomax et al. 1987). C-PC and linker polypeptides gene cluster in different cyanobacterial species transcribes different size of mRNAs (Grossman et al. 1993).

PBPs amino acids skeleton

We have selected 11 cyanobacterial species having both spherical as well as filamentous vegetative cells and collected the protein as well as nucleotide sequences from full genome sequence data base (Table. 1). The α and β subunits of PC with different amino acids composition has been shown in Figure 3. Chromophore attachment to the apo-protein of PBPs is catalyzed by a specific protein bilin lyase which is encoded by the genes (Zhou et al. 1992; Swanson et al. 1992). Chromophore attachment to apo-proteins forms enzyme substrate complex which is determined by the core of particular amino acid. Amino acid frame work may determine the structural integrity, stability and photo-convertibility of cyanobacteria. Cyanobacteria inhabiting diverse habitats have different nature of coloration due to difference in the structural integrity and composition of PBPs. Figure 4 represents the variation of all the 20 amino acids in which cysteine, tryptophan, histidine, and methionine have least occurrence in C-PC α whereas the other amino acid such as alanine, glycine, leucine, serine and threonine have frequent occurrence in C-PC β subunit. Kreil and Ouzounis (2001) reported the reduction of cystein and methionine due to the formation of disulphide bond and incorporation in the initiation codon.

Table 1. Selected cyanobacterial species along with species ID that are used to designate a particular species.

Species ID	Species
Sp1	*Nostoc* sp. PCC 7120
Sp2	*Microcystis aeruginosa* NIES-843
Sp3	*Synechococcus* sp. PCC 7002
Sp4	*Synechococcus elongatus* PCC 7942
Sp5	*Synechococcus elongatus* PCC 6301
Sp6	*Synechococcus* sp. JA-3-3Ab
Sp7	*Synechocystis* sp. PCC6803
Sp8	*Synechococcus* sp. CC9311
Sp9	*Synechococcus* sp. CC9902
Sp10	*Synechococcus* sp. WH8109
Sp11	*Synechococcus* sp. CC9605

Some other amino acids such as, aspartic acid, methionine and valine were found to be expressed in almost double amount. C-PE subunit has been divided into C-PE α_1 α_2 and C-PE β_1 β_2 (Fig. 5). The average fluctuation of both α and β subunits of amino acid occurs in leucine, serine, histidine and tryptophan. C-APC α was found to have cysteine, aspartic acid, phenylalanine in least amount than the amino acids alanine, aspartic acid, glycine, leucine and valine. In C-APC β subunit valine was more variable i.e., 8.88% than 6.59% in C-APC α subunit. Lysine, methionine, phenylalanine and proline have least changes in variability among the studied total phycobiliproteins (Fig. 6).

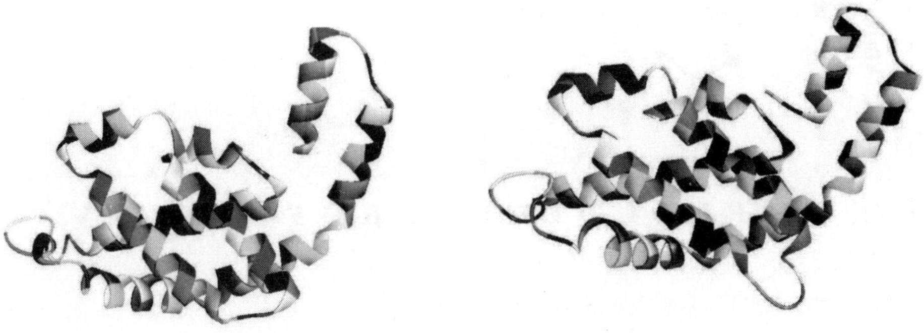

A B

Fig. 3. Three-dimensional structures of the α (A) and β (B) phycocyanin obtained from complete genome sequence of the cyanobacterium *Nostoc* sp. PCC 7120 which shows different amino acids residue in different colors.

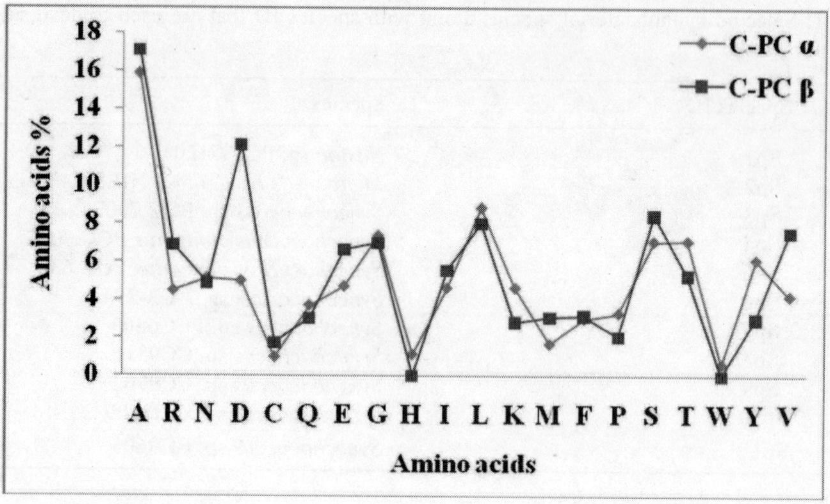

Fig. 4. Average amino acids (%) of phycobiliproteins subunits of C-PC α and C-PC β in all the studied cyanobacterial species.

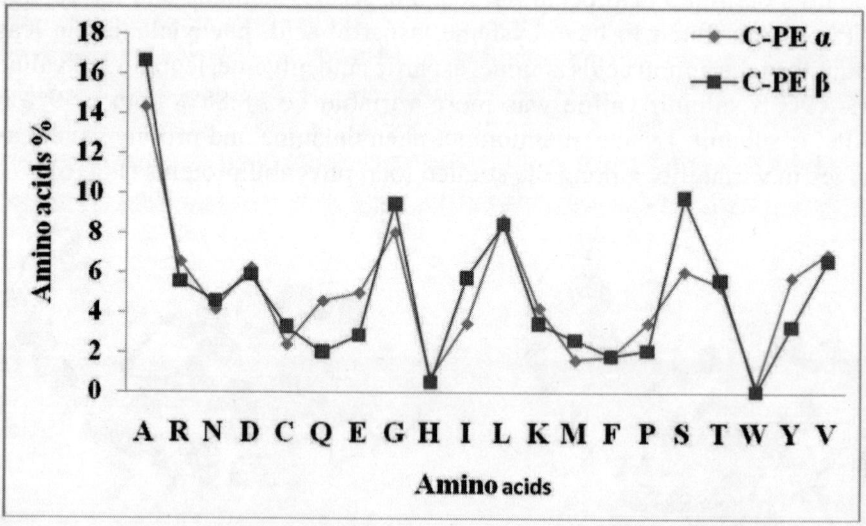

Fig. 5. Average amino acids (%) of phycobiliproteins subunits of C-PE α and C-PE β in all the studied cyanobacterial species.

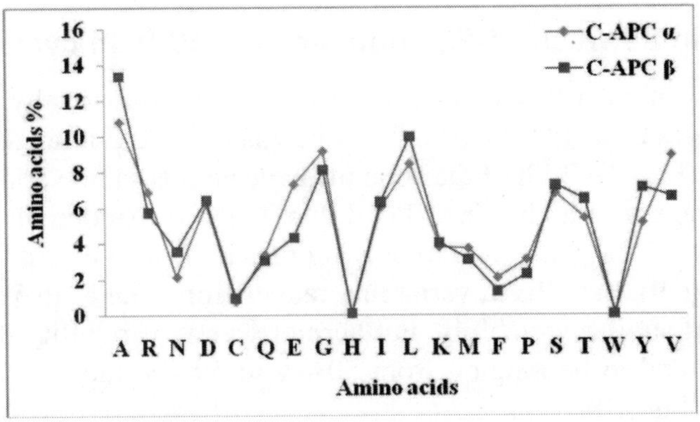

Fig. 6: Average amino acids (%) of phycobiliproteins subunits of C-APC α and C-APC β in all the studied cyanobacterial species.

Cyanobacterial nucleotide composition

For exploration of variability in amino acid constituents, G+C percentage variation was analyzed in total genome of all studied cyanobacteria (Fig. 7). It was observed that G+C ranges from 59.73 % to 48.98 %, whereas AT was ranging from 51.02 % to 40.27 %. Li and Watanabe (2002) also analyzed fifty strains of cyanobacteria and found 35.9 to 56.4 % G+C content. In a pioneer study, Sueoka (1961) reported that the genomic of GC ranged from 35% to 72% in bacteria. Lightfield et al. (2011) found the wide GC variation (16.6% to 74.9%) in bacteria. In eukaryotic species *S. cerevisiae* GC to AT mutation showed 2.9 fold increases but in opposite direction in mutation accumulation experiment (Lynch et al. 2008). GC and AT skew of various bacterial genomes predicts replication origin (Grigoriev 1998).

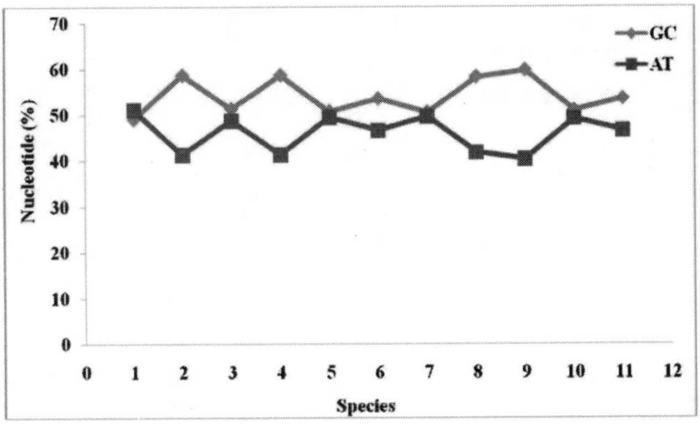

Fig. 7: Variable GC and AT nucleotides (%) in studied cyanobacterial species. In respect to amino acids variability we have analyzed the nucleotide content changes to explore the possible variability in amino acid constituents. The GC constituents in all subunit show significant variability in the studied cyanobacteria.

Relationship between PBPs subunits and GC % in cyanobacteria

G+C analysis of each PBPs have been summarised individually. In C-PC α variation ranges from 48.9 % to 58.7 % whereas in C-PC β it ranges from 49.2 % to 60.9 % (Fig. 8). Both of the subunits have almost same variation but the G+C percentage was higher in C-PC β than C-PC α. Among all the studied cyanobacteria, C-PE was found in only four species, such as, Sp8, Sp9, Sp10 and Sp11 (Fig. 9). In C-PE α, variability ranges from 53.4 % to 59.1 %. C-PE β also showed similar variability. Furthermore, G+C variability in C-APC α and β was found to be ranging from 50.4% to 59.4% and 47.2% to 60.1% respectively (Fig. 10).

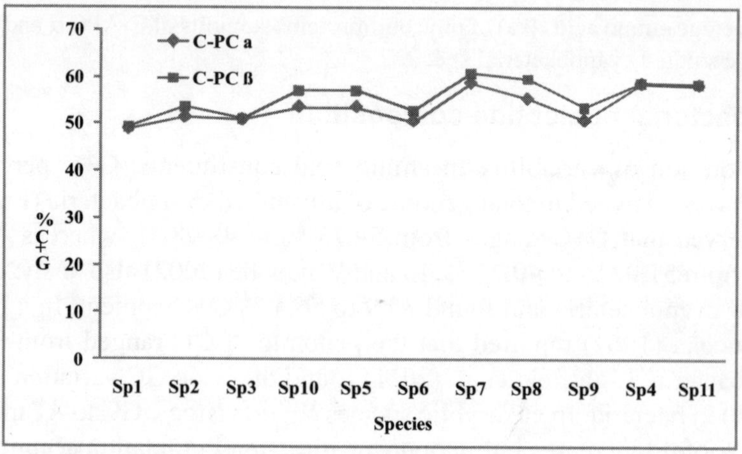

Fig. 8: Average G+C (%) for the genes of phycobiliproteins subunits C-PC α and C-PC β in all studied cyanobacterial species.

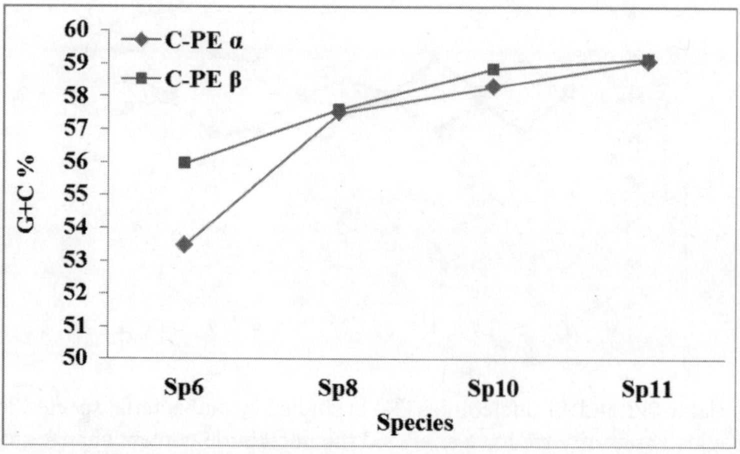

Fig. 9: Average G+C (%) for the genes of phycobiliproteins subunits C-PE α and C-PE β in only selected cyanobacterial species.

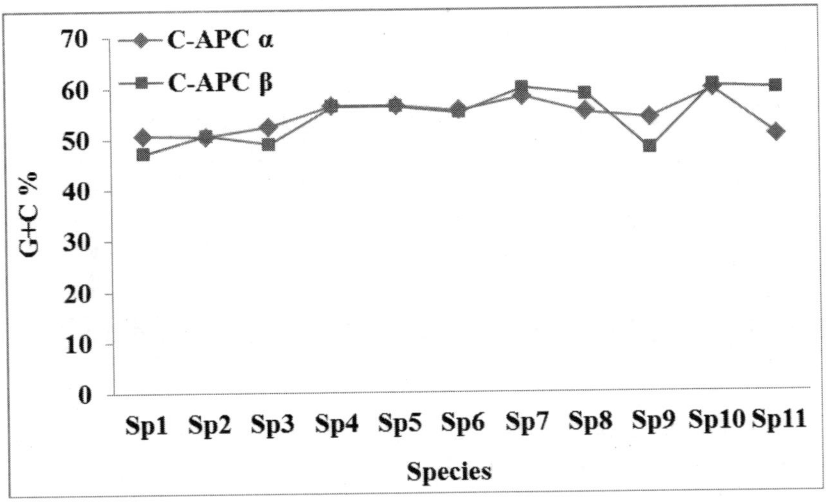

Fig. 10: Average G+C (%) for the genes of phycobiliproteins subunits C-APC α and C-APC β in all the studied cyanobacterial species.

Amino acids codons variability in PBPs subunits at GC_3 position

The third position of nucleotide was found to play an important role in the variation of total amino acids with specific codons. G3+C3 (GC_3) expressed 31 codons covering all 20 amino acids constituents. The variability in codons significantly alters the total number of amino acids at synonymous site. All three subunits of PBPs and their partner α and β subunits showed lot of variation in all amino acids constituents (Fig 11, 12 and 13). In C-PC α subunit codon GCC (Alanine) was less expressed than TAC (Tyrosine) and ACC (Threonine). Codons GAC (Aspartic acid), GGC (Glycine), ATC (Ileucine), CTG (Leucine), AAC (Asparagine) and AGC (Serine) were moderately expressed in both subunits of C-PC, whereas, other codons such as, ATG (Methionine), CGC (Arginine) and TCC (Serine) were also fairly expressed in β-subunit of C-PC, and all other codons were less expressed. The codon TAC (Tyrosine) codon was more expressed also in C-PE α than in C-PE β subunit. At the third codon level C-APC subunit also showed considerable variation.

Codon GGC (Glycine) and ACC (Threonine) were found to be more expressed in α and β subunits of C-APC respectively. G3+C3 coded codons analysis showed that GCC and TAC codon are highly expressed in α subunit and ATG, GGC, CGC codons in β subunit.

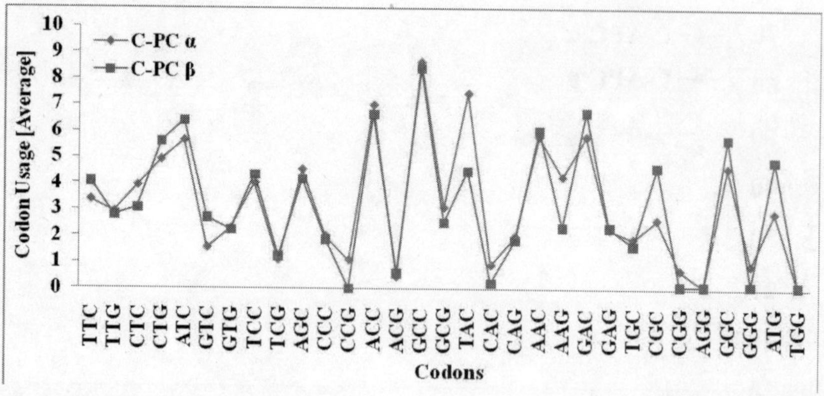

Fig. 11: GC_3 position codons variability in phycobiliproteins subunits C-PC α and C-PC β in all the studied cyanobacteria.

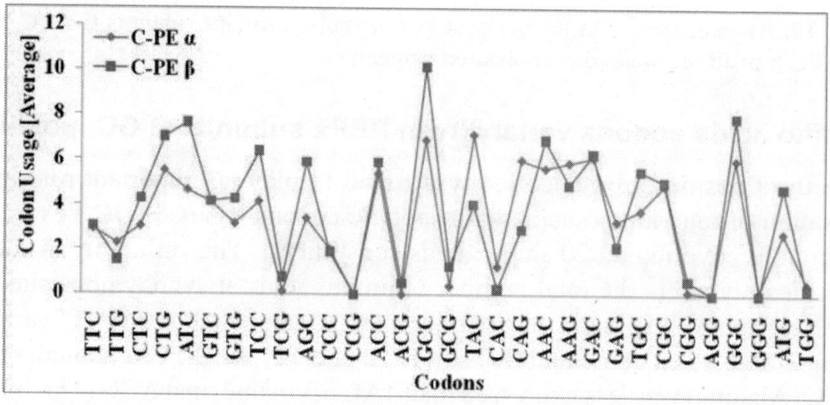

Fig. 12: GC_3 position codons variability in phycobiliproteins subunits C-PE α and C-PE β in all the studied cyanobacteria.

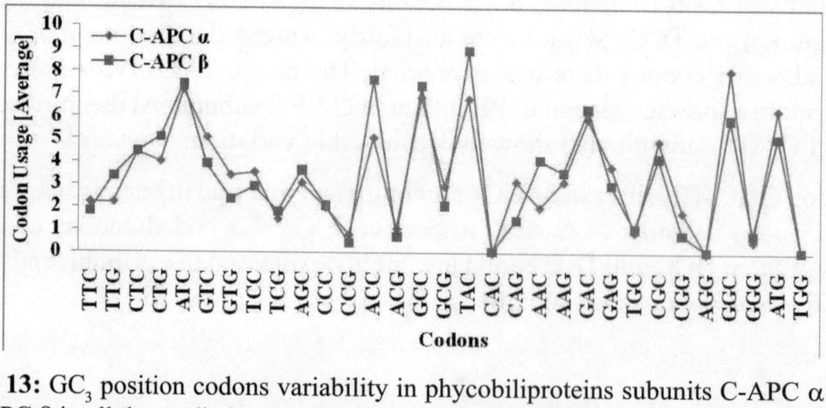

Fig. 13: GC_3 position codons variability in phycobiliproteins subunits C-APC α and C-APC β in all the studied cyanobacteria.

Gupta and Ghosh, (2003) also found the same G+C variation at the first, second and third codon levels. The result of G+C level showed the susceptible site of variation which is directly associated with the amino acid variation in the PBPs subunit. Increase in GC levels was found to have positive correlation with amino acids (GARP), while decreasing level of GC was found to be negatively correlated with amino acids (FYMINK) in bacteria (Banerjee et al. 2005). The GC constituents do not reflect the codon usage but vice-verse codon usage reflects particular GC constituents (Chen et al. 2003).

Conclusion

The analyses of codon expression of particular amino acids in specific gene segment of nucleotide will help to improve our understanding for the structural biochemical as well as evolutionary phenomena of cyanobacterial proteins. The specific variability in PBP subunit resulted due to more nucleotide variation on the third position than the first and second with specific codon expression. Moreover, the significant variability indicates the evolutionary forces originated from regional bias and mutational bias of gene specific nucleotide content and relatively expressed codons of particular amino acids.

Acknowledegments

Vinod K. Kannaujiya is thankful to CSIR for financial support in the form of CSIR-SRF. The work was also partially supported by the project (No. SR/ WOS-A/LS-140/2011), sanctioned to Richa by Department of Science & Technology, New Delhi.

References

Adir N, Dines M, Klartag M, McGregor A and Melamed-Frank M (2006) Assembly and disassembly of phycobilisomes. In: Microbiology Monographs: Inclusions in Prokaryotes. 2: 47-77.

Adir N (2005) Elucidation of the molecular structures of components of the phycobilisome: reconstructing a giant. Photosynth Res 85: 15-32.

Adzhubei AA, Adzhubei IA, Krasheninnikov IA and Neidle S (1996) Non-random usage of 'degenerate' codons is related to protein three-dimensional structure. FEBS Lett 399: 78-82.

Andrei Grigoriev (1998) Analyzing genomes with cumulative skew diagrams. Nucleic Acids Res 26: 2286-2290.

Banerjee T, Gupta SK and Ghosh TC (2005) Role of mutational bias and natural selection on genome-wide nucleotide bias in prokaryotic organisms. Bio Systems 81: 11-18.

Bharanidharan D, Bhargavi GR, Uthanumallian K and Gautham N (2004) Correlations between nucleotide frequencies and amino acid composition in 115 bacterial species. Biochem Biophys Res Commun 315: 1097-1103.

Brunak S and Engelbrecht J (1996) Protein structure and the sequential structure of mRNA: á Helix and β sheet signals at the nucleotide level. Proteins 25: 237-252.

Bulmer M (1991) The selection mutation drift theory of synonymous codon usage. Genetics 129: 897-907.

Cai YA, Murphy JT, Wedemayer GJ and Glazer AN (2001) Recombinant phycobiliproteins. Anal Biochem 290: 186-204.

Chen L-L and Zhang C-T (2003) Seven GC-rich microbial genomes adopt similar codon usage cyanobacterium *Synechococcus* sp. PCC 7002. Ph. D. thesis, The Pennsylvania State University, University Park, PA.

David L, Marx A and Adir N (2011) High-resolution crystal structures of trimeric and rod phycocyanin J Mol Biol 405: 201-213.

de Miranda AB, Alvarez-Valin F, Jabbari K, Degrave WM and Bernardi G (2000) Gene expression, amino acid conservation, and hydrophobicity are the main factors shaping codon preferences in *Mycobacterium tuberculosis* and *Mycobacterium leprae*. J Mol Evol 50: 45-55.

Dong H, Nilsson L and Kurland CG (1996) Co-variation of tRNA abundance and codon usage in *Escherichia coli* at different growth rate. J Mol Biol 260: 649-663.

dos Reis M and Wernisch L (2009) Estimating translational selection in eukaryotic genomes. Mol Biol Evol 26: 451-461.

Drew HR, Wing RM, Takano T, Broka C, Tanaka S, Itakura K and Dickerson RE (1981) Structure of a B-DNA dodecamer: Conformation and dynamics. Proc Natl Acad Sci U S A. 78:2179-2183.

Duret L, Arndt PF (2008) The impact of recombination on nucleotide substitutions in the human genome. PLoS Genet 4:e1000071

Fairchild CD and Glazer AN (1994) Oligomeric structure, enzyme kinetics, and substrate specificity of the phycocyanin α subunit phycocyanobilin lyase. J BiolChem 269: 8686-8694.

Glazer AN (1985) Light harvesting by phycobilisomes. Annu Rev Biochem 14: 47-77.

Grantham RC, Gautier C, Gouy M, Jacobzone M and Mercier R (1981) Codon catalog usage is a genome strategy modulated for gene expressivity. Nucleic Acids Res 9: 43-79.

Grossman A, Bhaya D, Apt K and Kohoe D (1995) Light-harvesting complexes in oxygenic photosynthesis: diversity, control and evolution. Annu Rev Genet 29: 231-288.

Grossman AR, Schaefer MR, Chiang GG and Collier JL (1993) The phycobilisomes a light harvesting complexresponsive to environtal conditions. Microbiol Rev 57: 725-749.

Guan X, Qin S, Zhao F, Zhang X and Tang X (2007) Phycobilisomes linker family in cyanobacterial genomes: divergence and evolution. Int J Biol Sci 3: 434-455.

Guisez Y, Robbens J, Remaut E and Fiers W (1993)Folding of the MS2 Coat Protein in *Escherichia coli* is modulated by translational pauses resulting from mRNA secondary structure and codons usage: A hypothesis. J Theor Biol 162: 243-252.

Gupta SK and Ghosh TC (2003) Reinvestigation on the causes of genomic GC variation between the orthologous genes of *Mycobacterium tuberculosis* and *Mycobacterium leprae*. Biochem Biophys Res Commun 303: 65-68.

Harrison RJ and Charlesworth B (2011) Biased gene conversion affects patterns of codon usage and amino acid usage in the *Saccharomyces sensu stricto* group of Yeasts. Mol Biol Evol 28: 117-129.

Hershberg R and Petrov DA (2008) Selection on codon bias. Annu Rev Genet 42:287-99.

Ikemura T (1985) Codon usage and tRNA content in unicellular and multicellular organisms. Mol Biol Evol 2:13-34

Knight RD, Freeland SJ and Landweber LF (2001) A simple model based on mutation and selection explains trends in codon and amino-acid usage and GC composition within and across genomes. Genome Biol 2: 1-13.

Kondo K, Mullineaux CW and Ikeuchi M (2009). Distinct roles of CpcG1-phycobilisome and CpcG2-phycobilisome in state transitions in a cyanobacterium *Synechocystis* sp. PCC 6803. Photosynth Res 99: 217-225.

Kondo K, Ochiai Y, Katayama M and Ikeuchi M (2007) The membrane-associated CpcG2-phycobilisome in *Synechocystis*: a new photosystem I antenna. Plant Physiol 144: 1200-1210.

Kreil DP, Ouzounis CA (2001) Identification of thermophilic species by the amino acid compositions deduced from their genomes. Nucleic Acids Res 29: 1608-1615.

Lightfield J, Fram NR and Ely B (2011) Across bacterial phyla, distantly-related genomes with similar genomic GC content have similar patterns of amino acid usage. PLoS ONE 6: e17677.

Li R and Watanabe MM (2002) DNA base composition of planktonic species of *Anabaena* (Cyanobacteria) and its taxonomic value. J Gen Appl Microbiol 48: 77-82.

Liu LN, Chen XL, Zhang YZ and Zhou BC (2005) Characterization, structure and function of linker polypeptides in phycobilisomes of cyanobacteria and red algae: an overview. Biochim Biophys Acta 1708: 133-142.

Lobry JR (1997) Influence of genomic G+C content on average amino-acid composition of proteins from 59 bacterial species. Gene 205: 309-316.

Lomax TL, Conley PB, Schiling J and Grossman AR (1987) Isolation and characterization of light-regulated phycobilisome linker polypeptide genes and their transcription as a polycistronic as a polycistronic mRNA. J Bacteriol 169: 2675-2684.

Lynch M, Sung W, Morris K, Coffey N, Landry CR, Dopman EB, Dickinson WJ, Okamoto K, Kulkarni S, Hartl DL and Thomas WK (2008) A genome wide view of the spectrum of spontaneous mutations in yeast. Proc Natl Acad Sci USA 105: 9272-9277.

MacColl R, Lee JJ and Berns DS (1971) Protein aggregation in C- Phycocyanin. Studies at very low concentrations with photoelectric scanner of the ultracentrifuge. Biochem J 122: 421-426.

Musto H, Cacciò S, Rodríguez-Maseda H and Bernardi G (1997) Compositional Constraints in the Extremely GC-poor Genome of *Plasmodium falciparum*. Mem Inst Oswaldo Cruz 92: 835-841.

Nayaa H, Romeroa H, Carelsb N, Zavalaa A and Mustoa H (2001) Translational selection shapes codon usage in the GC-rich genome of *Chlamydomonas reinhardtii*. FEBS Lett 501: 127-130.

Okamoto K, Kulkarni S, Hartl DL and Thomas WK (2008) A genome wide view of the patterns regardless of their phylogenetic lineages. Biochem Biophys Res Commun 306: 310-317.

Porter TD (1995) Correlation between codon usage, regional genomic nucleotide composition, and amino acid composition in the cytochrome P-450 gene superfamily. Biochim Biophys Acta 1261: 394-400.

Robinson M, Lilley R, Little S, Emtage JS, Yarranton G, Stephens P, Millican A, Eaton M and Humphreys G (1984) Codon usage can affect efficiency of translation of genes in *Escherichia coli*. Nucleic Acids Res 12: 6663-6671.

Singer GAC and Hickey DA (2000) Nucleotide bias causes a genome wide bias in the amino acid composition of proteins. Mol Biol Evol 17: 1581-1588.

Sinha RP and Häder D-P (1996) Photobiology and ecophysiology of rice field cyanobacteria. Photochem Photobiol Sci 64: 887-896.

Sinha RP, Lebert M, Kumar A, Kumar HD and Häder D-P (1995a) Spectroscopic and biochemical analyses of UV effect on phycobiliproteins of *Anabena* sp. and *Nostoc carmium*. Bot Acta 108: 87-92.

Sinha RP, Lebert M, Kumar A, Kumar HD and Häder DP (1995b) Disintegration of phycobilisomes in a rice field cyanobacterium *Nostoc* sp. following UV irradiation. Biochem Mol Biol Intern 37: 697-706.

Sinha RP, Kumar A, Tyagi MB and Häder D-P (2005) Ultraviolet-B-induced destruction of phycobiliproteins in cyanobacteria. Physiol Mol Biol Plants 11: 313-319.

Spencer CC, Deloukas P, Hunt S, Mullikin J, Myers S, Silverman B, Donnelly P, Bentley D and McVean G (2006) The influence of recombination on human genetic diversity. PLoS Genet 2: e148.

Sorensen MA, Kurland CG and Pedersen S (1989) Codon usage determines translation rate in *Escherichia coli*. J Mol Biol 207: 365-377.

Sueoka N (1961) Correlation between base composition of deoxyribonucleic acid and amino acid composition and protein. Proc Natl Acad Sci USA 47: 1141-1149.

Swanson RV, Zhou J, Leary JA, Williams T, de Lorimier R, Bryant DA and Glazer AN (1992) Characterization of phycocyanin produced by cpcE and cpcF mutants and identification of an intergenic suppressor of the defect in bilin attachment. J Biol Chem 267: 16146-16154.

Tandeau de Marsac N, Mazel D, Caupano V, Dameravl T and Houmard J (1990) Genetic analysis of the cyanobacterial light harvesting complex. In: The molecular biology of membrane-bound complexes in Phototrophic bacteria (Eds. G Drews), Plenum Publishing Corp, New York. pp. 143-153.

Thanaraj TA and Argos P (1996a) Protein secondary structural types are differentially coded on messenger RNA. Protein Sci 5: 1973-1983.

Thanaraj TA and Argos P (1996b) Ribosome mediated translational pause and protein domain organization. Protein Sci 5: 1594-1612.

Thomas JC and Passaquet C (1999) Characterization of a phycoerythrin without á-subunits from a unicellular red alga. J Biol Chem 274: 2472-2482.

Wada A (1992) Compliance of genetic code with base-composition deflecting pressure. Adv Biophys 28: 135-158.

Xia X (1998) How optimized is the translational machinery in *Escherichia coli*, *Salmonella typhimurium* and *Saccharomyces cerevisiae*? Genetics 149: 37-44.

Yu MH and Glazer AN (1982) Cyanobacterial phycobilisomes: Role of the linker polypeptides in the assembly of phycocyanin. J Biol Chem 257: 3429-3433.

Zhou J (1992) *Mutational analysis of the genes encoding phycobilisome components in the cyanobacterium Synechococcus sp. PCC 7002*. Ph. D. thesis, The Pennsylvania State University, University Park, PA.

Zhou J, Gasparich GE, Stirewalt VL, de Lorimier R and Bryant DA (1992) The *cpc*E and *cpc*F genes of *Synechococcus* sp. PCC 7002. Contruction and phenotypic characterization of interposon mutants. J Biol Chem 267: 16138-16145.

Zhang J, Wang M, Liu W, Zhou Jian-hua, Chen Hao-tai, Ma Li-na, Ding Yao-zhong, Yuan-xing Gu and Yong-sheng Liu (2011) Analysis of codon usage and nucleotide composition bias in polioviruses. Virol J 146: 1-8.

7

A Review on Toxicology, Biosynthesis and Detection of Cyanobacterial Toxins

Shweta Yadav and Rajeshwar P. Sinha

Abstract

Cyanobacteria, the Gram-negative photosynthetic prokaryotes are able to produce a wide range of secondary metabolites. Some of these metabolites are of general concern due to their toxicity and known as cyanobacterial toxins. These cyanotoxins are reported widely due to human and animal health hazard. On the basis of their toxicological target, these cyanotoxins can be classified into five different groups: hepatotoxins, neurotoxins, cytotoxins, dermatotoxins and irritant toxins (lipopolysaccharides). Hepatotoxins constitute a family of cyclic peptide variants and are classified as microcystins and nodularins. These hepatotoxic peptides possess a high degree of cellular specificity to the hepatocytes with similar mode of action and act as potent inhibitors of protein phosphatases 1 and 2A (PP1 and PP2A). Cyanobacterial neurotoxins are classified into two groups on the basis of their structure; (i) the alkaloids and (ii) the lipopeptides. They mainly interrupt normal signaling between neurons and muscles in several ways resulting in death due to paralysis of respiratory muscles. Another class of cyanotoxins includes cytotoxins represented by cylindrospermopsin, responsible for irreversible protein synthesis inhibition, cellular necrosis, DNA fragmentation. On the other hand lyngbyatoxins and aplysiatoxins are reported to act as dermatoxins causing severe primary contact irritation, dermatitis, blisters etc in mammals. Cyanobacterial lipopolysaccharides are recognized as irritant toxins responsible for pathological effects in humans, from gastro-intestinal illness, cutaneous signs and symptoms.

This chapter summarizes the different cyanotoxins, their structure, toxicology and additionally summarizes their detection, biosynthesis pathways and ecological significance.

Keywords: Cyanobacteria; Cyanotoxins; Toxicity; Polyketide synthase

Introduction

Cyanobacteria are an ancient group of photosynthetic prokaryotic organisms that are thought to be the first organisms to carry out oxygenic photosynthesis. Cyanobacteria are ubiquitous in nature and widely distributed in aquatic and terrestrial environments. Cyanobacteria represent important components of both aquatic and terrestrial ecosystems and possess a central position in the nutrient cycling due to their unique capacity to fix atmospheric nitrogen thereby enhancing productivity in a variety of environments. N_2-fixing cyanobacteria form a prominent component of microbial populations in wetland soils, especially in rice-fields, where they significantly contribute as a natural biofertilizer (Singh 1961; Roger and Watanabe 1986; Sinha et al. 2008). Some other important habitats include polluted lakes, ponds, water reservoirs, hypersaline lakes, tree barks, rocks and stones, deserts, thermal springs, estuaries etc. In warm, nutrient-rich water bodies, planktonic cyanobacteria grow quickly and cover the water surface resulting in the formation of waterblooms. Some authors attribute this adaptability, in part, to the ability of cyanobacteria to produce diverse secondary metabolites (Gademann and Portmann 2008; Rastogi and Sinha 2009; Yadav et al. 2011).

Cyanobacteria are known to produce a variety of toxic metabolites (cyanotoxins) that can be classified on the basis of their biological effects, toxicological targets and bioassay (Codd et al. 2005; Wiegand and Pflugmacher 2005; Dittmann et al. 2012). These compounds are extremely potent toxins which can kill a variety of organisms. Several genera of cyanobacteria such as *Microcystis, Anabaena, Nostoc, Oscillatoria, Cylindrospermum, Cylindrospermopsis, Aphanizomenon* and *Nodularia* are known to produce toxins. On the basis of the bioassay methods, the cyanotoxins are grouped into two categories; cytotoxins and biotoxins (Carmichael 1992). Another class of cyanobacterial poisons is mainly produced by marine cyanobacteria and cause inflammatory activity of skin due to contact and gastrointestinal inflammation upon the ingestion of contaminated water. Based on their toxicological targets cyanobacterial toxins can be classified into five groups: hepatotoxins, neurotoxins, cytotoxins, dermatotoxins and irritant toxins (Table 1) (Carmichael 1992; Wiegand and Pflugmacher 2005). Details of all the toxin groups are given below separately.

Table 1: Toxins produced by cyanobacteria (adapted from Yadav et al. 2011)

Cyanotoxin	Producer	Chemical nature	Mode of action	LD_{50}
Hepatotoxins				
Microcystins	*Anabaena, Anabaenopsis, Hapalosiphon, Microcystis, Nostoc, Oscillatoria, Planktothrix*	Cyclic heptapeptides	Inhibition of protein phosphatases (PP1 and PP2A)	50-160 µg/kg
Nodularin	*Nodularia*	Cyclic pentapeptides	Inhibition of protein phosphatase (PP1 and PP2A)	50 µg/kg
Neurotoxins				
Anatoxin-a	*Anabaena, Aphanizomenon, Cylindrospermum, Oscillatoria, Phormidium, Rhaphidiopsis*	Alkaloids	Binds irreversibly to the nicotinic acetylcholine receptors	375 µg/kg
Anatoxin-a (s)	*Anabaena*	Guanidine methyl phosphate ester	Inhibits acetylcholinesterase	200-250 µg/kg
Saxitoxins	*Aphanizomenon, Anabaena, Cylindrospermopsis, Lyngbya, Planktothrix*	Carbamate alkaloids channels in	Binds and blocks thesodium neural cells	10 µg/kg
Cytotoxins				
Cylindros permopsins	*Anabaena, Aphanizomenon, Cylindrospermopsis, Raphidiopsis, Umezakia*	Guanidine alkaloids biosynthesis	Inhibitor of protein and genotoxic	2.1 mg/kg
Dermatotoxins				
Lyngbyatoxin-a	*Lyngbya, Oscillatoria, Schizothrix*	Alkaloids	Cause erythema (dermatitis), blisters and necrosis in mammals; potent tumor promoters	—
Aplysiatoxins	*Lyngbya, Oscillatoria, Schizothrix*	Alkaloids	Inflammatory agents, protein kinase C activators	0.3 mg/kg
Endotoxins				
Lipopolysa-ccharides	All cyanobacteria	Lipopolysaccharide	Inflammatory agent, gastrointestinal irritants	45-190 mg/kg

Hepatotoxins

The hepatotoxins are the most common and widely reported cyanotoxins that have been responsible for killing of animals in virtually every part of the world (Falconer 1989; Runnegar et al. 1991; Carmichael 1992; Sivonen 1996, 1998; Haider et al. 2003; Ferrão-Filho and Kozlowsky-Suzuki 2011). These toxins mainly damage the hepatocytes of the liver thus owe their name. Death occurs in few hours to few days after initial exposure due to intrahepatic haemorrhage and hypovolemic shock. These are transported preferentially into the hepatocytes and cause upto 100 % increase in liver weight in small animals under laboratory conditions. Hepatotoxins constitute a family of cyclic peptide variants and are classified as microcystins (heptapeptides) and nodularins (pentapeptides) on the basis of number of amino acids.

Microcystin

Microcystins (MCs) are mainly produced by *Microcystis aeruginosa*, but are also produced by members of several cyanobacterial genera including *Anabaena, Planktothrix* (*Oscillatoria*), *Anabaenopsis*, *Nostoc* and *Hapalosiphon* and are most frequently associated with hepatotoxicity in mammals and other vertebrates (Dawson 1998). MCs are cyclic heptapeptides (909-1067 Da) having the general structure cyclo (-D-Ala^1-X^2- D-Me Asp^3-Z^4-$Adda^5$-D-Glu^6-$Mdha^7$) in which X and Z are variable L-amino acids, D-Me Asp is D-erythro-β-methylaspartic acid, Mdha is N-methyldehydroalanine and Adda is a novel amino acid (2S, 3S, 8S, 9S)- 3-amino-9-methoxy-2, 6, 8-trimethyl-10 phenyldeca-4, 6-dienoic acid. About 90 MC isoforms varying by degree of methylation, hydroxylation, epimerization, peptide sequence and toxicity have been identified (Rinehart et al. 1994; Codd et al. 1999; Sivonen and Jones 1999; McElhiney and Lawton 2005; Welker and von Dohren 2006). The most common and toxic variant MC-LR contains amino acids leucine (L) and arginine (R) at positions 2 and 4 respectively (Fig. 1). Biosynthesis and extra-cellular concentration of MC have been reported to vary under diverse conditions of light, temperature and phosphate which might have implication for the toxicity of cyanobacterial blooms in an aquatic ecosystem. The LD_{50} (i.p. mouse) for the most toxic MCs has been reported to be 50-160 μg/kg body weight.

The target organ for microcystins is primarily the liver (Dawson 1998). Several authors found microcystin in animal's organs such as liver, muscle, kidney, brain and skin. In case of fish and birds, that are intensively studied and the highest concentrations were found in liver or hepatopancreas (Bury et al. 1998; Karjalainen et al. 2003; Krienitz et al. 2003; Lewin et al. 2003; Chen and Xie 2005; Ibelings et al. 2005). Radiolabeled technique also proved the liver as microcystin target. Robinson and coworkers (1991) administered tritiated H_3-Microcystin-LR into mice and rats and found that more than 50% of the radiolabeled MC-LR is present in the liver. These results indicate that the main transportation mechanism of MCs into the vertebrate's liver cell is through bile acid carrier transport system (Dawson 1998).

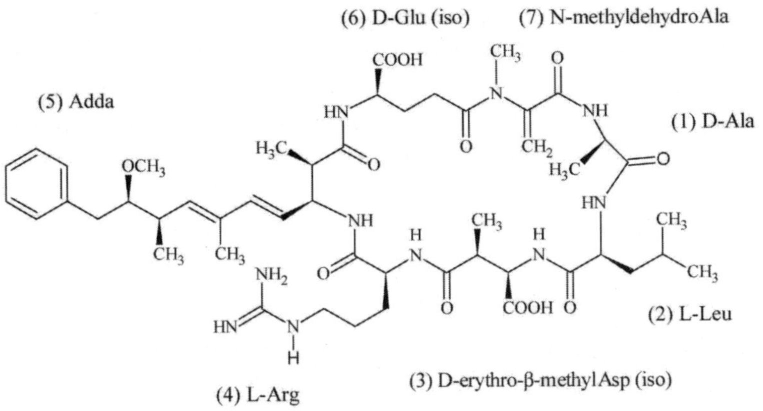

Fig. 1: Structure of microcystin-LR

Death of animals is mostly due to the severe liver damage which starts with cytoskeletal disorganization and may lead to cell blebbing, cellular disruption, lipid peroxidation, loss of membrane integrity, DNA damage, apoptosis, necrosis, intrahepatic bleeding, and eventually death by hemorrhagic shock (Wickstrom et al. 1995; Fischer and Dietrich 2000).

Hepatocytes, the functional cells of the liver are the primary target for MCs (Meriluoto et al. 1990). MCs possess a high degree of cellular specificity to the hepatocytes (Eriksson et al. 1990) and are transported into the liver through a selective transport system (Runnegar et al. 1991). Intermediate filaments and microfilaments are basically protein polymers of cytoskeleton and are the components most affected by MCs. As a result of shrinkage of cytoskeleton filament, the finger-like projections through which hepatocytes interact with neighbouring cells withdraw, leading to the loss of well organized network of hepatocytes (Carmichael 1994). MCs are known to increase the phosphorylation of cytosolic and cytoskeletal proteins in isolated hepatocytes (Eriksson et al. 1990; Nishiwaki-Matsushima et al. 1992; Ohta et al. 1994). It has been reported that both *in vivo* and *in vitro*, MCs acts as potent inhibitors of protein phosphatases 1 and 2A (PP1 and PP2A) (Eriksson et al. 1990; Matsushima et al. 1990; Ohta et al. 1994; Runnegar et al. 1995; Campos and Vasconcelos 2010).

PP1 and 2A are the most abundant phosphatases in mammalian tissues (Toivola et al. 1994) and are involved in dephosphorylation of serine/threonine residues on a huge variety of regulatory and structural proteins (Barford 1996). The mechanism of action of MC-LR resembles to that of okadaic acid, a polyether fatty acid from marine sponges, both of which are known to inhibit PP1 and PP2A. Overall alteration in cytoskeletal organization (intermediate filaments and microfilaments), due to excessive protein-phosphorylation are correlated with MC-LR treatment (Eriksson et al. 1990; Carmichael 1994). The balance between the activities of these two groups of enzymes (PP1 and 2A) powerfully influences cell structure and function, intermediary metabolism, cell division and cancer cell growth (Cohen 1989; Toivola et al. 1994). Since it is known that the activity of kinases such as MAP-kinase is regulated by the activity of PP2A therefore inhibition of PP2A by MC could result in the kinase activation resulting in unchecked cell proliferation that could lead to tumor promotion (Falconer 1991; Nishiwaki-Matsushima et al. 1992; Toivola et al. 1994).

Repeated consumption of low levels of MC have been shown to cause increased liver weight, hepato-histological damage (Heinze 1999), liver tumors (Herfindal and Selheim 2006; Li et al. 2009; Martinez Hernandez et al. 2009) and kidney damage (Milutinovic et al. 2003).

Nodularin

Another class of hepatotoxins includes nodularins (NODLN), which are cyclic pentapeptides (824 Da) produced by species of *Nodularia.* NODLNs have a generalized structure of cyclo (-D-MeAsp^1-L-Arg^2-Adda^3-D-Glu^4-Mdhb^5) where Mdhb is 2-(methylamino)-2-dehydrobutyric acid. Structure of NODLN is closely related to that of the MCs having Mdhb(5) instead of Mdha(7) but lacking D-Ala(1) and X(2) (Fig. 2) (Lehtimäki et al. 1997). The smaller and less diverse pentapeptide family of nodularins comprises about 10 different variants (Rinehart et al. 1988; Sivonen and Jones 1999). NODLN-Har, a structural variant of NODLN was isolated from *Nodularia* PCC7804, having a homoarginine in place of arginine and two dihydronodularins have been chemically synthesized (Namikoshi et al 1993). It has an LD_{50} of 60 µg/kg body weight (Ressom et al. 1994).

These closely related peptides are potent protein phosphatase inhibitors with a primary effect on hepatocytes and its mode of action is very similar to that of the MCs (Ohta et al. 1992), except that NODLN does not bind covalently to PP1 or 2A (Falconer and Yeung 1992; Runnegar et al. 1993). NODLN is also report to cause liver cancer in male rat F-344 (Ohta et al. 1994; Fujiki and Suganuma 2009). It is however, interesting to note that toxins, by themselves do not initiate a cell progression towards becoming cancerous. Instead they can only promote the development of carcinogenic alrterations in the cells when triggered by such chemicals (Nishiwaki-Matsushima et al. 1992).

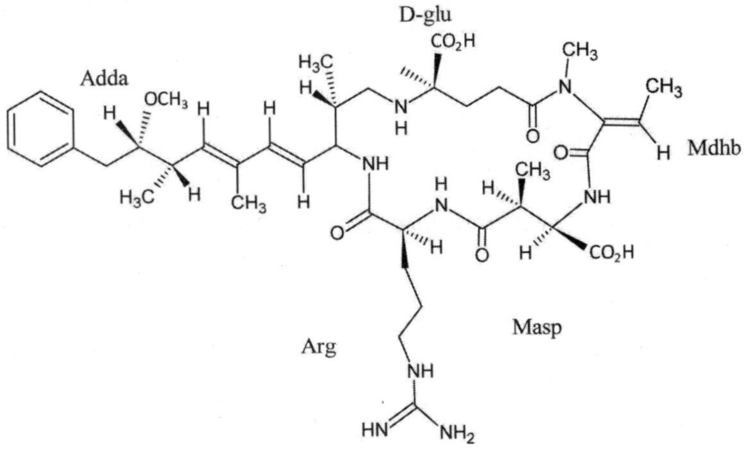

Fig. 2: Structure of nodularin

Neurotoxins

Cyanobacterial neurotoxins can be classified into two groups on the basis of their structure; (i) the alkaloids, including anatoxin-a and its analog homoanatoxin-a, anatoxin-a(s), and paralytic shellfish poisoning (PSP) toxins e.g., saxitoxin, gonyautoxin, and their derivatives, and (ii) the lipopeptides, including antillatoxins A and B, kalkitoxin, and the jamaicamides. Neurotoxins disrupt normal signaling between neurons and muscles in several ways, leading to death by causing paralysis of respiratory muscles followed by suffocation.

Alkaloid toxins have been reported from species of *Anabaena*, *Aphanizomenon*, *Microcystis*, *Cylindrospermopsis* and *Planktothrix* (Sivonen and Jones 1999). One of the most extreme cases of cyanobacterial poisoning is represented by blooms of *Anabaena flos-aque*, which produce anatoxins.

Anatoxin-a

Anatoxin-a produced by a toxic strain of *Anabaena flos-aquae* was the first alkaloid toxin to be isolated and characterized (Fig. 3a) (Carmichael et al. 1985). Anatoxin-a has also been isolated from toxic strains of other *Anabaena* species and from *Oscillatoria, Aphanizomenon* and *Microcystis* (Codd et al. 1999; Sivonen and Jones 1999). It is a potent postsynaptic neuromuscular blocker with a LD_{50} (i.p. mouse) of 200-375 µg/kg body weight (Fitzgeorge et al. 1994; Ressom et al. 1994).

Homoanatoxin-a

A related compound, homoanatoxin-a (Fig. 3b), has been isolated from a strain of *Phormidium formosum* (Skulberg et al. 1992). It is a potent postsynaptic neuromuscular blocker with a LD_{50} (i.p.) value of 375 μg/kg body weight.

Anatoxin-a and homoanatoxin-a cause potent neuromuscular blockade, which manifests as staggering, gasping and muscle fasciculation in affected animals; death results from paralysis of respiratory muscles (Skulberg et al. 1992; Ressom et al. 1994; Codd et al. 1999). Anatoxin-a binds to neuronal nicotinic acetylcholine receptors at neuromuscular junctions, mimicking the effect of acetylcholine. This causes an influx of Na^+ and subsequently, depolarisation, which results in the opening of voltage sensitive Ca^{++} and Na^+ channels. In contrast to acetylcholine, anatoxin-a is not susceptible to enzymatic hydrolysis by acetylcholinesterases and this leads to exhaustion and overexcitation of muscle cells. It also causes muscle membrane depolarization and desensitization resulting in neuromuscular blockade. Symptoms of anatoxin-a toxicity hence include muscle fasciculation, loss of coordination, staggering, gasping, convulsion and subsequent acute intoxication death by respiratory paralysis (Metcalf and Codd 2005). Homoanatoxin-a is a methyl derivate of anatoxin-a with slightly lower potency. Its toxicity is based on an enhanced Ca^{++} flux in the cholinergic nerve terminals. Comparable to anatoxin-a, homoanatoxin-a acts very quickly and may cause death due to respiratory arrest within minutes (van Apeldoorn et al. 2007).

Anatoxin-a(s)

A structurally unrelated toxin, anatoxin-a(s) has been found in some *Anabaena* species (Sivonen and Jones 1999). Anatoxin-a(s) (Fig. 3c) irreversibly inhibits acetylcholinesterase and is reported to be 10 times more toxic to mice than anatoxin-a. Its intra-peritoneal LD_{50} in mice is 20 μg/kg body weights with a survival time of 10-30 min.

As these neurotoxic symptoms are very similar to those of anatoxin-a, anatoxin-a(s) was initially considered to be an anatoxin-a homologue. Anatoxin-a(s) resembles in its structure and blocking the acetylcholine regenerating enzyme in a manner similar to with synthetic organophosphorus insecticides such as malathion and parathion. In contrast to anatoxin-a, anatoxin-a(s) is a potent irreversible inhibitor of acetylcholinesterase (Mahmood and Carmichael 1986).

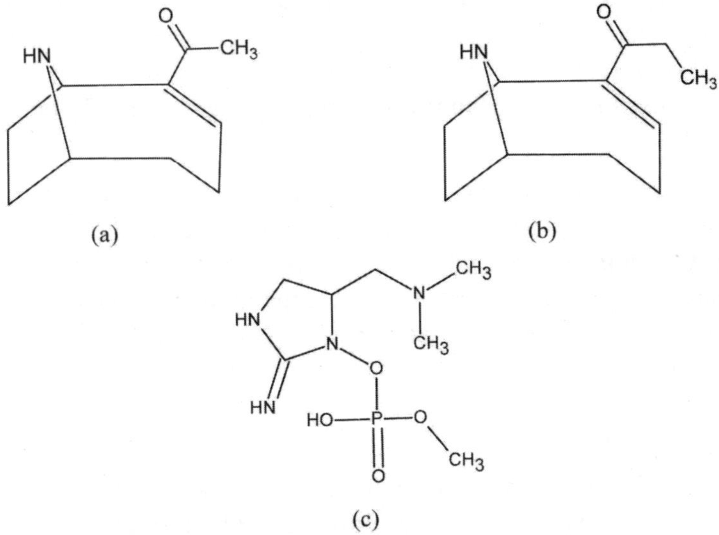

Fig. 3: Structure of (a) anatoxin-a (b) homoanatoxin-a and (c) anatoxin-a(s)

Saxitoxins

The other group of cyanobacterial alkaloid neurotoxins is the saxitoxins. Saxitoxins (Fig. 4) are also known as paralytic shellfish poisons (PSPs) because they were initially isolated from shellfish, which concentrated the toxin from marine dinoflagellates (Gorham and Carmichael 1988). As a consequence of the consumption of contaminated shellfish, PSPs become available for higher trophic levels and have been associated with animal, livestock and human mortalities (Kuiper-Goodman et al. 1999; Landsberg 2002). As a consequence of the consumption of contaminated shellfish, PSPs become available for higher trophic levels and have been associated with animal, livestock and human mortalities (Kuiper-Goodman et al. 1999; Landsberg 2002). About 21 variants of saxitoxins which differ among others predominantly in the substitution of five variable positions; have been recognized from different cyanobacterial species belonging to *Anabaena, Aphanizomenon, Lyngbya and Cylindrospermopsis* sp. (Lagos et al. 1999; Sivonen and Jones 1999;

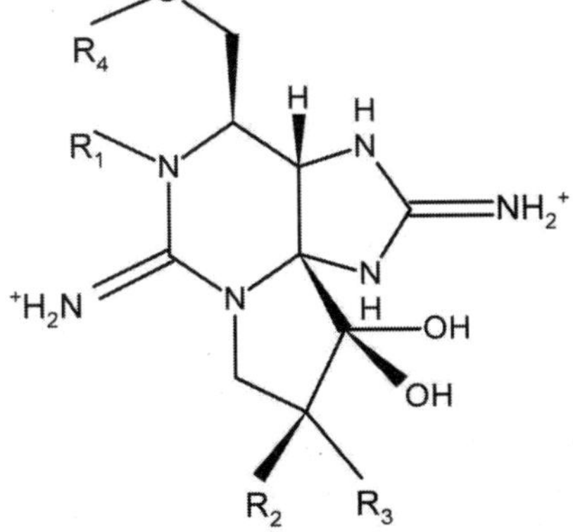

Fig. 4: Structure of saxitoxin

Sivonen 2000; van Apeldoorn et al. 2007). These are a group of alkaloid tricyclic compounds that are either non-sulphated (saxitoxins and neosaxitoxin), single sulphated (gonyautoxins) or doubly sulphated (C-toxins) (Wiese et al. 2010). Toxicity of saxitoxin has been investigated in mice with a LD_{50} between 10-30 µg/kg body weights.

The toxicity of PSPs, saxitoxin and neo-saxitoxins is attributable to a generalised blockade of sodium channels in the excitable membranes in nerve axons (Kuiper- Goodman et al. 1999). This results in numbness, paralysis and death in mammals via respiratory arrest by blocking the channel opening and stopping the inflow of sodium which leads to muscle paralysis and death by respiratory arrest (Cestèle and Catterall 2000; Llewellyn 2006; Humpage 2008).

Lipopeptide neurotoxins

These lipopeptides (antillatoxins, jamaicamides and kalkitoxin) were initially described as being acutely ichthyotoxic (antillatoxin A LC_{50}=0.1 µM, antillatoxin B LC_{50}=1 µM, kalkitoxin LC_{50}=0.7µM, jamaicamides A, B, and C LC_{100} >5 ppm) but later identified as neurotoxic (Berman et al. 1999; Wu et al. 2000; Edwards et al. 2004). Kalkitoxin and jamaicamide C are potent brine shrimp toxins with a LC_{50} of 0.17 µM and LC_{25}=10 ppm, respectively (Wu et al. 2000; Edwards et al. 2004). Kalkitoxin also inhibits cell division in sea urchin embryos (IC_{50} ≈0.025 µM). These compounds have been verified through experimental studies, however, it remains unclear to what extent they affect aquatic organisms in the environment and detailed studies on these neurotoxins are lacking.

Cytotoxins

Cylindrospermopsin

Cylindrospermopsin (CYN), a toxic alkaloidal polyketide, produced by a number of fresh water cyanobacterial species. It comprises a tricyclic guanidine combined with hydroxymethyl uracil and is a highly water-soluble molecule (Fig. 5). It has been found that an integral pyrimidine ring is essential for CYN toxicity, as both, substitution of a proton by chlorine or oxidation of a carbon change CYN to a non-toxic form (Banker et

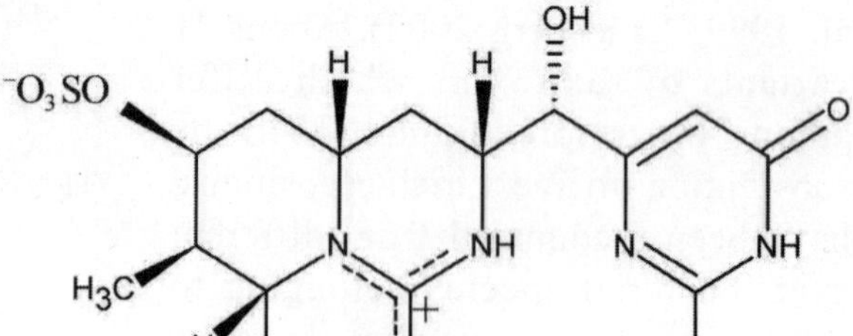

Fig. 5: Structure of cylindrospermopsin

al. 2001). Deoxy-cylindrospermopsin that may also be obtained from *C. raciborskii* is not as much toxic as CYN.

It is mainly produced by *Cylindrospermopsis raciborskii*, however some other cyanobacterial species (*Umezakia natans*, *Anabaena bergii*, *Anabaena lapponica*, *Raphidiopsis curvata* and some *Aphanizomenon* sp.) have also been reported to produce cylindrospermopsin (Falconer 2005; Spoof et al. 2006). It has been reported that CYN and its analog deoxycylindrospermopsin (DeoxyCYN) does not stall protein phosphatases but they are significant and permanent inhibitor of protein biosynthesis (Froscio et al. 2003; Neumann et al. 2007).

The toxin was first isolated and characterized by Ohtani et al. (1992). Cylindrospermopsin is also classified as a hepatotoxic alkaloid because principal organ affected is the liver but toxicological studies have shown that other organs such as the kidney are also affected (Terao et al. 1994; Falconer et al. 1999). However, its structure and mechanism of toxicity are different from MCs. Therefore cylindrospermopsin and its analogs have also been generally classified as cytotoxins (i.e., substances toxic to cells). The mouse (i.p.) LD_{50} for cylindrospermopsin is reported as 2.1 mg/kg body weight (Ohtani et al. 1992).

The mechanism for cellular uptake is unknown, but it does require the bile acid transport system (Chong et al. 2002). Diffusion was suggested as a possible uptake mechanism (Chong et al. 2002) however, the charged nature of cylindrospermopsin makes this route unlikely without the help from a pore or transporter molecule. Cylindrospermopsin exposure mainly results into irreversible protein synthesis inhibition, cellular necrosis, atrophy, DNA fragmentation, tumor initiation and mutagenesis (Terao et al. 1994; Froscio et al. 2003; Wiegand and Pflugmacher 2005; van Apeldoorn et al. 2007). In consequence of the inhibition of protein synthesis, cylindrospermopsin primarily injures tissues presenting a high synthesis/turnover of proteins, i.e. liver and kidney, intestine, and the immune system including spleen and thymus. Poisonings may lead to decreased functionality of these organs and subsequently even death, depending on the duration and dose of exposure (Falconer 2005).

Dermal toxins/Dermatoxins/Tumor promoters

The marine filamentous cyanobacterium *Lyngbya majuscula* has been reported to cause severe primary contact irritation in swimmers for over four decades (Capper et al. 2005; Smith et al. 2008). The toxins responsible for these effects have been identified as aplysiatoxin and debromoaplysiatoxin (Moore et al. 1984). Lyngbyatoxin A, also found in *L. majuscula*, is another inflammatory agent (Moore 1996). Lyngbyatoxins A (LA), B and C are indole alkaloids in nature while aplysiatoxins are phenolic bis-lactones consisting of aplysiatoxin (AT), debromoaplysiatoxin (DAT) and their less toxic anhydro variants. Debromoaplysiatoxin has also been isolated from other marine cyanobacteria such as *Schizothrix* and *Oscillatoria* (Moore et al. 1984). These potent inflammatory toxins operate through mechanisms similar to those of phorbol esters; they are potent tumour promoters, activating protein kinase C (Moore et al. 1984). Lyngbyatoxins and aplysiatoxins act as both dermatoxins, (Ito et al. 2002) and tumor promoters (Fujiki et al. 1983). Topical application with LA, AT, or DAT, on the other hand, resulted in erythema (dermatitis), blisters and necrosis in mammals.

Cyanobacterial lipopolysaccharides/Irritant toxins

In general lipopolysaccharide (LPS) is known as a common constituent of the cell wall of Gram-negative bacteria as well as most of the cyanobacteria, consist of an internal acylated glycolipid, core polysaccharides, and an outer polysaccharide chain (Wiegand and Pflugmacher 2005). Cyanobacterial LPSs are well recognized with a range of pathological effects in humans, from gastro-intestinal illness, cutaneous signs and symptoms, allergy, respiratory disease, headache and fever. Several genera of cyanobacteria including *Microcystis*, *Anabaena*, *Spirulina* and *Oscillatoria* are known to produce toxic lipopolysaccharides (LPS). Cyanobacterial lipopolysaccharides (LPS) are generally classified as endotoxins and are located within or at the surface of the outer cell layer. These are comprised of a carbohydrate polymer, a core oligosaccharide and an acylated glycolipid (≈ lipid A), the latter segment is linked to toxicity in enteric, heterotrophic Gram-negative bacteria (Stewart et al. 2006). Cyanobacterial LPS differs from LPS of enteric bacteria by having a larger variety of long chain unsaturated fatty acids and hydroxy fatty acids and lack of phosphate. Glucosamine seems to be the backbone sugar, in which amounts of 2-keto 3-deoxyoctonate (KDO), galactoses, and heptoses also varies (Martin et al. 1989).

Cyanobacterial LPS are reported to be responsible for a variety of health effects, from skin rashes to gastrointestinal, respiratory and allergic reactions. Inhalation of LPS might result in cyanobacteria related adverse effects known as "flue-like symptoms," characterized by cough, chills and sore throat. LPS are also known to cause fever in mammals and to be involved in "septic shock syndrome" which may worsen toxicant-induced liver damage (Choi and Kim 1998) by the release of inflammatory mediators such as tumor necrosis factor-a (TNF-a), interferon-g, interleukin (IL) 1 and 6, leukotrienes, prostanoids, and nitric oxide (Hewett and Roth 1993). LPS can also inflict the biotransformation enzyme systems, directly or mediated by factors of the immune reaction and also affect the detoxifying system of diverse organisms (Wiegand and Pflugmacher 2005). Being a common component of the cell wall of cyanobacteria, determination of contribution of these LPS to cyanobacterial toxicity or the modulation of their toxic effects is necessary.

Biosynthesis of Cyantoxins

In last few years, the biosynthesis pathways of the four major cyanotoxins: microcystin, nodularin, saxitoxin and cylindrospermopsin, have been genetically and biochemically enlightened.

MCs are synthesized non-ribosomally via a giant enzyme complex MC synthetase which includes nonribosomal peptide synthetase (NRPS), polyketide synthase (PKS), mixed NRPS/PKS and some additional tailoring enzymes. Nonribosomal peptide synthetases (NRPSs) are involved in the synthesis of linear, cyclic, and branched-cyclic peptides. Modular PKSs (PKS type I) are multifunctional megasynthases organized into repeated functional units. PKSs assemble acyl coenzyme A monomers by using the core domains ketosynthase, acyltransferase, and acyl carrier protein (Christiansen et al. 2003). These are encoded by *mcy* gene cluster (55 kbp) containing 10 genes in *Microcystis* sp. (Fig. 6) of which six large open reading frames (ORFs) (*mcy* A-E and G) are of a mixed NRPS/PKS nature together with four small ORFs (*mcy* F and H-J) with putative precursor and MC tailoring function (Tillett et al. 2000; Pearson and Neilan 2008).

For MC synthcsis, a cyanobacterium must have genes that are related to MC production prior to other factors associated to MC production (Fig. 6). The production of secondary metabolites like MCs in microorganisms a mechanism, thio-template nonribosomal peptide synthesis is there. These MC synthetase genes not only exist in all toxic strains of *Microcystis*, but also in MC-producing strains of *Anabaena*, *Planktothrix*, and *Nostoc* (Neilan et al. 1999; Christiansen et al. 2003). It has been reported that biosynthesis and extracellular concentration of microcystin may be enhanced under different conditions of light, temperature and phosphate, which might have implications for the toxicity of cyanobacterial blooms in an aquatic ecosystem (Kaebernick et al. 2000).

The proposed pathway for nodularin biosynthesis is similar to that for microcystin. The gene cluster *ndaS* responsible for nodularin biosynthesis was sequenced and characterized from *Nodularia spumigena* NSOR10 (Moffitt and Neilan 2004). The 48 kb region of the genome made of nine ORFs (*ndaA-I*) transcribed from a bidirectional regulatory promoter region (Fig. 6) (Pearson et al. 2010).

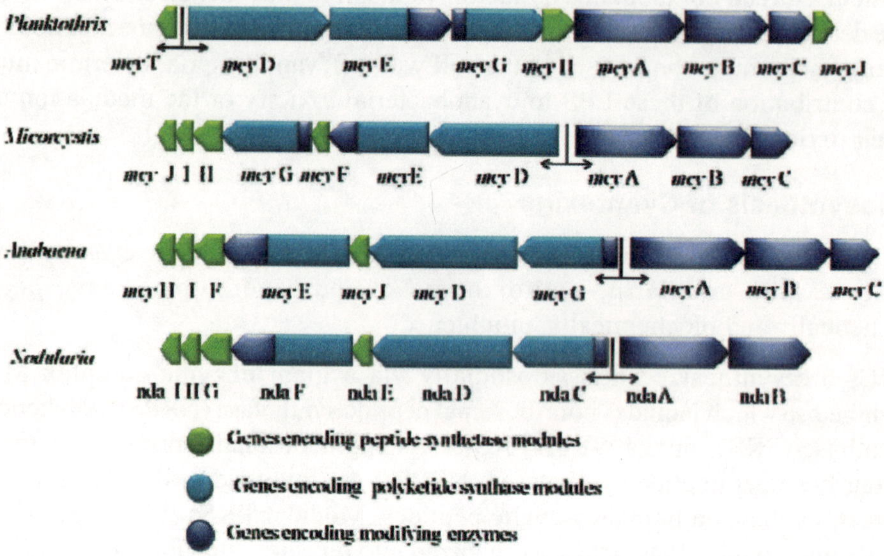

Fig. 6: Microcystin and nodularin gene cluster (modified from Kurmayer and Christiansen 2009).

The gene cluster putatively responsible for the biosynthesis of saxitoxin was identified in *Cylindrospermopsis raciborskii* T3 by Kellmann and co-workers (*sxt*) applying a reverse genetic approach. The *sxt* gene cluster (Fig. 7) is encoded by more than 35 kb and comparative sequence analysis assigns 30 catalytic functions to 26 proteins (Pearson et al 2010).

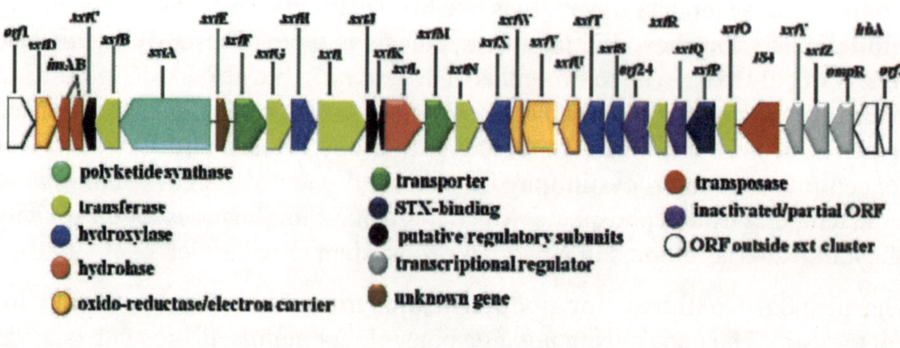

Fig. 7: *sxt* gene cluster from *C. raciborskii* T3 (adapted and modified from Kellmann et al. 2008).

Recently gene cluster for cylindrospermopsin biosynthesis (*cyr*) was sequenced from *C. raciborskii* AWT205 (Fig. 8) (Mihali et al. 2008). Biosynthesis, regulation and export of the cylindrospermopsin is encoded by 43 kb gene cluster consist of 15 ORFs.

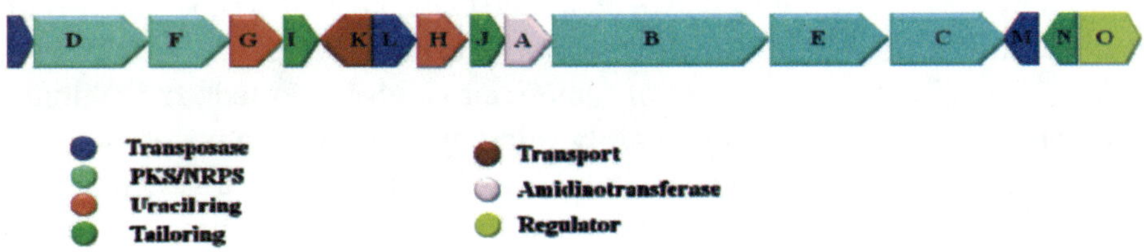

Fig. 8: Cylindrospermopsin gene cluster from *C. raciborskii* AWT205 (adapted and modified from Mihali et al. 2008).

Biosynthesis is initiated via an amidinotransfer onto glycine followed by five polyketide extensions and subsequent reductions, rings are formed via Michael additions in a step-wise manner. The uracil ring is formed by a novel pyrimidine biosynthesis mechanism and tailoring reactions, including sulfation and hydroxylation that complete biosynthesis (Pearson et al 2010).

Ecological Role of Cyanotoxins

Biosynthesis of cyanotoxins is an energetically dependent process but their functions are unclear. Study of the conditions that regulate cyanotoxin production may help to reveal the potential functions of these molecules. Effect of different environmental conditions such as light, temperature and varying nutrients (like phosphorus, nitrogen and iron) on cyanotoxin production have been investigated (Sivonen 1990; Kaebernick et al. 2000; Kaebernick and Neilan 2001; Dolman et al. 2012). *Anabaena, Microcystis, Aphanizomenon, Oscillatoria, Nodularia* and *Cylindrospermopsis* are the most common cyanobacteria associated with toxicity (Hitzfeld et al. 2000; Lakshmana Rao et al. 2002).

Cyanotoxins exhibit high toxicity to vertebrates (including mammals) also have a dramatic influence on many groups of organisms (Wiegand and Pflugmacher 2005). Cyanobacterial blooms may represent a hazard to aquatic organisms, when cells collapse, releasing cyanotoxins to the water. Cyanotoxins seem to be more toxic to mammals than to aquatic organisms (Ibelings and Havens 2008). Cyanobacteria produce potent toxins such as microcystin-LR, saxitoxins and anatoxin-a, which cause injury in vertebrates even at very low doses (micrograms per kg body weight) (Falconer 1998; Codd et al. 1999; Kuiper-Goodman et al. 1999).

Besides animal toxicity, allelopathic effects of the cyanobacterial toxins on algal and growth and photosynthesis of other algae and plants have also been reported by several workers (Ikawa et al. 2001; Papke et al. 1997; Singh et al. 2001). Growth of several aquatic plants (*Ceratophyllum*, *Myriophyllum*, *Cladophora*, *Elodea*, *Lemna* and *Phragmites* etc) is also reported to be inhibited in the presence of cyanotoxins (mainly microcystins) (Casanova et al. 1999; Romanoska-Duda and Tarczynska 2002). In addition to aquatic system, effect of cyanobacterial toxins on terrestrial plants has also attained research interests due to contamination via spray irrigation (Codd et al. 1999; McElhiney et al. 2001; Hamvas et al. 2003).

Using a variety of molecular methods to monitor the expression of genes and proteins involved in cyanotoxin biosynthesis is particularly appealing as a way to help uncover the ecological relevance of cyanotoxins.

Detection methods for cyanotoxins

Several methods have been developed for the detection and analysis of cyanobacterial toxins (Carmichael 1992; Toivola et al 1994; Harada 2004; Diehnelt et al 2005; Aráoz et al 2008; Pearson and Neilan 2008). Bioassay, chemical assay and immunoassay are main techniques of detection for the cyanotoxins.

Classical methods for screening of waterblooms and laboratory culture or cell extracts for toxicity include mouse bioassay. This is probably the typical first test for screening of toxicity which is inexpensive and can detect, within a few hours, the qualitative and quantitative characteristics of the toxin present. Hepatotoxin and neurotoxin can be distinguished on the basis of signs of poisoning.

Biochemical methods for detection of toxins include estimation of protein phsophatase and cholinesterase activity, as the activities of these enzymes are severely inhibited by hepatotoxin and anatoxin respectively (An and Carmichael 1994; Henriksen et al 1997). The extent of the inhibition of protein phsophatase by hepatotoxins can therefore be used as a measure of toxin concentrations. The basis of this type of procedure is the measurement of phosphate release from a phosphorylated protein in the presence of a phosphatase enzyme preparation and an inhibitor such as microcystin. Phosphatase inhibition by cyanobacterial hepatotoxins such as microcystins and nodularin can be determined using ^{32}P radiolabelled substrates (MacKintosh et al. 1990;

Yoshizawa et al. 1990; Honkanen et al. 1990, 1994; Ash et al. 1995; Holmes and Boland 1993) and measuring the radiolabelled phosphate produced.

The most promising method of detection for cyanotoxins seems to be some type of immunoassay. For the detection of minute quantities of cyanotoxins in the sample, methods such as enzyme linked immunosorbant assay (ELISA) and radioimmunoassay (RIA) are frequently used (Sivonen 1998). These immunoassay based method has also been used successfully to detect low level toxin amounts in drinking water and also in animal tissue (Chu et al. 1989).

NMR and mass spectroscopic methods such as FABMS/CID/ LC-MS are reported to be very useful for detection and determination of structure of cyanobacterial toxins because of their sensitivity and speed of analysis (Harada 2004; Diehnelt et al. 2005). Mass Spectrometry (MS) is another analytical tool for measuring molecular mass of a toxin sample, and may be utilized for analysis of peptides, drug discovery and water quality (Ashcroft 2004; Beresovsky et al 2006). This instrument is often coupled with chromatography apparatus to separate compounds for more accurate resolution of components by mass spectrometry. LC-MS is very useful for detection and determination of structure of cyanotoxin because of their sensitivity and speed of analysis (Harada 2004; Diehnelt et al 2005).

Various molecular techniques mostly based on PCR are also used for screening, identification and characterization of toxicity and genes responsible for toxin production (Pearson and Neilan 2008). Recently multiplex PCR has been routinely employed for the monitoring of *mcy* genes (Hotto et al 2007; Saker et al 2007; Pearson and Neilan 2008; Valério et al 2010).

Conclusion

Cyanobacteria are common members of the plankton community of marine, brackish, and freshwaters throughout the world. Cyanotoxins are responsible for many cases of intoxication involving animals such as cattle, swine, dogs, fish, birds and bats. Humans may also be exposed to cyanobacterial toxins via different routes. Physiological and biological effects in aquatic plants due to cyanotoxins exposure may represent a risk to all parts of the food chain including aquatic animals following bioaccumulation and biotransfer of these toxins to animals through the food chain. Thus studying cyanotoxins has both toxicological and economical importance. Determination of taxonomic groups in cyanobacteria responsible for production of toxins has been investigated in last few years. For many of these toxins, their role in the producer cells, their evolution and natural function as well as their impact and fate in the aquatic environment needs further research. Furthermore, more information is needed on the biosynthesis of these cyanotoxins.

References

An J and Carmichael WW (1994) Use of a colorimetric protein phosphatase inhibition assay and enzyme linked immunosorbent assay for the study of microcystins and nodularins. Toxicon 32: 1495-1507.

Aráoz R, Guérineau V, Rippka R, Palibroda N, Herdman M, Laprevote O, von Döhren H, Tandeau de Marsac N and Erhard M (2008) MALDI-TOF-MS detection of the low molecular weight neurotoxins anatoxin-a and homoanatoxin-a on lyophilized and fresh filaments of axenic *Oscillatoria* strains. Toxicon 51: 1308-1315.

Ash C, MacKintosh C and MacKintosh R (1995) Development of a colorimetric protein phosphorylation assay for detecting cyanobacterial toxins. Water Sci Technol 31: 47-49.

Ashcroft AE (2004) An introduction to mass spectrometry. The University of Leeds, Leeds.

Banker R, Carmeli S,Werman M, Teltsch B, Porat R and Sukenik A (2001) Uracil moiety is required for toxicity of the cyanobacterial hepatotoxin cylindrospermopsin. J Toxicol Environ Health 62: 281-288.

Barford D (1996) Molecular mechanisms of the protein serine/threonine phosphatases. Trends Biochem Sci 21: 407-412.

Beresovsky D, Hadas O, Livne A, Sukenik A, Kaplan A and Carmeli S (2006) Toxins and biologically active secondary metabolites of *Microcystis* sp. isolated from lake Kinneret. Israel J chem 46: 79-87.

Berman FW, Gerwick WH and Murray TF (1999) Antillatoxin and kalkitoxin, ichthyotoxins from the tropical cyanobacterium *Lyngbya majuscula*, induce distinct temporal patterns of NMDA receptor-mediated neurotoxicity. Toxicon 37: 1645-1648.

Bury NR, Codd GA, Wendelaar Bonga SE and Flik G (1998) Fatty acids from the cyanobacterium *Microcystis aeruginosa* with potent inhibitory effects of fish gill Na^+/K^+-ATPase activity. J Exp Biol 201: 81-89.

Campos A and Vasconcelos VM (2010) Molecular mechanisms of microcystin toxicity in animals. Int J Mol Sci 11: 268-287.

Capper A, Tibbetts IR, O'Neil JM and Shaw GM (2005) The fate of *Lyngbya majuscula* toxins in three potential consumers. J Chem Ecol 31: 1595-1606.

Carmichael WW (1992) Cyanobacteria secondary metabolites – The cyanotoxins. J Appl Bacteriol 72: 445-459.

Carmichael WW (1994) The toxins of cyanobacteria. Sci Am 270: 64-70.

Carmichael WW, Jones CLA, Mahmood NA and Theiss WC (1985) Algal toxins and water-based diseases. CRC Crit Rev Environ Control 15: 275-313.

Casanova MT, Burch MD, Brock MA and Bond PM (1999) Does toxic *Microcystis aeruginosa* affect aquatic plant establishment? Environ Toxicol 14: 97-109.

Cestèle S and Catterall WA (2000) Molecular mechanisms of neurotoxin action on voltage-gated sodium channels. Biochimie 82: 883-892.

Chen J and Xie P (2005) Seasonal dynamics of the hepatotoxic microcystins in various organs of four freshwater bivalves from the large eutrophic lake Taihu of subtropical China and the risk for human consumption. Environ Toxicol 20: 572-584.

Choi SH and Kim SG (1998) Lipopolysaccharide inhibition of rat hepatic microsomal epoxide hydrolase and glutathione S-transferase gene expression irrespective of nuclear factor-kB activation. Biochem Pharmacol 56: 1427-1436.

Chong MWK, Wong PKS, Shaw GR and Seawright AA (2002) Toxicity and uptake mechanism of cylindrospermopsin and lophyrotomin in primary rat hepatocytes. Toxicon 40: 205-211.

Christiansen G, Fastner J, Erhard M, Borner T and Dittmann E (2003) Microcystin Biosynthesis in *Planktothrix*: genes, evolution, and manipulation. J Bacteriol 185: 564-572.

Chu FS, Huang X, Wei RD and Carmichael WW (1989) production and characterization of antibodies against microcystins. Appl Environ Microbiol 55: 1928-1933.

Codd GA, Bell SG, Kaya K, Ward CJ, Beattie KA and Metcalf JS (1999) Cyanobacterial toxins, exposure routes and human health. Eur J Phycol 34: 405-415.

Codd GA, Morrison LF and Metcalf JS (2005) Cyanobacterial toxins: risk management for health protection. Toxicol Appl Pharmacol 203: 264-272.

Cohen P (1989) The structure and regulation of protein phosphatases. Ann Rev Biochem 58: 453-508.

Dawson RM (1998) The toxicology of microcystins. Toxicon 36: 953-962.

Diehnelt CW, Peterman SM and Budde WL (2005) Liquid chromatography-tandem mass spectrometry and accurate m/z measurements of cyclic peptide cyanobacteria toxins. Trend Anal Chem 24: 622-634.

Dittmann E, Fewer DP and Neilan BA (2012) Cyanobacterial toxins: biosynthetic routes and evolutionary roots. FEMS Microb Doi: 10.1111/1574-6976.12000.

Dolman AM, Rücker J, Pick FR, Fastner J, Rohrlack T, Mischke U and Wiedner C (2012) Cyanobacteria and Cyanotoxins: The Influence of Nitrogen versus Phosphorus. doi:10.1371/journal.pone.0038757.

Edwards DJ, Marquez BL, Nogle LM, McPhail K, Goeger DE, Roberts MA and Gerwick WH (2004) Structure and biosynthesis of the jamaicamides, new mixed polyketidepeptide neurotoxins from the marine cyanobacterium *Lyngbya majuscula.* Chem Biol 11: 817-833.

Eriksson JE, Toivola D, Meriluoto JAO, Karaki H, Han Y-G and Hartshorne D (1990) Hepatocyte deformation induced by cyanobacterial toxins reflect inhibition of protein phosphatase. Biochem Biophys Res Commun 173: 1347-1353.

Falconer IR (1989) Tumor promotion and liver injuries caused consumption by cyanobacteria. Envirol. Toxicol. Water Qual 6: 177-184.

Falconer IR (1991) Tumour promotion and liver injury caused by oral consumption of cyanobacteria. Environ. Toxicol. Water Qual 6: 177-184.

Falconer IR (1998) Algal toxins and human health. In: *The Handbook of Environmental Chemistry* 5 Part C (Ed. Hrubec J), Springer-Verlag, Berlin, Germany. pp. 53-82.

Falconer IR (2005) Cyanobacterial toxins of drinking water supplies: Cylindrospermopsins and microcystins. CRC Press, Boca Raton, Florida.

Falconer IR and Yeung DS (1992) Cytoskeletal changes in hepatocytes induced by *Microcystis* toxins and their relation to hyperphosphorylation of cell proteins. Chemico-biol Interactions 81: 181-196.

Falconer IR, Hardy SJ, Humpage AR, Froscio SM, Tozer GJ and Hawkins PR (1999) Hepatic and renal toxicity of the blue-green alga (cyanobacterium) *Cylindrospermopsis raciborskii* in male Swiss albino mice. Environ Toxicol 14: 143-150.

Ferrão-Filho ADS and Kozlowsky-Suzuki B (2011) Cyanotoxins: Bioaccumulation and effects on aquatic animals. Mar Drugs 9: 2729-2772.

Fischer WJ and Dietrich DR (2000) Pathological and biochemical characterization of microcystin-induced hepatopancreas and kidney damage in carp (*Cyprinus carpio*). Toxicol Appl Pharmacol 164: 73-81.

Fitzgeorge R, Clark S and Keevil C (1994) Routes of intoxication. In: *Detection Methods for Cyanobacterial Toxins* (Eds. Codd GA, Jefferies TM, Keevil CW and Potter P), Royal Society of Chemistry, Cambridge. pp. 69-74.

Froscio SM, Humpage AR, Burcham PC and Falconer IR (2003) Cylindrospermopsin-induced protein synthesis inhibition and its dissociation from acute toxicity in mouse hepatocytes. Environ Toxicol 18: 243-251.

Fujiki H and Suganuma M (2009) Carcinogenic aspects of protein phosphatase 1 and 2A inhibitors. Prog Mol Subcell Biol 46: 221-254.

Fujiki H, Sugimura T and Moore RE (1983) New classes of environmental tumor promoters: indole alkaloids and polyacetates. Environ Health Perspect 50: 85-90.

Gademann K and Portmann C (2008) Secondary metabolites from cyanobacteria: complex structures and powerful bioactivities. Curr Org Chem 12: 326-341.

Gorham PR and Carmichael WW (1988) Hazards of freshwater blue-green algae (Cyanobacteria). In: *Algae and Human Affairs* (Eds. Lembi CA and Waaland JR), Cambridge University Press, Cambridge. pp. 403-432.

Haider S, Naithani V, Viswanathan PN and Kakkar P (2003) Cyanobacterial toxins: a growing environmental concern. Chemosphere 52: 1-21.

Hamvas M, Máthe C, Molnár E, Vasas G, Grigorszky I and Borbély G (2003) Microcystin-LR alters the growth, anthocyanin content and single stranded DNase activities in *Sinapis alba* L. seedlings. Aquat Toxicol 62:1-9.

Hamvas M, Máthe C, Molnár E, Vasas G, Grigorszky I and Borbély G (2003) Microcystin-LR alters the growth, anthocyanin content and single stranded DNase activities in *Sinapis alba* L. seedlings. Aquat Toxicol 62: 1-9.

Harada KI (2004) Production of secondary metabolites by freshwater cyanobacteria. Chem Pharm Bull 52: 889-899.

Heinze R (1999) Toxicology of the cyanobacterial toxin microcystin-LR to rats after 28 days intake with the drinking water. Environ Toxicol 14: 57-60.

Henriksen P, Carmichael WW, An J and Moestrup Ø (1997) Detection of an anatoxin-a(s)-like anticholinesterase in natural blooms and cultures of cyanobacteria/blue-green algae from Danish lakes and in the stomach contents of poisoned birds. Toxicon 35: 901-913.

Herfindal L and Selheim F (2006) Microcystin produces disparate effects on liver cells in a dose dependent manner. Mini-Reviews Med Chem 6: 279-285.

Hewett JA and Roth RA (1993) Hepatic and extrahepatic pathobiology of bacterial lipopolysaccharides. Pharmacol Rev 45: 381-411.

Hitzfeld BC, Lampert CS, Spaeth N, Mountfort D, Kaspar H and Dietrich DR (2000) Toxin production in cyanobacterial mats from ponds on the McMurdo Ice Shelf. Antarctica. Toxicon 38: 1731-1748.

Holmes CFB and Boland MP (1993) Inhibitors of protein phosphatase -1 and -2A; two of the major serine/threonine protein phosphatases involved in cellular regulation. Current Opin Struct Biol 3: 934-943.

Honkanen RE, Codispoti BA, Tse K and Boynton AL (1994) Characterization of natural toxins with inhibitory activity against serine/threonine protein phosphatases. Toxicon 32: 339-350.

Honkanen RE, Zwiller J, Moore RE, Daily SL, Khatra BS, Dukelow M and Boynton AL (1990) Characterization of microcystin-LR, a potent inhibitor of type 1 and type 2A protein phosphatases. J Biol Chem 265: 19401-19404.

Hotto AM, Satchwell MF and Boyer GL (2007) Molecular characterization of potential microcystin- producing cyanobacteria in Lake Ontario embayments and nearshore waters. Appl Environ Microbiol 73: 4570-4578.

Humpage A (2008) Toxin types, toxicokinetics and toxicodynamics. In: *Cyanobacterial Harmful Algal Blooms: State of the Science and Research Needs* (Ed. Hudnell HK), Adv Exp Med Assoc, Springer Press, New York, USA. pp. 383-416.

Ibelings BW and Havens KE (2008). Cyanobacterial toxins: a qualitative meta-analysis of concentrations, dosage and effects in freshwater, estuarine and marine biota. In: *Cyanobacterial Harmful Algal Blooms: State of the Science and Research Needs* (Ed. Hudnell HK), Springer, New York. pp. 675-732.

Ibelings BW, Bruning K, de Jonge J, Wolfstein K, Pires LMD, Postma J and Burger T (2005) Distribution of microcystins in a lake food web: No evidence for biomagnifications. Microb Ecol 49: 487-500.

Ikawa M, Sasner JJ and Haney JF (2001) Activity of cyanobacterial and algal odor compounds found in lake waters on green alga *Chlorella pyrenoidosa* growth. Hydrobiologia 443: 19-22.

Ito E, Takai A, Kondo F, Masui H, Imanishi S, and Harada K (2002) Comparison of protein phosphatase inhibitory activity and apparent toxicity of microcystins and related compounds. Toxicon 40: 1017-1025.

Kaebernick M and Neilan BA (2001) Ecological and molecular investigations of cyanotoxin production. FEMS Microbiol Ecol 35: 1-9.

Kaebernick M, Neilan BA, Borner T and Dittmann E (2000) Light and the transcriptional response of the microcystin biosynthesis gene cluster. Appl Environ Microbiol 66: 3387:3392.

Karjalainen M, Reinikainen M, Lindvall F, Spoof L and Meriluoto JAO (2003) Uptake and accumulation of dissolved, radiolabeled nodularin in Baltic Sea Zooplankton. Environ Toxicol 18: 52-60.

Kellmann R, Mihali TK, Jeon YJ, Pickford R, Pomati F and Neilan BA (2008) Biosynthetic intermediate analysis and functional homology reveal a saxitoxin gene cluster in cyanobacteria. Appl Environ Microbiol 74: 4044-4053.

Krienitz L, Ballot A, Kotut K, Wiegand C, Pqtz S, Metcalf JS, Codd GA and Pflugmacher S (2003) Contribution of hot spring cyanobacteria to the mysterious deaths of Lesser Flamingoes at Lake Bogoria, Kenya. FEMS Microbiol Ecol 43: 141-148.

Kuiper-Goodman T, Falconer I and Fitzgerald J (1999) Human health aspects. In: *Toxic Cyunohacteria in Water* (Eds. Chorus I and Bartram J), Who and E & FN Spon, London. pp. 113-154.

Kurmayer R and Christiansen G (2009) The genetic basis of toxin production in cyanobacteria. Freshwater Rev 2: 31-50.

Lagos N, Onodera H, Zagatto PA, Andrinolo SM, Azevedo SMF and Oshima Y (1999) The first evidence of paralytic shellfish toxins in the freshwater cyanobacterium *Cylindrospermopsis raciborskii*, isolated from Brazil. Toxicon 37: 1359-1373.

Lakshmana Rao PV, Bhattacharya R, Gupta N, Parida MM, Bhaskar ASB and Dubey R (2002) Involvement of caspase and reactive oxygen species in cyanobacterial toxin anatoxin-a induced cytotoxicity and apoptosis in rat thymocytes and Vero cells. Arch Toxicol 76: 227-235.

Landsberg JH (2002) The effects of harmful algal blooms on aquatic organisms. Rev Fish Sci 10: 113-390.

Lehtimäki J, Moisander P, Sivonen K and Kononen K (1997) Growth, nitrogen fixation and nodularin production by two Baltic Sea cyanobacteria. Appl Environ Microbiol 63: 1647-1656.

Lewin WC, Kamjunke N and Mehner T (2003) Phosphorus uptake by *Microcystis* during passage through fish guts. Limnol Oceanogr 48: 2392-2396.

Li H, Xie P, Li G, Hao L and Xiong Q (2009) *In vivo* study on the effects of microcystin extracts on the expression profiles of proto-oncogenes (*c-fos*, *c-jun* and *c-myc*) in liver, kidney and testis of male Wistar rats injected i.v. with toxins. Toxicon 53: 169-175.

Llewellyn LE (2006) Saxitoxin, a toxic marine natural product that targets a multitude of receptors. Nat Prod Rep 23: 200-222.

MacKintosh C, Beattie KA, Klumpp S, Cohen P and Codd GA (1990) Cyanobacterial microcystin-LR is a potent and specific inhibitor of protein phosphatases 1 and 2A from both mammals and higher plants. FEBS Lett 264: 187-192.

Mahmood NA and Carmichael WW (1986) The pharmacology of anatoxin-a(s), a neurotoxin produced by the freshwater cyanobacterium *Anabaena flos-aquae* NRC 525-17. Toxicon 24: 425-434.

Martin C, Codd GA, Siegelman HW and Weckesser J (1989) Lipopolysaccharides and polysaccharides of the cell envelope of toxic *Microcystis aeruginosa* strains. Arch Microbiol 152: 90-94.

Martinez Hernandez J, Lopez-Rodas V and Costas E (2009) Microcystins from tap water could be a risk factor for liver and colorectal cancer: A risk intensified by global change. Med Hypothe 72: 539-540.

Matsushima R, Yoshizawa S, Watanabe MF, Harada KI, Furusawa M, Carmichael WW and Fujiki H (1990) *In vitro* and *in vivo* effects of protein phosphatase inhibitors, microcystins and nodularin, on mouse skin and fibroblasts. Biochem Biophys Res Commun 171: 867-874.

McElhiney J and Lawton LA (2005) Detection of the cyanobacterial hepatotoxins microcystins. Toxicol Appl Pharmacol 203: 219-230.

Meriluoto J, Nygard SE, Dahlem AM and Eriksson JE (1990) Synthesis, organotropism and hepatocellular uptake of tritium labeled epimers of dihydromicrocystin-LR, a cyanobacterial peptide toxin analog. Toxicon 28: 1439-1446.

Metcalf JS and Codd GA (2003) Analysis of cyanobacterial toxins by immunological methods. Chem Res Toxicol 16: 103-112.

Mihali TK, Kellmann R, Muenchhoff J, Barrow KD and Neilan BA (2008) Characterization of the gene cluster responsible for cylindrospermopsin biosynthesis. Appl Environ Microbiol 74: 716-722.

Milutinovic A, Zivin M, Zorc-Pleskovic R, Sedmak B and Suput D (2003) Nephrotoxic effects of chronic administration of microcystins -LR and -YR. Toxicon 42: 281-288.

Moffitt MC and Neilan BA (2004) Characterization of the nodularin synthetase gene cluster and proposed theory of the evolution of cyanobacterial hepatotoxins. Appl Environ Microbiol 70: 6353-6362.

Moore RE (1996) Cyclic peptides and depsipeptides from cyanobacteria: A review. J Ind Microbiol 16: 134-143.

Moore RE, Cheuk C and Patterson GML (1984) Hapalindoles: new alkaloids from the blue-green alga *Hapalosiphon fontinalis*. J Am Chem Soc 106: 6456-6457.

Namikoshi M, Carmichael WW, Sakai R, Jares-Erijman EA, Kaup AM and Rinehart KL (1993) 9-Deazaadenosine and its 5V-a-dglucopyranoside isolated from the cyanobacterium *Anabaena affinis* strain VS-1. J Am Chem Soc 115: 2504-2505.

Neilan BA, Dittmann E, Rouhiainen L, Bass RA, Schaub V, Sivonen K and Börner T (1999) Nonribosomal peptide synthesis and toxigenicity of cyanobacteria. J Bact 181: 4089-4097.

Neumann C, Bain P and Shaw G (2007) Studies of the comparative *in vitro* toxicology of the cyanobacterial metabolite deoxycylindrospermopsin. J Toxicol Environ Health 70: 1679-1686.

Nishiwaki-Matsushima R, Ohta T, Nishiwaki S, Suganuma M, Kohyama K, Ishikawa T, Carmichael WW and Fujiki H (1992) Liver tumor promotion by the cyanobacterial cyclic peptide toxin microcystin-LR. J Cancer Res Clin Oncol 118: 420-424.

Ohta T, Nishiwaki R, Yatsunami J, Komori A, Suganuma M and Fujiki H (1992) Hyperphosphorylation of cytokeratins 8 and 18 by microcystin-LR, a new liver tumour promoter, in primary cultured rat hepatocytes. Carcinogenesis 13: 2443- 2447.

Ohta T, Sueoka E, Iida N Komori A, Suganuma M, Nishiwaki R, Tatematsu M, Kim SJ, Carmichael WW and Fujiki H (1994) Nodularin, a potent inhibitor of protein phosphatases 1 and 2A, is a new environmental carcinogen in male F344 rat liver. Cancer Res 24: 6402-6406.

Ohtani I, Moore RE and Runnegar MTC (1992) Cylindrospermopsin: a potent hepatotoxin from the blue-green algae *Cylindrospermopsis raciborskii.* J Am Chem Soc 114: 7942-7944.

Papke U, Gross EM and Francke W (1997) Isolation, identification and determination of the absolute configuration of fischerellin B. A new algicide from the freshwater cyanobacterium *Fischerella muscicola* (Thuret.). Tetrahedron Lett 38: 379-382.

Pearson L, Mihali T, Moffitt M, Kellmann R and Neilan B (2010) On the chemistry, toxicology and genetics of the cyanobacterial toxins, microcystin, nodularin, saxitoxin and cylindrospermopsin. Mar Drugs 8: 1650-1680.

Pearson LA and Neilan BA (2008) The molecular genetics of cyanobacterial toxicity as a basis for monitoring water quality and public health risk. Curr Opin Biotechnol 19: 281-288.

Rastogi RP and Sinha RP (2009) Biotechnological and industrial significance of cyanobacterial secondary metabolites. Biotechnol Adv 27: 521-539.

Ressom R, San Soong F, Fitzgerald J, Turczynowicz L, El Saadi O, Roder D, Maynard T and Falconer I (1994) Health effects of toxic cyanobacteria (Blue-green algae), Australian Government Publishing Service, Canberra. pp. 27-69.

Rinehart KL, Harada K, Namikoshi M, Chen C, Harris CEJ, Munroe MHG, Blunt JW, Mulligan PE, Beasley VR, Dahlem AM and Carmichael WW (1988) Nodularin, microcystin and the configuration of ADDA. J Am Chem Soc 110: 8557-8558.

Rinehart KL, Namikoshi M and Choi BW (1994) Structure and biosynthesis of toxins from blue-green algae (cyanobacteria). J Appl Phycol 6: 159-176.

Robinson NA, Pace JG, Matson CF, Miura GA and Lawrence WB (1991) Tissue distribution, excretion and hepatic biotransformation of microcystin-LR in mice. J Pharmacol Exp Ther 256: 176-182.

Roger PA and Watanabe I (1986) Technologies for using biological nitrogen fixation in wetland rice: potentialities, current usage and limiting factors. Fert Res 9: 39-77.

Romanoska-Duda Z and Tarczynska M (2002) The influence of microcystin-LR and hepatotoxic cyanobacterial extract on the water plant *Spirodela oligorrhiza.* Environ Toxicol 17: 434-440.

Runnegar M, Berndt N, Kong SM, Lee EY and Zhang L (1995) *In vivo* and *in vitro* binding of microcystin to protein phosphatases 1 and 2A. Biochem Biophys Res Commun 216: 162-169.

Runnegar MT, Kong S and Berndt N (1993) Protein phosphatase inhibition and *in vivo* hepatotoxicity of microcystins. Am J Physiol 265: 224-230.

Runnegar MTC, Gerdes RG and Falconer IR (1991) The uptake of cyanobacterial hepatotoxin microcystin by isolated rat hepatocytes. Toxicon 29: 43-51.

Saker ML, Vale M, Kramer D and Vasconcelos VM (2007) Molecular techniques for the early warning of toxic cyanobacteria blooms in fresh water lakes and rivers. Appl Microbiol Biotechnol 75: 441-449.

Singh DP, Tyagi MB, Kumar A, Thakur JK and Kumar A (2001) Antialgal activity of a hepatoxin-producing cyanobacterim *Microcystis aeruginosa.* World J Microbiol Biotechnol 17: 15-22.

Singh RÍ (1961) The Role of Blue-Green Algae in Nitrogen Economy of Indian Agriculture, Indian Council of Agricultural Research, New Delhi.

Sinha RP, Kumari S and Rastogi RP (2008) Impacts of ultraviolet-B radiation on cyanobacteria: Photoprotection and repair. J Sci Res 52: 125-142.

Sivonen K (1990) Effects of light, temperature, nitrate, orthophosphate, and bacteria on growth of and hepatotoxin production by *Oscillatoria agardhii* strains. Appl Environ Microbiol 56: 2658-2666.

Sivonen K (1996) Cyanobacterial toxins and toxin production. Phycologia 35: 12-24.

Sivonen K (1998) Toxins produced by cyanobacteria. In: *Mycotoxins and Phycotoxins* (Eds. Miraglia M, Edmond HV, Brera C and Gilbert J), Development in Chemistry, Toxicology and Food Safety, USA. pp. 547-568.

Sivonen K (2000) Freshwater cyanobacterial neurotoxins: ecobiology, chemistry and detection. In: *Seafood and Freshwater Toxins* (Ed. Botana LM), Marcel Dekker Inc, New York, USA. pp. 567-582.

Sivonen K and Jones G (1999) Cyanobacterial toxins. In: *Toxic Cyanobacteria in Water: A Guide to their Public Health Consequences, Monitoring and Management* (Eds. Chorus I and Bartram J), Spon Press, London. pp. 41-111.

Skulberg OM, Carmichael WW, Andersen RA, Matsunuga S, Moore RE and Skulberg R (1992) Investigations of a neurotoxic *Oscillatorian* strain (cyanophyceae) and its toxin. Isolation and characterization of homoanatoxin-a. Environ Toxicol Chem 11: 321-329.

Smith JL, Boyer GL and Zimba PV (2008) A review of cyanobacterial odorous and bioactive metabolites: impacts and management alternatives in aquaculture. Aquaculture 280: 5-20.

Spoof L, Berg KA, Rapala J, Lahti K, Lepistö L, Metcalf JS, Codd GA and Meriluoto J (2006) First observation of cylindrospermopsin in *Anabaena lapponica* isolated from the boreal environment (Finland). Environ Toxicol 21: 552-560.

Stewart I, Schluter PJ and Shaw GR (2006) Cyanobacterial lipopolysaccharides and human health: A review. Environmental Health: A Global Access Science Source 5: 7.

Terao K, Ohmori S, Igarashi K, Ohtani I, Watanabe MF, Harada KI, Ito E and Watanabe M (1994) Electron microscopic studies on experimental poisoning in mice induced by cylindrospermopsin isolated from blue-green alga *Umezakia natans*. Toxicon 32: 833-843.

Tillett D, Dittmann E, Erhard M, von Döhren H, Börner T and Neilan BA (2000) Structural organization of microcystin biosynthesis in *Microcystis aeruginosa* PCC7806: an integrated peptide polyketide synthetase system. Chem Biol 7: 753-764.

Toivola DM, Eriksson JE and Brautigan DL (1994) Identification of protein phosphatase 2A as the primary target for microcystin-LR in rat liver homogenates. FEBS Lett 344: 175-180.

Valério E, Chambel L, Paulino S, Faria N, Pereira P and Tenreiro R (2010) Multiplex PCR for detection of microcystins-producing cyanobacteria from freshwater samples. Environ Toxicol 25: 251-260.

van Apeldoorn ME, Van Egmond HP, Speijers GJA and Bakker GJI (2007) Toxins of cyanobacteria. Mol Nutr Food Res. 51:7-60.

Welker M and Von Döhren H (2006) Diversity of cyanobacterial peptides-nature's own nonribosomal combinatorial biosynthesis. FEMS Microbiol Rev 30: 530-563.

Wickstrom ML, Khan SA, Haschek WM, Wyman JF, Eriksson JE, Saeffer DJ and Beasley VR (1995) Alterations in microtubules, intermediate filaments, and microfilaments induced by microcystin- LR in cultured cells. Toxicol Pathol 23: 326-337.

Wiegand C and Pflugmacher S (2005) Ecotoxicological effects of selected cyanobacterial secondary metabolites: a short review. Toxicol Appl Pharmacol 203: 201-218.

Wiese M, D'Agostino PM, Mihali TK, Moffitt MC and Neilan BA (2010) Neurotoxic alkaloids: saxitoxin and its analogs. Mar Drugs 8: 2185-2211.

Wu M, Okino T, Nogle LM, Marquez BL, Williamson RT, Sitachitta N, Berman FW, Murray TF, McGough K, Jacobs R, Colsen K, Asano T, Yokokawa F, Shioiri T and Gerwick WH (2000) Structure, synthesis, and biological properties of kalkitoxin, a novel neurotoxin from the marine cyanobacterium *Lyngbya majuscula*. J Am Chem Soc 122: 12041-12042.

Yadav S, Sinha RP, Tyagi MB and Kumar A (2011) Cyanobacterial secondary metabolites. Int J Pharm Bio Sci 2: B-144-B-167.

Yoshizawa S, Matsushima R, Watanabe MF, Harada K-I, Ichihara A, Carmichael WW and Fujiki H (1990) Inhibition of protein phosphatases by microcystin and nodularin associated with hepatotoxicity. J Cancer Res Clin Oncol 116: 609-614.

8

Recent Advances in Antimicrobial Potentials of Himalayan Plants

Gaurav Raj Dwivedi, Vigyasa Singh, Mahendra Pandurang Darokar and Rakesh Pandey

Abstract

The extensive use of antibiotics has raised serious health concern throughout the world due to development of resistance in microbes towards many classes of antibiotics. The emergence of multidrug resistance (MDR) in microbes and lack of novel classes of antimicrobial agents have raised an immediate need to identify new agents with novel mode of action. In addition to novel agents, inhibitors of efflux pumps are being now viewed as one of the most viable management strategy for treating MDR infections. In this concern, plants have always been a source of inspiration for novel drugs. In the development of new drugs, the role of plants is very crucial because their compounds may work as natural blueprint or may be directly used as phytomedicine. Plants either contain antimicrobial compounds or possess compounds that have no antimicrobial activity of its own, but possess the ability to enhance activity of ineffective antibiotics. Limited numbers of plants are reported for their antimicrobial properties even after Indian Himalayan region harbors 10000 angiosperms, 44 species of gymnosperms and 600 species of pteridophytes, more than 1737 species of bryophytes, 983 species lichens. In this book chapter authors try to concise up-to-date information regarding the MDR, efflux pumps and Indian Himalayan plants as antimicrobial agents.

Keywords: Antimicrobial; Indian Himalayan Plants; MDR; Novel drugs; Phytomedicine

Introduction

About 100,000 tons of antibiotics are manufactured annually worldwide but due to the limited number of core chemical structures of antibiotics and drug resistant microbes peoples are not in safe hands (Projan 2003; Piddock 2006a; Maurya et al. 2013). Recent estimate revealed that in the US, even single drug resistant bacterial strain may contribute to more deaths than HIV (Brown and Wright 2005; Chatelain and Ioset 2011). In microbes, single drug resistant, multidrug resistant (MDR), extensively drug resistant (XDR) and superbug strains presented a serious challenge to modern medicine (Kumarasamy et al. 2010; Raghvendra et al. 2011). In context to multidrug resistance, all living cells use special types of membrane transport proteins, often referred to as 'molecular vacuum cleaners', to keep their interiors free from noxious agents of cellular metabolism as well as a variety of environmental toxins (Nikaido 2009). Such proteins are known as MDR efflux pumps, which might handle structurally diverse compounds (Lomovskaya and Bostian 2006).

Out of 352382 flowering plants, more than 50000 flowering plants were referred as medicinal plants (Pushpangadan 1995; Schippmann et al. 2002; Negi et al. 2010). India is one of the world's twelve mega plant biodiversities consist of 20000 species of angiosperms, 64 gymnosperms, 1200 pteridophytes, 2850 bryophytes, 2021 lichens, 15,500 fungi and 6500 algae (Joppa et al. 2011). India is rich in its own flora including endemic plant species like 5725 angiosperms, 10 gymnosperms, 193 pteridophytes, 678 bryophytes, 260 liverworts, 466 lichens, 3500 fungi, and 1924 algae (Sanjappa 2005). This proportion of medicinal plants is the highest number of plants known for their medical value in any country of the world for the existing flora (Sharma et al. 1979; Negi et al. 2010).

Indian Himalayan region provides immense plant diversity which is evident from more than 10000 species of angiosperms, 44 species of gymnosperms, 600 species of pteridophytes, 1737 species of bryophytes, 983 species of Lichens (Mishra 1985; Singh and Hajra 1996; Kala 2004; Bhat et al. 2013). On the regional scale, the maximum species of medicinal plants have been reported from Uttarakhand, followed by Sikkim and North Bengal (Kala and Mathur 2002; Bhat et al. 2013). Much work has been carried out on ethnomedicinal aspects of plants (Negi et al. 1993; Bhat et al. 2013). Even hundreds of plant species have been screened for antimicrobial properties, but the vast majority of these have not been adequately evaluated. In concern of MDR microbes, more attention and interest for antimicrobial activity of natural products is highly needed (Taylor et al. 1996).

Here in the present book chapter, the literature being reviewed concerning microbial resistance and will highlight the plants of Himalayan region for promising new antimicrobial agents. The majority of agents discussed are putative inhibitors of microbes. The present book chapter also explores the problems posed microbes possessing efflux mechanisms as a mechanism of resistance towards clinically relevant drugs.

Drug Resistance

Acquired or intrinsic mechanism of reduction in effectiveness of a drug is known as drug resistance. This process is well concerned in case of microbes where the growing number of antibiotic resistant pathogens presents a serious challenge to modern medicine (Tenover 2006). This antimicrobial resistance can be mediated by various mechanisms such as enzymatic inactivation of the drug, alteration of the target and decreased intracellular concentration of the antimicrobials which is mediated by either decreased influx or increased efflux or a combination of both (Kohanski et al. 2010).

Multi drug resistance (MDR)

Multidrug resistance is the phenomenon where development of resistance to two or more than two structurally unrelated drugs (Zgurskaya and Nikaido 2000). The multidrug resistance phenomenon is often associated with the over-expression of the transporters that recognize and efficiently expel from the cells a broad range of structurally unrelated compounds (Gupta et al. 2011). Analysis of the available genome sequences of various bacteria revealed that known and putative drug efflux transporters constitute from 6% to 18% of all transporters (Paulsen et al. 1998).

Efflux pumps

Efflux pumps are transporter proteins of cytoplasmic membrane in combination with cell wall which are involved in the extrusion of toxic substrates (including all classes of clinically relevant antibiotics) from cells into the external environment (Levy 2002). Efflux proteins are prevalently found in bacteria as well as in eukaryotic organisms (Nikaido 2009). These efflux pumps may be specific for one substrate or may transport a range of structurally dissimilar compounds (including antibiotics of multiple classes).

In the prokaryotic kingdom there are five major families of efflux transporter: MFS (major facilitator super-family), MATE (multidrug and toxic efflux), RND (resistance-nodulation-division), SMR (small multidrug resistance) and ABC (ATP binding cassette) (Piddock 2006b). All these systems utilize the proton motive force as an energy source, apart from the ABC transporters, which utilizes ATP hydrolysis to drive the export of substrates.

Proton driven antiporters

In prokaryotes, the extrusion of drugs and other cytotoxic compounds is conferred largely by pumps in which the drug efflux process is coupled to the influx of a proton (H^+) (Pao et al. 1998). These pumps are termed as proton antiporters and can be categorized into a number of families which include: the MFS, the SMR and the RND family. These families are mainly differentiated on the basis of hydrophobic motifs in which SMR and RND have 4 and 12-helix respectively while MFS has 12 and 14-helix hydrophobic motifs (Dwivedi 2012). These transporters are shown utilizing the proton-motive force generated by respiration to expel antibiotics and other drugs into the periplasmic space between the inner and outer membranes. Major facilitator and RND transporters can act in concert with a membrane fusion protein (MFP)/periplasmic adaptor protein (PAP) and an outer membrane factor (OMF)/ outer membrane protein (OMP) which translocates the drugs across both the inner and outer membranes, expelling them from the cell (Nikaido 1996).

Sodium and ATP driven transport

In the stressed microbial systems, even though proton-driven antiport is the major mechanism of drug efflux, other important mechanisms have also been described. A novel group of transporter proteins belong to MATE family; the common theme shared by this group of proteins is that drug efflux is coupled to sodium ion influx (He et al. 2010). In addition to above groups, ABC group of transporters present ubiquitously in all kind of cells and provide the predominant mechanism of efflux in eukaryotes (Higgins 1992). ABC transporter proteins catalyses the efflux of drugs/toxin/compounds by utilizing the energy released upon the hydrolysis of ATP.

Why antimicrobials from plants?

Infectious diseases are one of the major causes of morbidity and mortality in human beings. Various causative agents such as bacteria, viruses, parasites and fungi have affected human health and economy by emerging more than 335 diseases since 1940 (Jones et al. 2008) It is critical time to discover and invent new agents of greater therapeutic profile to fight against drug resistance, emerging and re-emerging infectious diseases (Erturk et al. 2006).

The exploitation of plants by human for the treatment of various diseases has been in practice for a very long time. Herbal drugs constitute a major part in all the traditional system of medicines (Higa et al. 1994; *Singh* and *Lahiri 2010*). According to a survey by NCI, USA, among the 877 small-molecule new chemical entities introduced as drugs worldwide during 1981–2002 were mainly inspired by natural products (Newman et al. 2003). Plant species still serves as a rich source of many novel biologically active compounds such as terpenoids, alkaloids, glycosteroids, flavonoids, polyphenols etc. (Heinrich and Gibbons 2001).

Plant preparations with lower adverse reactions and reduced cost are encouraging the public and national health care institutions to consider plant medicines as alternatives to synthetic drugs (Upadhyay et al. 2012). Screening of plant products for antimicrobial properties has gained much importance because World Health Organization (WHO) is keenly interested in the development and utilization of medicinal plant resources in the traditional system of medicine in the developing countries (Goud et al. 2005; Dwivedi 2012).

Himalayan Diversity as Antimicrobial Agents

The entire Himalayan ranges can be divided into three major botanical regions, the western, the central and the eastern Himalayas (Samant et al. 1998). The western region is divided into western Himalayas and north western Himalayas. The central Himalayas confirm to the mountainous ranges of Nepal. The eastern Himalayas include Darjeeling, Sikkim, Bhutan and Arunachal Pradesh. The presence of a large number of native (31%), endemic (15.5%) and threatened (14%) plant species manifest the unique diversity of plants in the Himalayan region (Samant et al. 2001). Seventy one genera and 5 plant families (Tetracentraceae, Hamamelidaceae, Circaesteraceae, Butomaceae and Stachyuraceae) are endemic to Himalayan region (Bhat et al. 2013). The largest family of flowering plants in the hotspot is the Orchidaceae, with 750 species, and a large number of orchids, many representing rather young endemic species, The Eastern Himalaya are also a center of diversity for several widely distributed plant taxa, such as *Rhododendron*, *Primula*, and *Pedicularis* (Samant et al. 2001; http://www.eoearth.org/article/Biological_ diversity_in_ the_ Himalayas %20/).

In the Himalaya, a zone of permanent rock and ice begins at about 5,500-6,000 meters and vascular plants occurring at these highest elevations. Cushion plants found at more than 6,100 meters, while *Ermania himalayensis*, was found at 6,300 meters. The rich floristic diversity of Himalayas is a potential source of many drug yielding, aromatic and oil yielding plants (Malhotra and Basu 1984; Bhat et al. 2013). Many of these plants yield alkaloids, glycoside alkaloids, resins and other secondary metabolites which have been used for medicine (Hammer et al. 1999; Dwivedi 2012). For smooth reading purposes these antimicrobial plants are classified as angiosperms, gymnosperms, bryophytes, pteridophytes and lichens.

Angiosperms as Antimicrobials

The medicinal values of Himalayan angiosperms (flowering plants) are already referred in Charaka Samhita and Sushruta Samhita and other ancient literatures (Singh and Lahiri 2010). Regional populations have developed their own health care system based on locally available plants and these are named as Nepali jaributi, Lepcha and Tibetan medicinal system (Zysk 1998). Several plants from different families of angiosperms have been reported for antimicrobial properties and are documented in Table1.

About 80,000 plant species are utilized by the different system of Indian medicine, but only numerous plants are reported for their antimicrobial activities (Zysk 1996; Prajapati et al. 2006). In context of MDR, this is the demand of time to screen unexplored medicinal plants as antimicrobial agents (Polombo and Semple 2001; Upadhyay et al. 2012; Dwivedi 2012).

Gymnosperms as Antimicrobials

Like angiosperms only few gymnosperms are reported for antimicrobial activities. *Cycas* and *Gingo* were found as sources of novel antimicrobial peptides (Kobayashi and Matsumoto 1965; Hung et al. 2000; Sati and Joshi 2011). The leaves of *Sabina chinensis, Cedrus deodara, Pinus bungeana, Platgcladus orientalis, Torryeya grandissh* contain the essential oil with reported antimicrobial activity (Dorman and Deans 2000). *Ables balsamea Abies religiosa, Pinus brutia* and *Pinus roxburghii* of family pinaceae are one of the major sources of antimicrobial agents turpentine obtained from their resin (Kizil et al. 2002).

Pteridophytes as Antimicrobials

Pteridophytes are known for their medicinal value to human for 2000 years ago (Banerjee and Sen 1980). In comparison to the angiosperms, pteridophytes were neglected in modern chemotherapy and researches as antimicrobials (Dixit 1974). Some researchers did antimicrobial activity of several vascular plants which included some pteridophytes of which 17 species had antimicrobial activity (Hayes 1947; Spencer et al. 1947; Nickell 1959; Dhawan et al. 1977).

Adiantum group are reported as good source of antimicrobial substances (Singh et al. 2008). The antimicrobial properties of *Adiantum caudatum, A. philippense,*

Table 1: List of Himalayan Angiosperms with reported antimicrobial activities

Sl No.	Himalayan Plant Species	Family	Parts used	Reported Molecules/ compounds	References
1.	*Abelia grandiflora*	Caprifoliaceae	Leaf, Stem	Water Extracts	Madsen and Pates 1952
2.	*Acacia catechu*	Mimosaceae	Stem	80% Ethanolic extract	Patel et al. 2009
3.	*Aconitum heterophyllum*	Ranunculaceae	Root, tuber	Alkaloid, Ethanolic extract	Pandya et al. 1990
4.	*Acorus calamus*	Araceae	Leaf ,Rhizome	Asarone	Kar and Jain 1971
5.	*Aegle marmelous*	Rutaceae	Fruit, leaf	Rutacin	Ahmad et al. 1998
6.	*Aesculus hippocastanum*	Hippocastanaceae	Seed, tree bark, bud, flowers	protein 1 (Ah-AMP1),	Fant et al. 1999
7.	*Ageratum conyzoides*	Asteraceae	Root	Ageconyfavones	Ming 1999
8.	*Aisandra butyracea*	Sapotaceae	Seed, pulp, bark	Triterpenoids	Latitha and Venkatraman 1991
9.	*Albizia lebbeck*	Mimosaceae	Seed, flower	Triterpenoids	Srinivasan et al. 2001
10.	*Alliaria petiolata*	Cruciferae	Leaf & root	Extracts	Chopra et al. 1986
11.	*Alpinia officinarum*	Zingiberaceae	Rhizome	Extracts	Ray and Majumdar 1976
12.	*Alstonia scholaris*	Apocynaceae	Leaf, bark	Alkaloids, ditamine, echitenine and echitamine	Khan et al. 2003
13.	*Ammannia baccifera*	Lythraceae	Roots, whole plants	4-Hydroxy-α-tetralone	Upadhyay et al. 2013
14.	*Ammannia* multiflora	Lythraceae	Roots, whole plants	4-Hydroxy-α-tetralone	Upadhyay et al. 2012
15.	*Andrographis paniculata*	Acanthaceae	Leaf	Andrographolide	Coon and Ernst 2004
16.	*Anemarrhena asphodeloides*	Asphodelaceae	Rhizome	Extracts	Yeung Him-Che 1985
17.	*Anthocephalus cadamba*	Rubiaceae	Bark, fruits	Extracts	Umachigi et al. 2007
18.	*Arisaema consanguineum*	Araceae	Root	Extracts	Bown D 1995
19.	*Arnebia euchroma*	Boraginaceae	Root	shikonin	Duke and Ayensu 1985
20.	*Artemisia maritima*	Compositae	Leaf	Essential oil	Nadkami1976
21.	*Artemisia sieversiana*	Compositae	Leaf	Essential oil	Chopra et al. 1986
22.	*Belamcanda chinensis*	Iridaceae	Root	Extracts	Bown D 1995
23.	*Berberis angulosa*	Berberidaceae	rhizome	berbeine	Duke and Ayensu 1985
24.	*Berberis aristata*	Berberidaceae	rhizome	berbeine	Duke and Ayensu 1985

Sl No.	Himalayan Plant Species	Family	Parts used	Reported Molecules/ compounds	References
25.	*Berberis chitria*	Berberidaceae	rhizome	berbeine	Duke and Ayensu 1985
26.	*Berberis jaeschkeana*	Berberidaceae	rhizome	berbeine	Duke and Ayensu 1985
27.	*Berberis sikkimensis*	Berberidaceae	rhizome	berbeine	Duke and Ayensu 1985
28.	*Bergenia ciliata*	Saxifragaceae	Root, leaf	Extracts	Manandhar 2002
29.	*Betula utilis*	Betulaceae	Bark	Extract	Kumaraswamy et al. 2008
30.	*Callicarpa macrophylla*	Verbenaceae	Leaf	sesquiterpenes, and triterpenes	Jones and Kinghorn 2008
31.	*Chenopodium ambrosioides*	Chenopodiaceae	Whole plant	Essential oil	Kumar et al. 2007
32.	*Cichorium intybus*	Asteraceae	Root, seeds	Extracts	Petrovic et al. 2004
33.	*Cimicifuga foetida*	Ranunculaceae	Root	Extract	Chopra *et al.* 1986
34.	*Cinnamomum tamala*	Lauraceae	Leaf	Eugenol and Oleoresins	Kapoor et al. 2008
35.	*Citrus medica*	Rutaceae	Fruit	Extracts	Samy 2005
36.	*Coptis teeta*	Ranunculaceae	root	Extracts	Duke and Ayensu 1985
37.	*Crassocephalum crepidioides*	Asteraceae	Leaves and stem	Extracts	Grubben and Denton 2004
38.	*Crocus sativus*	Iridaceae	Stigma	Extracts	Arora and Kaur 1999
39.	*Curcuma caesia*	Zingiberaceae	Rhizome	Cinnamate	Gupta et al. 1976
40.	*Curcuma zedoaria*	Zingiberaceae	Rhizome	P methoxy Cinnamate	Gupta et al. 1976
41.	*Cyperus rotundus*	Cyperaceae	Leaves	Extracts	Chopra et al. 1986
42.	*Eclipta prostrata*	Asteraceae	Shoot	alkaloids	Chopra et al. 1986
43.	*Elaeocarpus sphaericus*	Elaeocarpaceae	Fruit	alkaloid	Singh and Nath 1999
44.	*Eupatorium odoratum*	compositae	Leaf, root	Extracts	Iwu and Chiori 1985
45.	*Gaultheria fragrantissima*	Ericaceae	Leaf	Essential oil	Chopra et al. 1986
46.	*Gentiana kurroo*	Gentianaceae	Root	Root extracts	Chiej R 1984
47.	*Geranium nepalense*	Geraniaceae	Whole plants	Essential oil	Pande *et al.* 1984
48.	*Gmelina arborea*	Verbenaceae	Leaf	Extracts	Sharma *et al.* 2001
49.	*Hedychium spicatum*	Zingiberaceae	Rhizome	Limonene, linalool	Kapoor et al.1971
50.	*Heracleum nepalense*	Apiaceae	Root	methanol extract	Dash et al. 2005
51.	*Holarrhena antidycntrica*	Apocynaceae	Bark, seed, leaves	Terpenes	Ballal et al. 2000
52.	*Houttuynia cordata*	Saururaceae	Whole plants	Extracts	Grieve 1984

Sl No.	Himalayan Plant Species	Family	Parts used	Reported Molecules/ compounds	References
53.	*Hypericum japonicum*	Clusiaceae	Whole plants	Flavonoids	Dhar et al. 1973
54.	*Hypericum perforatum*	Clusiaceae	Whole plants	Flavonoids	Reichling et al. 2001
55.	*Hyptis suaveolens*	Lamiaceae	Whole Plants	1-limonene, sesquiterpenes	Aswal et al. 1996
56.	*Impatiens balsamina*	Balsaminaceae	Flower	Extracts	Chopra et al. 1986
57.	*Indigofera glandulosa*	Fabaceae	Root and Leaves	Alkaloids and Flavonoides	Dhawan et al. 1977
58.	*Indigofera tinctoria*	Fabaceae	Root	Flavonoids and alkaloids	Dhawan et al. 1977
59.	*Inula racemosa*	Compositeae	Root and whole plants	Essential oil,isoalantolactone	Lokhande et al. 2007
60.	*Ipomoea muricata*	Convolvulaceae	seed	lysergol	Maurya et al., 2013
61.	*Jasminum officinale*	Oleaceae	Whole plants	Lupeol, jasminol	Aswal et al. 1996
62.	*Juniperus communis*	Cupressaceae	Fruit and oil	Diterpene ketone	Pepeljnjak et al. 2005
63.	*Litsea glutisnosa*	Lauraceae	Fruit,Bark extracts	Terpenes	Mandal et al. 2000
64.	*Lycium barbarum*	Solanaceae	Root, bark	Extracts	Bown 1995
65.	*Lycopodium clavatum*	Lycopodiaceae	Whole plants	alkaloids	Dhar et al. 1968
66.	*Mahonia napaulensis*	Berberidaceae	rhizome	berberine	Duke and Ayensu 1985
67.	*Moringa oleifera*	Moringaceae	Seed, Root, bark, flowers and fruits	Terpenes	Eilert et al. 1981
68.	*Murraya koenigii*	Rutaceae	Stem Bark	Alkaloids	Goutam and Purohit 1974
69.	*Myrica esculenta*	Moringaceae	Bark	Extracts	Ahmad and Beg 2001
70.	*Myrsine semiserrata*	Myrsinaceae	Fruit	Embelic acid	Chopra et al. 1986
71.	*Nardostachys jatamansi*	Valerianaceae	Rhizome (AB)	Jatamanshic acid	Kumar et al. 2006
72.	*Nepeta clarkei*	Lamiaceae	Leaf	Actinidine	Saxena and Mathela, 1996
73.	*Nepeta leucophylla*	Lamiaceae	Leaf	Iridodial b-monoenol acetate	Saxena and Mathela, 1996
74.	*Ougenia dalbergoides*	Papilionaceae	Core Wood		Chopra et al. 1986
75.	*Paedaria foetida*	Rubiaceae	Stem	Extract	Chopra et al. 1986
76.	*Paeonia suffruticosa*	Paeoniaceae	Bark and Leaf	Monoterpenes	Ren-Bo et al. 2006
77.	*Panax pseudoginseng notoginseng*	Araliaceae	Root	Root Extracts	Kelmanson et al. 2000
78.	*Phellodendron amurense*	Rutaceae	Bark	Extract	Grieve 1984

Sl No.	Himalayan Plant Species	Family	Parts used	Reported Molecules/ compounds	References
79.	*Physalis peruviane*	Solanaceae	Leaf andFruit (AB)	Withanolides	Sethi et al. 1974
80.	*Phytolacca acinosa*	Phytolaccaceae	Root	Root extracts	Bown 1995
81.	*Picrorhiza kurroa*	Scrophulariaceae	Root	Glucosides,glycosides	Vohora et al. 1972
82.	*Piper betle*	Piperaceae	Leaf	Chavicol	Nair and Chanda 2008
83.	*Piper cubeba*	Piperaceae	Fruit	Linoleic acid	Chopra et al. 1986
84.	*Piper longum*	Piperaceae	Fruit and Root	Piperine	Chopra et al. 1986
85.	*Plantago lanceolata*	Plantaginaceae	Leaf	Mucilage, tannin and silic acid	Chopra et al. 1986
86.	*Prunis cerasoides*	Rosaceae	Bark and fruit	Flavanone Glycoside	Ray and Majumdar 1976
87.	*Prunus amygdalus*	Rosaceae	Kernel	Glycoside	Neogi et al. 2008
88.	*Pterospermum acerifolium*	Sterculiaceae	Leaf, bark, flower	Extracts	Thatoi et al.2008
89.	*Punica granatum*	Punicaceae	Fruit,flower	pelletierines	Bown 1995
90.	*Rauvolfia serpentina*	Apocynaceae	Root, Leaf	Alkaloids	Ahmad et al. 2002
91.	*Rheum australe*	Polygonaceae	leaf	Oxalic acid	Chopra et al. 1986
92.	*Rhus chinensis*	Anacardiaceae	Fruit	Extracts	Chopra et al. 1986
93.	Salix caprea	Salicaceae	Bark, young leaf	Flavonoides	Malterud *et al.*1985
94.	*Saussurea lappa*	Asteraceae	Root	Extracts	Ray and Majumdar 1976
95.	*Saxifraga stolonifera*	Saxifragaceae	Leaf, stem	Extract	Nickell 1959
96.	*Sida spinosa*	Malvaceae	Whole plant	Extracts	Ray and Majumdar 1976
97.	*Smilax china*	Liliaceae	Tubers	Steroidal saponins	Song et al. 1999
98.	*Swertia chirayita*	Gentianaceae	Whole plant	Extracts	Awasthi et al. 2005
99.	*Tamarindus indica*	Leguminosae	Leaf, fruit, bark, flowers		Extracts Ray and Majumdar 1976
100.	*Tragia involucrata*	Euphorbiaceae	Whole plant	Vinyl hexylether	Samy et al. 2004
101.	*Vetiveria zizanoides*	Poaceae	Root	Terpenes and essential oil	Gupta et al., 2012
102.	*Vitex negundo*	Verbenaceae	Leaf	Extracts	Duke and Ayensu 1985
103.	*Woodfordia fruticosa*	Lythraceae	Whole plant	Extracts	Das et al. 2007
104.	*Xanthium strumarium*	Compositae	fruits	Extract	Cerdeiras et al. 2007
105.	*Zanthoxylum armatum*	Rutaceae	Fruits, bark	Essential oil	Mehta *et al.* 1981
106.	*Zingiber officinale*	Zingiberaceae	Rhizome	Sesquiterpenes & Zinzerol	Azu et al. 2007

A. venustum, A. incisum are well established (Dixit 1974; Sharma and Vyas 1985; Jain 1991; Singh et al. 2008). *Adiantum capillus-veneris,* which is extensively used in folk medicine and its leaves essential oil had good antibacterial activity (Banerjee 1974; Victor et al. 2003). *Dryopteris* group namely *Dryopteris cochleata, Dryopteris odontoloma, Dryopteris wallichiana,* species were also reported for antimicrobial activity (Anonymous 1952; Jain 1991; Parihar 2006). In addition to above groups, *Botrychium lanuginosum, Botrychium ternatum, Glaphyropteridopsis erubescens, Microsorium membranaceum, Lepisorus clathratus, Polypodium irioides, Pyrrosia mannii, Phymatodes ebenipes* and *Equisetum arvense* deserve special mention in antimicrobial activity (Anonymous 1986; Shimpi and Dutta 2004; Radulovi et al. 2006). At last we can say that only nominal pteridophytes are reported as antimicrobial but maximum are unexplored and future research should be focused to find the active antimicrobial principle and their possible mode of action.

Bryophytes as Antimicrobials

The bryophytes "amphibians of plant kingdom" extensively used in herbal medicines in China, India, and among Native Americans since ancient times (McCleary et al. 1960; Wu, 1982; Frahm 2004; Singh and Lahiri 2010). Bryophytes used traditionally for the treatment of heart problems, inflammation, fever, liver ailments urinary, digestive problems, female problems, lung disease, skin problems and ringworm infections (Banarjee 1974; Pant et al. 1986; Frahm 2004). Out of 2850 bryophytes reported in India 1737 bryophytes are reported from Himalayan region, but their ethnomedical uses are limited. Isoflavonoids, flavonoids, and bioflavonoids are the key players for antimicrobial activity of bryophytes (Anhut et al. 1984; Basile et al. 1999; Gupta and Singh 1971). In some cases of bryophytes, terpenoids, volatile constituents and phenolics played important role for antimicrobial activity (Saritas et al. 2001; Alam 2012).

Several mosses and liverworts extracts have antibacterial, antifungal, and antiviral activity such as *Marchantia polymorpha* or *Marchantia palmate, Sphagnum portoricense, S. strictum, Conocephalum* extracts against tested against bacterial and viral pathogens (Van Hoof et al. 1981; Mewari and Kumar 2008). *Plagiochasma appendiculatum, Thuidium delicatulum, Thuidium cymbifolium, Bryum cellulare, Bryum argentium* and *Racomitrium crispulum Hypnum cupressiforme Eurynchium praelongum, Rhynchostegium* spp., *Fissidens grandifrons* and different species of *Hygroamblystegium* had marked antibacterial and antifungal effects (Saxena and Harinder 2004; Singh et al. 2007; Bodade et al. 2008) published a paper on antimicrobial activity of some Indian mosses, which included 15 moss species out of them 12 were from

Himalayan region namely *Sphagnum junghuhnianum, Brachythecium populeum, Brachythecium rutabulum, Macrothamnium submacrocarpum, Trachypodopsis serrulata, Cratoneuron filicinum, Mnium marginatum, , Entodon pulchellus, Entodon cf rubicundus*, Jeag., *Barbula arcuata* Griff., *Thuidium cymbifolium* and *Anoectangium stracheyanum*. Ethanolic extract of 12 moss species were found good antibacterial and antifungal activity. The active antimicrobial constituents have been identified from several bryophytes, such as polygodial (*Porella)*, norpiguisone (*Conocephalum conicum)*, and lunularin (*Lunularia cruciata)* but maximum are unexplored (Bodade et al. 2008).

Lichens as Antimicrobials

In Asian countries, lichens have been used in folk and alternative medicines for centuries (Richardson 1988; Saklani and Upreti 1992). The Himalayan region has roughly 983 species of lichens (235 genera), which comprise nearly 50% of the species occurring in the Indian sub-continent (Sharma et al. 2012). Oak forests are known for their rich lichen flora and Himalayas has always be the main area for lichen collection (Upreti 1996).

Biological activities of lichens are generally correlated to their special ecological circumstances (Lawrey 1986; Rikkinen 1995; Singh et al. 2000; Kumar and Upreti 2001). Initially, lichen extracts were screened for antibacterial activity in 1950s (Stoll et al. 1950) now the study elaborated. Now a days study concentrated on isolation and characterization of active principle of lichen and more detailed biological evaluation (Shahi et al. 2000; Ray et al. 2003; Gulluce et al. 2006).

Lichens are composite organisms consisting of a fungus (the mycobiont) and a photosynthetic partner (the photobiont or phycobiont) growing together in a symbiotic relationship but fungus (the mycobiont) partner is mainly responsible for the production of secondary metabolites (Lawrey 1989). Lichens provide a promising source of bioactive molecules and more than 800 compounds are reported in different lichens while many more still need to be characterized (Boustie and Grube 2005). Lichens are known to produce a large number of secondary metabolites, viz. phenolic compounds (orcinol and b-orcinol derivatives), dibenzofuranes and usnic acids (e.g. usnic acid), depsides (e.g. barbatic acid), depsidones (e.g. salazinic acid), depsones (e.g. picrolichenic acid), lactones (e.g. protolichesterinic acid, nephrosterinic acid), quinones (e.g. parietin), pulvinic acid derivatives (e.g. vulpinic acid), polyphenolic substances, methyl β-orcinolcarboxylate (MBO) and ethyl haematommate (EH) (Lauterwein et al. 1995; Huneck and Yoshimura, 1996; Boustie and Grube 2005; Yoshimura et al. 2007; Gupta et al. 2007; 2012; Sharma et al. 2012).

Now many lichen species are known for their secondary metabolites such as *Usnea articulata, U hirta, U filipendula* (usnic acid & salazinic acid), *Hypogymnia apinnata* (atranorin), *Letharia columbiana* (vulpinic acid), *Lobaria pulmonaria* (stictic acid, constictic acid & norstictic acid), *Alectoria sarmentosa* (Alectosarmentin), *Hypogymnia tubulosa* (3-hydroxyphysodic acid), *Xanthoparmelia pokornyi* (gyrophoric and stenosporic acid), *Parmelia tinctorum, Parmelia sulcata* (salazinic acid), *Pseudevernia furfuracea* (physodic acid, chloroatranorin, atranorin, and olivetoric acid), *Cetraria aculeata* (usnic acid and protolichesterinic acid) (Gollapudi et al. 1994; Turk et al. 2003; Yilmaz et al. 2004; Candan et al. 2006; 2007; Renzaka and Sigler 2007).

Efflux pump inhibitors

In stressed conditions, microbes over expressed multidrug resistance efflux pumps; however, in case of bacteria where they exert their strongest effect in intrinsic and acquired antibiotic resistance, reaching as much as several orders of magnitude change in bacterial susceptibility (Viveiros et al. 2005; Poole 2007). Bacterial efflux pumps could be utilized as a drug target to combat serious and problematic infections (Lomovskaya and Watkins, 2001). These transporter proteins have no close homologs in mammalian system and become essential in the presence of noxious agents/antibiotics (Lomovskaya and Watkins 2001; Lomovskaya and Bostian 2006).

The approach of inhibiting/modulating the efflux pumps can be highly useful in combating the multidrug resistance in microbes. Although the exact mechanism of inhibiting/modulating efflux pump is not well understood, it might bind to the antibiotic binding site or may disrupt the overall activity of the efflux pumps (Pages and Amaral, 2009). Knowledge of the structure and function at molecular level of the MDR efflux pump and its individual components is important for the intelligent design of potential combinational agents for the therapy of MDR microbial infections. These efflux pump inhibitors (EPIs)/ efflux pump modulators are compounds which potentiate the activity of an antibiotic (first/primary drug) against a resistant strain. Many molecules/compounds clearly demonstrate drug resistance reversal/modulatory properties, although the mechanistic processes are poorly understood (Table2). However, these molecules may function as (i) by prevention the degradation or modification of the primary drug (an antibiotic), (ii) by the accumulation and retention of the primary drug (an antibiotic) by inhibiting the efflux pumps, (iii) inhibition of the intrinsic repair pathway or tolerance mechanism of cells to the primary drug (an antibiotic) (iv) act like antibiotics which target a similar or different pathway that is inhibited by the first antibiotic drug (Pages and Amaral 2009; Dwivedi 2012).

Table 2: Known drug efflux pump inhibitors

Name of Inhibitors	Possible mode of inhibition	References
Quinoline derivatives (Alkoxy- and thioalkoxy-quinolines including piperidinoethyl chains)	Competitive inhibitors of the antibiotic flux	Mahamoud et al. 2007
Tetracycline analogue (Tigecycline)	Competitive inhibitors of the antibiotic efflux	Marquez 2005
Aminoglycoside analogue	Competitive inhibitors of the antibiotic efflux	Marquez 2005
Carbonyl cyanide *m*-chlorophenyl hydrazone (CCCP)	Collapses the membrane energy	Kaatz et al. 2003
Verapamil	Voltage-dependent Ca2+ channel blockers	Marquez 2005
Phenothiazine	Inhibits transport of molecules such as those involved in generating cellular energy from hydrolysis of ATP	Amaral et al. 2004
Ouabain	Inhibit transport of such as those involved in generating cellular energy from hydrolysis of ATP	Clausen, 2003
Reserpine	ATPase inhibitors with dual mode of action also inhibit the different efflux pumps	Khan et al. 2006
Piperine	Inhibits the different efflux pump	Khan et al. 2006
Arylpiperidines and Phenylpiperidine	Modification of the lipopolysaccharide structure	Bohnert and Kern 2005
5'-methoxyhydnocarpin-D (5-MHC-D)	Inhibits the different efflux pump	Stermitz et al. 2007
Chalcone	Inhibits the different efflux pump	Stavri et al. 2007
Stilbene	Inhibits the different efflux pump	Stavri et al. 2007
Catechin gallates	Inhibits the different efflux pump	Stavri et al. 2007
Carnosic acid	Inhibits the different efflux pump	Stavri et al. 2007
Lysergol	Inhibit ABC efflux pumps	Maurya et al. 2013
Lysergol derivatives	Inhibit ABC efflux pumps	Maurya et al. 2013
Niaziridin	Inhibit ABC efflux pumps	Dwivedi 2012
4hydroxy α-tetralone	Inhibit ABC efflux pumps	Dwivedi 2012; Upadhyay et al. 2012

The concept of using compounds/molecules that inhibit resistance in microbes which may be employed with a conventional antibiotic is well proven in case of augmentin which is the combination of amoxicillin (a beta lactam antibiotic) and an inhibitor of beta lactamases, clavulanic acid (Sa´nchez-Navarro 2005). Clavulanic acid greatly increases the stability of amoxicillin against degradation by beta lactamases and the product is highly useful in the treatment of patients with community-acquired pneumonia or sinusitis due to b-lactamase producing pathogens (Sa´nchez-Navarro 2005; Nikaido, 2009). During this decade, various compounds have been designed and synthesized to inhibit the activity microbial efflux pumps and listed in Table 2.

Advances in Antimicrobial Research

A. In-vitro Bioassays

Several *in-vitro* bioassays have been developed to evaluate antimicrobial activity in chemical compounds. In most of these methods, microbes are cultured in various types of broth and agar based media. Handling of microbes is recommended in a BSL-1 to 4 (Biosafety Level-1-4) facilities to check its dissemination and to protect workers and the environment. Moreover, creating a BSL-3 and BSL-4 facility involve a large amount of money, expertise, and dedicated trained work force.

1. Agar Diffusion

This is a disc method, which gives a preliminary indication whether a chemical compound is active or not. Disc diffusion assay are carried out by modified Kirby-Bauer method (CLSI). Although this is not a quantitative evaluation but grading can be done by visual observation of the zone of inhibition. The high polar chemical compounds do not diffuse properly in this medium and hence cannot be studied.

2. Macro Agar Dilution

This method gives quantitative evaluation of activity and minimum inhibitory concentration (MIC) determination by using known concentrations of extracts or compounds in an agar medium. The minimum inhibitory concentration (MIC) values are determined by 2-fold serial dilution broth assay.

3. Micro Broth Dilution

This method is most recommended, widely used, rapid, performed in 96 well plate formats and determine MIC values by 2-fold serial dilution broth assay for all the microbes (CLSI).

4. Combination studies

The combination studies are performed by broth checkerboard method as well as fix dose combination of plant compounds with known antibiotics. The MIC for each combination is recorded as the last dilution without any turbidity as per CLSI guidelines. Results are recorded in terms of fold reduction and fractional inhibitory index. The interaction of each combination was based on the fractional inhibitory concentration (FIC) and fractional inhibitory concentration index (FICI). As classically defined, FICI of <0.5 represented synergism, where as FICI of 0.5 to 4.0 represented typical additive action, while FIC index of >4.0 represented antagonism (Hsieh et al. 1993).

5. Radiorespirometry

This is also a quantitative growth inhibition method in which MIC can be calculated by using various concentrations of test sample and measuring $^{14}CO_2$ production (Siddiqi 1996). This method is mainly utilized for the detection for compounds having anti TB activity. In this method, the growth of *Mycobacterium* can be determined within one week by the extent of oxidation of 14C-palmitic acid in a liquid Middlebrook 7H12 medium to $^{14}CO_2$, which is measured quantitatively in Bactec 460 TB system. In the recent past, some non-radiometric automated systems have also been developed, where oxygen consumption or carbon dioxide production is determined to quantify the growth-inhibition done by the test compound (Diaz-Infantes et al.2000; Sanders et al. 2004)

6. Reporter Gene Assays

This assay is designed to simplify the quantification and enabling easy determination of growth/inhibition kinetics. Reporter genes may be used for rapid determination of bacterial viability by measuring expression of an introduced fluorescent or luminescent protein (Cho et al. 2002).

B. In-vivo Assays

These assays are performed in animals susceptible to particular microbial strain. Mice models are the most commonly used. The initial infection is created by different ways in mice with virulent strains of microbes. After attaining a proper stage of infection (at active phase or plateau phase) the treatment is started based on the experimental design. Different experiments are performed for different periods of time (short-term/medium-term).

In spite of all above methods this is the need of time to develop methods for early detection of MDR strains from clinical specimens so that the appropriate antibiotic therapy can be initiated to have better clinical outcome.

Overall this book chapter will be useful for those readers who are interested in antimicrobials from plants and as well as those working in the field of putting into reuse of the old antibiotics which were considered to be the medical miracle during the period of second world war until late 1980s.

Conclusion

Probably, the time has come not to fight on different types of drug resistance but to find alternatives to avoid and avert it. In this regard, these plants based compounds will be a promising candidate as antimicrobials or efflux pump inhibitors which may be useful as: (i) alternatives of antibiotics with novel mode of action (ii) lowering the dose of antibiotics; (iii) reducing the drug resistance development frequency and (iv) by increasing the efficacy of antibiotics against multidrug resistant microbes. This chapter may be also helpful in the development of inexpensive and dose economic antimicrobial drugs/ combinations from very common and widely distributed Himalayan plants.

Acknowledgement

The authors wish to thank director CSIR-CIMAP for providing the facility for preparation of this book chapter. We also thank all of our colleagues for their assistance in completion of this study.

References

Ahmad I and Beg AZ (2001) Antimicrobial and phytochemical studies on 45 Indian medicinal plants against multi-drug resistant human pathogens. J Ethanopharm 74: 113-123.

Ahmad I, Mehmood Z and Mehmood I (1998) Screening of some Indian medicinal plants for their antimicrobial properties. J Ethanopharm 62: 183–193.

Ahmed S, Amin MN, Anjum A and Haque ME (2002) The *in vitro* antibacterial activity of Rauvolfia serpentina and its tissue culture. J Nat Prod Med 6: 45-49.

Alam A (2012) Some Indian bryophytes known for their biologically active compounds Int J Appl Biol Pharma Tech 3: 239-246.

Amaral L and Kristiansen JE (2000) Phenothiazines: an alternative to conventional therapy for the initial management of suspected multidrug resistant tuberculosis: A call for studies. Int J Antimicrob Agents 14: 173–176.

Amaral L, Viveiros M and Molnar J (2004) Antimicrobial activity of phenothiazines. *In-Vivo* 18: 725-731.

Anhut S, Zinsmeister HD, Mues R, Barz W, Mackenbrock, K, Khter J and Markham KR (1984) The first identification of isoflavones from a Bryophyte. Phytochem 23: 1073-1075.

Anonymous (1948) The wealth of India. Vol. I. Publications and Information Directorate, C.S.I.R., New Delhi.

Anonymous (1952) The Wealth of India: Raw Materials. III. Publications and Information Directorate, C.S.I.R., New Delhi.

Anonymous (1986) Useful plants of India. Publications and Information Directorate, C.S.I.R., New Delhi.

Arora DS and Kaur J (1999) Antimicrobial activity of spices. Int J Antimicrob Ag 12: 257-262.

Aswal BS, Goel AK, Kulshrestha DK, Mehrotra BN and Patnaik GK (1996) Screening of Indian plants for biological activity. Ind J Exp Biol 34: 444-467.

Awasthi AK, Bisht GS and Anroop B (2005) *In vitro* antimicrobial activity of *Swertia chirata.* Ind J Nat Prod 21: 27–29.

Azu NC, Onyeagba RA, Nworie O and Kalu J (2007) Antibacterial Activity Of *Allium cepa* (Onions) And *Zingiber officinale* (Ginger) On *Staphylococcus aureus* And *Pseudomonas aeruginosa* Isolated From High Vaginal Swab. The Internet J Trop Med 2: 1540-2681.

Ballal M, Sreejan D, Bhat K K, Shirwaikar A and Shivananda PG (2000) Antibacterial activity of *Holarrhena antidysenterica* against the enteric pathogens. Ind J Pharm 32: 392-393.

Banerjee RD (1974) Studies on antibiotic activity of bryophytes and pteridophytes. Ph.D. Thesis University of Kalyani, Kalyani, India.

Banerjee RD and Sen SP (1980) Antibiotic Activity of Pteridophytes. Eco Bot 34: 284-298.

Basile A, Giordono S, Lopez-Saez JA and Cobianchi RC (1999) Antibacterial Activity of pure flavonoids isolated from mosses. Phytochem 52: 1479-1482.

Bhat JA, Kumar M and Bussmann RW (2013) Ecological status and traditional knowledge of medicinal plants in Kedarnath Wildlife Sanctuary of Garhwal Himalaya, India. J Ethnobiol Ethnomed 9: 1.

Bodade RG, Borkar PS, Arfeen MS and Khobragade CN (2008) *In vitro* Screening of Bryophytes for Antimicrobial Activity. J Med Plan 7: 23-28.

Bohnert JA and Kern WV (2005) Selected arylpiperazines are capable of reversing multidrug resistance in *Escherichia coli* over expressing RND efflux pumps. Antimicrob Agents Chemother 49: 849–852.

Boustie J and Grube M (2005) Lichens-a promising source of bioactive secondary metabolites. Pla Gen Res 3: 273–287.

Bown D (1995) Encyclopaedia of Herbs and their Uses. Dorling Kindersley, London.

Brown ED and Wright GD (2005) New targets and screening approaches in antimicrobial drug discovery. Chem Rev 105: 759-774.

Candan M, Yilmaz M, Tay T and Kivanc M (2006) Antimicrobial activity of extracts of the lichen *Xanthoparmelia pokornyi* and its gyrophoric and stenosporic acid constituents. Zeit Natur 61: 319-323.

Candan M, Yilmaz M, Tay T, Erdem M and Türk AO (2007) Antimicrobial activity of extracts of the lichen *Parmelia sulcata* and its salazinic acid constituent. Zeit Natur 62: 619-621.

Chatelain E and Ioset J-R (2011) Drug discovery and development for neglected diseases: the DNDi model. Drug Dis Develop Ther 5: 175-181.

Chiej R (1984) Encyclopaedia of Medicinal Plants. Macdonald & Co, London, 274.

Cho SH, Chang CL and Franzblau SG (2002) Interscience conference on antimicrobial agents and chemotherapy; 42nd, American Society for Microbiology, San Diego.

Chopra RN, Nayar SL and Chopra IC (1986) Glossary of Indian Medicinal Plants (Including the Supplement). Council of Scientific and Industrial Research, New Delhi.

Clausen T (2003) Na+-K+ pump regulation and skeletal muscle contractility. Physiol Rev 83: 1269-1324.

Clinical Laboratory Standards Institute (CLSI). (2006a) Performance Standards for Antimicrobial Disk Susceptibility Tests. Approved Standard, 9th ed. Document M2-A9. CLSI, Wayne, PA.

Clinical Laboratory Standards Institute. (2006b) Methods for Dilution Antimicrobial Susceptibility Tests for Bacteria that Grow Aerobically. 7th ed. Approved Standard, M7-A7. CLSI, Wayne, PA.

Coon JT and Ernst E (2004) Andrographis paniculata in the treatment of upper respiratory tract infections: a systematic review of safety and efficacy. Plant Medi 70: 293-298.

Das PK, Goswami S,Chinniah A, Panda N, Banerjee S, Sahu NP and Achari V (2007) *Woodfordia fruticosa*: Traditional uses and recent findings. J Ethnopharm 110: 189-199.

Dash S, Nath LK, Bhise S and Bhuyan N (2005) Antioxidant and antimicrobial activities of *Heracleum nepalense* D Don root. Trop J Pharm Res 4: 341-347.

Dhar, ML, Dhar MM, Dhawan BN, Mehrotra BN and Ray C (1968) Screening of Indian plants for biological activity: part I. Ind J Exp Biol 6: 232-247.

Dhar, ML, Dhar MM, Dhawan BN, Mehrotra BN, Srimal RC and Tandon JS (1973) Screening of Indian plants for biological activity: part IV. Ind J Exp Biol 11: 43-54.

Dhawan BN, Patnaik G K, Rastogi RP, Singh KK and Tandon JS (1977) Screening of Indian plants for biological activity: Part VI. Ind J Exp Biol 15: 208-219.

Diaz-Infantes MS, Ruiz-Serrano MJ, Martinez-Sanchez L, Ortega A and Bouza E (2000) Evaluation of the MB/BacT mycobacterium detection system for susceptibility testing of *Mycobacterium tuberculosis*. J Clin Microbiol 2000: 1988–1999.

Dixit RD (1974) Ferns a much neglected group of medicinal plants. J Ind Med Res 9: 59-68.

Dorman HJD and Deans SG (2000) Antimicrobial agents from plants: antibacterial activity of plant volatile oils. J App Microbiol 88: 308-316.

Duke JA and Ayensu ES (1985) Medicinal Plants of China. Reference Publications. New York.

Dwivedi GR (2012) Studies on efflux pumps for high throughput screening of phytomolecules combating bacterial multidrug resistant infections. PhD Thesis. Dr. R. M. L. Avadh University, Faizabad, India.

Eilert U, Walters B and Nahrstedt A (1981) The antibiotic principle of seeds of *Moringa oleifera* and *Moringa stenopatala*. Planta Medica 42: 55-61.

Erturk O, Kati H, Yayli N and Demirbag Z (2006) Antimicrobial properties of *Silene multifida* (Adams) Rohrb. Plant extract. Turk J Biol 30: 17-21.

Fant F, Vranken WF and Borremans FAM (1999) The three dimensional solution structure of *Aesculus hippocastanum* antimicrobial protein 1 determined by H-1 nuclear magnetic resonance. Prot Str Funct Genet 37: 388–403.

Frahm JP (2004) New frontiers in bryology and lichenology: Recent developments of commercial products from bryophytes. Bryologist 107: 277-283.

Ghione M, Parrello D and Grasso L (1988) Usnic acid revisited its activity on oral flora. Chemioterapia 7: 302-305.

Gollapudi SR, Telikepalli H, Jampani HB, Mirhom YW, Drake SD, Bhattiprolu KR, Velde VD and Mitscher LA (1994) Alectosarmentin, a new antimicrobial dibenzofuranoid lactol from the lichen, Alectoria sarmentosa. J Nat Prod 57: 934-938.

Goud PSP, Murthy KSR, Pillaiah T and Babu GVAK (2005) Screening for Antibacterial and Antifungal activity of some medicinal plants of Nallamalais, Andhra pradesh, Ind. J. Eco. Tax Bot 29: 704- 708.

Goutam MP and Purohit RM (1974) Antimicrobial activity of the essential oil of the leaves of *Murraya koenigii.* Ind J Pharm 36: 11.

Grieve M (1971) *A modern herbal.* Dover Publications, New York.

Grubben GJH and Denton OA (2004) Plant Resources of Tropical Africa 2. Vegetables. PROTA Foundation, Wageningen.

Gulluce M, Aslan A, Sokmen M, Sahin F, Adiguzel A, Agar G and Sokmen A (2006) Screening the antioxidant and antimicrobial properties of the lichens *Parmelia saxatilis, Platismatia glauca, Ramalina pollinaria, Ramalina polymorpha and Umbilicaria nylanderiana.* Phytomed 13: 515–521.

Gupta KG and Singh B (1971) Occurrence of antibacterial activity in moss extracts. Res Bull Pun Uni 22: 237-239.

Gupta SK, Banarjee AB and Achari B (1976) Isolation of ethyl p methoxy cinnamate, the major antifungal principle of *Cucuma zedoaria*. Llodia 39: 218.

Gupta S, Mishra M, Sen N, Parihar R and Dwivedi GR et al. (2011) DbMDR: a relational database for multidrug resistance genes as potential drug targets. Chem Biol Drug Des 78: 734-738.

Gupta S, Dwivedi GR, Shrivastava SK and Darokar MP (2012) Antimycobacterial activity of fractions and isolated compounds from *Vetiveria zizanoides*. Med Chem Res 21:1283–1289.

Gupta VK, Darokar MP, Saikia D, Pal A, Fatima A and Khanuja SPS (2007) Antimycobacterial Activity of Lichens. Pharma Biol 45: 200-204.

Gupta VK, Verma S, Gupta S, Singh A, Pal A, Srivastava SK, Srivastava PK, Singh SC and Darokar MP (2012) Membrane-damaging potential of natural L-(-)-usnic acid in *Staphylococcus aureus*. Eur J Clin Microbiol Infect Dis 31: 3375-3383.

Hammer KA, Carson CF and Riley TV (1999) Antimicrobial activity of essential oils and other plant extracts. J App Microbiol 86: 985-990.

Hayes L (1947) Survey of higher plants for presence of antibacterial substances. Bot Gaz 108: 408-414.

He X, Szewczyk P, Karyakin A, Evin M and Hong WX et al. (2010) Structure of a cation-bound multidrug and toxic compound extrusion transporter. Nature 467: 991-994.

Heinrich M and Gibbons S (2001) Ethnopharmacology in drug discovery: an analysis of its role and potential contribution. J Phar Pharmacol 53: 425–432.

Higa T, Anka J, Kitamara A, Koyama T, Akshashi M and Uchida T (1994) Bioactive compounds from marine sponges. Pure App Hemat 66: 2227-2236.

Higgins CF (1992) ABC transporters: from microorganisms to man. Annu Rev Cell Biol 8: 67-113.

Hsieh MH, Yu CM, Yu VL and Chow JW (1993) Synergy assessed by checkerboard a critical analysis. Diag Microbiol Infect Dis 16: 343-349.

Huneck S, Yoshimura I (1996) Identification of Lichen Substances. Springer, Berlin.

Hung X, Xie WJ and Gong ZZ (2000) Characteristics and antifungal activity of a chitin binding protein from Ginkgo biloba. FEBS Letter 478: 123-126.

Iwu MM and Chiori CO (1985) Antimicrobial activity of *Eupatorium odoratum* extracts Fitoterapia 55: 354-356.

Jain SK (1991) Dictionary of Indian folk medicine and ethnobotany. Deep Publications, New Delhi.

Jones KE, Patel N, Levy MA, Storeygard A and Balk D et al. (2008) Global trends in emerging infectious diseases. Nature 451: 990-993.

Jones WP and Kinghorn AD (2008) Biologically Active Natural Products of the Genus *Callicarpa* . Curr Bioa Comp 4: 15-32.

Joppa LN, Roberts DL and Pimm SL (2011) How many species of flowering plants are there? Proc R Soc B 278: 554–559.

Kaatz GW, Moudgal VV, Seo SM and Kristiansen JE (2003) Phenothiazines and thioxanthenes inhibit multidrug efflux pump activity in *Staphylococcus aureus*. Antimicrob Agents Chemother 47: 719-726.

Kala CP (2004) Revitalizing traditional herbal therapy by exploring medicinal plants: A case study of Uttaranchal State in India. In *Indigenous Knowledges: Transforming the Academy, Proceedings of an International Conference* Pennsylvania: Pennsylvania State University. pp.15-21.

Kala C and Mathur VB (2002) Patterns of plant species distribution in the trans- Himalayan region of Ladakh, India. J Veg Sci 13: 751-754.

Kapoor IPS, Singh B and Singh G (2008) Essential oil and oleoresins of C*innamomum tamala* (tejpat) as natural food preservatives for pineapple fruit juice. J Food Process Preserv 32: 719-728.

Kapoor LD, Sharma PC and Kapoor SL (1971) Medicinal plant wealth around religious shrines of India Bhadrinath area. A Biblio Ind Med 6: 87-94.

Kar A and Jain SR (1971) Antibacterial evaluation of some indigenous medicinal volatile oils. Plan Food Human Nutri J 20: 231-237.

Katiyar C, Gupta A, Kanjilal S, Katiyar S (2012) Drug discovery from plant sources: An integrated approach. Ayu 33: 10-19.

Kelmanson JE, Jäger AK and van-Staden J (2000) Zulu medicinal plants with antibacterial activity. J Ethnopharm 69: 241-246.

Khan IA, Mirza ZM, Kumar A, Verma V and Qazi GN (2006) Piperine, a phytochemical potentiator of ciprofloxacin against *Staphylococcus aureus*. Antimicrob Agents Chemother 50: 810-812.

Khan MR, Omoloso AD and Kihara M (2003) Antibacterial activity of *Alstonia scholaris* and *Leea tetramera* .Fitoterapia 74: 736-740.

Kizil M, Kizil G, Yavuz M and Aytekin C (2002) Antimicrobial activity of the tar obtained from the roots and stems of *Pinus brutia*. Pharm Biol 40: 135-138.

Kobayashi A and Matsumoto H (1965) Studies on methylazoxy methanol the aglycone of cycasin, Isolation, biological and chemical properties. Arch Biochem Biophy 110: 373–380.

Kohanski MA, Dwyer DJ and Collins JJ (2010) How antibiotics kill bacteria: from targets to networks? Nature Rev Microbiol 8: 423-435.

Kumar K and Upreti DK (2001) *Parmelia* spp. (lichen) in ancient medicinal plant lore of India. Eco Bot 55: 458–459.

Kumar R, Mishra AK, Dubey N K and Tripathi Y B (2007) Evaluation *of Chenopodium ambrosioides* oil as a potential source of antifungal, antiaflatoxigenic and antioxidant activity. Int J Food Microb 115: 159-64.

Kumar VP, Chauhan NS, Padh H and Rajani M (2006) Search for antibacterial and antifungal agents from selected Indian medicinal plants. J Ethnopharm 107: 182-188.

Kumarasamy KK, Toleman M A, Walsh TR, Bagaria J and Butt F et al. (2010) Emergence of a new antibiotic resistance mechanism in India, Pakistan, and the UK: a molecular, biological, and epidemiological study. The Lancet Infect Dis 10: 597- 602.

Kumaraswamy MV, Kavitha HU and Satish S (2008) Antibacterial Evaluation and Phytochemical Analysis of *Betula utilis* D. Don Against Some Human Pathogenic Bacteria. World J Agricult Sci 4: 661-664.

Lalitha T and Venkataraman LV (1991) Antifungal Activity and Mode of Action of Saponins from *Madhuca-Butyracea* Macb. Ind J Exp Biol 29: 558-562.

Lauterwein M, Oethinger M, Belsner K, Peters T and Marre R (1995) In vitro activities of the lichen secondary metabolites vulpinic acid, (+)-usnic acid, and (-)-usnic acid against aerobic and anaerobic microorganisms. Antimicrob Agen Chemother 39: 2541-2543.

Lawrey JD (1986) Biological role of lichen substances. Bryol 89:111–122.

Lawrey JD (1989) Lichen Secondary Compounds: Evidence for a Correspondence between Antiherbivore and Antimicrobial Function. Bryol 89: 326–328.

Levy SB (2002) Active efflux, a common mechanism for biocide and antibiotic resistance. J App Microbiol 92: 65-71.

Lokhande PD, Gawai KR, Kodam KM, Kuchekar BS, Chabukswar AR and Jagdale SC (2007) Antibacterial Activity of Isolated Constituents and Extract of Roots of *Inula racemosa.* Res J Medi Plant 1: 7-12.

Lomovskaya O and Bostian KA (2006) Practical applications and feasibility of efflux pump inhibitors in the clinic–a vision for applied use. Biochem Pharmacol 71: 910–918.

Madsen GC and Pates AL (1952) Occurrence of antimicrobial substances in chlorophyllose plants growing in Florida. Bot Gazette 113: 293-300.

Lomovskaya O and Watkins WJ (2001) Efflux pumps: their role in antibacterial drug discovery. Curr Med Chem 8: 1699-1711.

Mahamoud A, Chevalier J, Alibert-Franco S, Kern WV and Pagès JM (2007) Antibiotic efflux pumps in Gram-negative bacteria: the inhibitor response strategy. J Antimicrob Chemother 59: 1223–1229.

Malhotra CL and Basu D (1984) A preliminary survey of plants resources of medicinal and aromatic value from Almora. J Eco Tax Bot 5: 859-864.

Malterud KM, Bremnes TE, Faegri A, Moe T, Dugstad SEK, Anthonsen T and Henriksen LM (1985) Flavonoids from the wood of *Salix caprea* as inhibitors of wood-destroying fungi. J Nat Prod 48: 559-563.

Manandhar NP (2002) Plants and People of Nepal. Timber Press. USA.

Mandal SC, Kumar CK, Majumder A, Majumder R and Maity BC (2000) Antibacterial activity of Litsea glutinosa bark. Fitoterapia 71: 439-441.

Marquez B (2005) Bacterial efflux systems and efflux pumps inhibitors. Biochim 87: 1137–1147.

Maurya A, Dwivedi GR, Shrivastava SK and Darokar MP (2013) Antibacterial and synergy of clavine alkaloid lysergol and its derivatives against nalidixic acid resistant *Escherichia coli*. Chemi Biol Drug Des 81: 484-490.

McCleary JA, Sypherd PS and Walkington DL (1960) Mosses as a possible source of antibiotics. Science 131:108.

Mehta MB, Kharya MD, Srivastava R and Varma KC (1981) Antimicrobial and anthelmintic activities pf the essential oil of *Zanthoxylum alatum* Roxb. Ind Perfumer 25: 1-3.

Mewari N and Kumar P (2008) Antimicrobial Activity of Extracts of *Marchantia polymorpha.* Pharm Biol 46: 819–822.

Ming LC (1999) Ageratum conyzoides: A tropical source of medicinal and agricultural products. In: J. Janick (Ed), *Perspectives on new crops and new uses*. ASHS Press, Alexandria, VA pp. 469–473.

Mishra PL (1985) The medicinal plants of Himalaya. *Himala. J* Sci 4: 65-80.

Nadkami AK (1976) Artemisia maritime In: Indian materia medica Popular Parkashan, Bombay. pp. 142-143.

Nair R and Chanda S (2008) Antimicrobial activity of *Terminalia catappa*, *Manilkara zapota* and *Piper betel* leaf extract. Ind J Pharma Sci 70: 390-393.

Negi AS, Kumar JK, Luqman S, Saikia D and Khanuja SPS (2010) Antitubercular potential of plants: A brief account of some important molecules. Med Res Rev 30: 603–645.

Negi KS, Tiwari JK, Gaur RD and Pant KC (1993) Notes on ethnobotany of five districts of Garhwal Himalaya, Uttar Pradesh, India. Ethnobot 5: 73-81.

Neogi U, Saumya R, Mishra RK and Raju KC (2008) Lipid Content and *in vitro* Antimicrobial Activity of Oil Seeds of Some Indian Medicinal Plants. Curr Res Bacteri 1: 1-6.

Newman DJ, Cragg GM and Snader KM (2003) Natural products as sources of new drugs over the period 1981–2002. J Nat Prod 66: 1022–1037.

Nickell LG (1959) Antimicrobial activity of vascular plants. Eco Bot 13: 281-318.

Nikaido H (1996) Multidrug efflux pumps of Gram-negative bacteria. J Bacteriol 178: 5853–5859.

Nikaido H. (2009) Multidrug Resistance in Bacteria. Annu Rev Biochem78: 119–146.

Pages JM and Amaral L (2009) Mechanisms of drug efflux and strategies to combat them: Challenging the efflux pump of Gram-negative bacteria. Biochim Biophy Acta 1794: 826-833.

Pande PC, Joshi P and Pande HC (1984). Studies on the ethnobotany II. On some less known medicinal and economic ferns of Kumaun region of Western Himalaya. In: Current Research in Plant Science (Eds Sharma TA, Saini SS, Trivedi ML, Sharma M). Bishen Singh and Mahendra Pal Singh, Dehra Dun. pp. 135-137.

Pandya KK, Patel RB and Chakravarty BK (1990) Antibacterial activity of some Indian medicinal plants. Ind Drugs 27: 415-417.

Pant G, Tewari SD, Pargaien MC and Bisht LS (1986) Bryological activities in North- West Himalaya – II. A bryophyte foray in the Askot region of district Pithoragarh (Kumaun Himalayas). Bryol Times 39: 2-3.

Pao SS, Paulsen IT and Saier MH (1998) Major facilitator superfamily. Microbiol Mol Biol Rev 62: 1–34.

Parekh J and Chanda S (2006) In-vitro Antimicrobial Activities of Extracts of *Launea procumbens* Roxb. (Labiateae), *Vitis vinifera* L. (Vitaceae) and *Cyperus rotundus* L. (Cyperaceae). Afr J Biomed Res 9: 89-93.

Patel JD, Kumar V and Bhatt AS (2009) Antimicrobial screening and phytochemical analysis of the resin part of Acacia catechu. Pharm Biol 47: 34-37.

Parihar P (2006) Antibacterial effect of *Dryopteris cochleata* (Don) C. Chr. against various human as well as plant pathogenic bacteria. Ind Fern J 23: 15-18.

Paulsen IT, Sliwinski MK and Saier MH (1998) Microbial genome analyses: global comparisons of transport capabilities based on phylogenies, bioenergetics and substrate specificities. J Mol Biol 277: 57-592.

Pepeljnjak S, Kosalec I, Kalodera Z and Blazevic N (2005) Antimicrobial activity of *Juniper* berry essential oil (*Juniperus communis* L., Cupressaceae). Acta Pharmaceu 55: 417-422.

Petrovic J, Stanojkovic A, Comic Lj and Curcic S (2004) Antibacterial activity of *Cichorium intybus*. Fitoterapia 75: 737-739.

Piddock LJ (2006a) Multidrug-resistance efflux pumps not just for resistance. Nature Rev Microbiol 4: 629-636.

Piddock LJ (2006b) Clinically relevant chromosomally encoded multidrug resistance efflux pumps in bacteria. Clin Microbiol Rev 19: 382–402.

Polombo EA and Semple SJ (2001) Antibacterial activity of traditional Australian medicinal plants. J Ethnopharm 77: 15-17.

Poole K (2007) Efflux pumps as antimicrobial resistance mechanisms. Ann Med 39:162-76.

Prajapati ND, Purohit SS, Sharma AK and Kumar T (2006) *A Hand book of Medicinal Plants*. 3rd Edition. Agrobios Hindustan Printing Press, Jodhpur.

Projan S (2003) Why is big pharma getting out of antibacterial drug discovery? Curr Opin Microbiol 6: 427–430.

Pushpangadan P (1995) Ethnobiology in India: a status report, GOI New Delhi.

Radulovi N, Stojanovi G, Pali R (2006) Composition and antimicrobial activity of *Equisetum arvense* L. essential oil. Phyto Res 20: 85-88.

Raghvendra, Sharma V, Arya SG, Hedaytullah MD, Tyagi S et al. (2011) Pharmacological and biochemical aspects of new delhi metallo-beta-lactamase (NDM-1)-a superbug: an overview. Int J Preclin Pharma Res 2:17-21.

Ray PG and Majumdar SK (1976) Antimicrobial activity of some Indian plants. Eco Bot 30: 317-320.

Ray S, Sinhababu A and Mandal NC (2003) Antibacterial activity of three lichen specimen viz., *Usnea articulata,Ramalina jamesii* and *Parmelia tinctorum* from the Eastern Himalaya. J Hill Res 16: 66-69.

Reichling J, Weseler A and Saller R (2001) A current review of the antimicrobial activity of *Hypericum perforatum* L. Pharmacopsychiatry 34: 116-118.

Ren-Bo A, Hyun-Chul K, Sung-Hee L, GiI-Saeng J, Dong-Hwan S, Hyun P, Dong-Yeul K, Lee JH and Youn-Chul K (2006) A New Monoterpene Glycoside and Antibacterial Monoterpene Glycosides from *Paeonia suffruticosa.* Arch Pharm Res 29: 815-820.

Renzaka T and Sigler K (2007) Hirtusneanoside, an unsymmetrical dimeric tetrahydroxanthone from the lichen *Usnea hirta.* J Nat Prod 70: 1487-1491.

Richardson DHS (1988) Medicinal and other aspects of lichens. In: *Handbook of Lichenology*, (Eds. Galun M), CRC Press Boca Raton, FL 33487 United States. pp. 93–108.

Rikkinen J (1995) What's behind the pretty colours? A study on the photobiology of lichens. Bryobrothera 4: 1–239.

Saklani A and Upreti DK (1992) Folk uses of some lichens in Sikkim. J Ethnopharm. 37: 229–233.

Samant SS, Dhar U and Palni LMS (1998) Medicinal plants of Indian Himalaya: Diversity, distribution and potential values. Gyanodaya Prakashan, Nainital. pp. 163.

Samant SS, Dhar U and Rawal RS (2001) Diversity, distribution and indigenous uses of threatened medicinal plants of Askot wildlife sanctuary in west Himalaya:Conservation and management perspectives. In : Himalayan Medicinal Plants: Potential and Prospects (Eds. S. S. Samant, U. Dhar and L. M. S. Palni). Gyanodaya Prakasan, Nainital. pp. 167-185.

Samy PR (2005) Antimicrobial activity of some medicinal plants from India. Fitoterapia 76: 697-699.

Samy PR, Sarumathi M and Ignacimuthu S (2004) Antibacterial screening of Tragia involucrata L. (Euphorbiaceae)–a tribal used medicinal plant. In: B.K.H. Tan, B.H. Bay and Y.Z. Zhu, Editors, Novel compounds from Natural products in the New Millennium Potential and Challenges, World Scientific Publishing Company Private Limited, Singapore. pp. 112–122.

Sa´nchez-Navarro A (2005) New formulations of amoxicillin/clavulanic acid: a pharmacokinetic and pharmacodynamic review. Clin Pharmacokinet 44: 1097–1115.

Sanders CA, Nieda RR and Desmond EP (2004) Validation of the use of Middlebrook 7H10 agar, BACTEC MGIT 960, and BACTEC 460 12B media for testing the susceptibility of *Mycobacterium tuberculosis* to levofloxacin. J. Clin. Microbiol. 2004: 5225–5228.

Sanders DW, Weatherwax P and McClung LS (1945) Antibacterial substances from plants collected in Indiana. J Bact 49: 611-615.

Sanjappa M (2005) Plant diversity in India—status, conservation and challenges (Maheshwari P Medal Award Lecture). In: XXVIII Conference of Indian Botanical Society B.S.I. Dehradun. pp. 5–6.

Saritas Y, Sonwa MM, Iznaguen H, Konig WA, Muhle H and Mues R (2001) Volatile constituents in mosses (Musci). Phytochem 57: 443-457.

Sati SC and Joshi S (2011) Antibacterial Activities of *Ginkgo biloba* L. Leaf Extracts. Sci World J 11: 2241-2246.

Saxena J and Mathela CS (1996) Antifungal Activity of New Compounds from *Nepeta leucophylla* and *Nepeta clarkei.* App Environ Microbiol 62: 702–704.

Saxena K and Harinder S (2004) Uses of Bryophytes. Resonance 9: 56-65.

Schippmann U, Leaman DJ and Cunningham AB (2002) Impact of Cultivation and Gathering of Medicinal Plants on Biodiversity: Global Trends and Issues. In (FAO) Biodiveristy and the ecosystem approach in agriculture, forestry and fishries. Satellite event on the occasion of the Ninth regular session of the commission on genetic resources for food and agriculture. Rome 12–13 October 2002. Inter departmental working group on biological diversity for food and agriculture, Rome.

Sethi PD, Ravindran PC, Sharma KB and Subramanian SS (1974) Antibacterial activity of some C28 steroidal lactones. Ind J Pharm 36: 122-123.

Shahi SK, Shukla AC, Uperti DK and Dikshit A (2000) Use of lichens as antifungal drugs against superficial fungal infections. J Med Arom Plant Sci 22: 169–172.

Sharma BC, Kalikotay S and Rai B (2012) Assessment of antimicrobial activity of extracts of few common lichens of Darjeeling hills. Ind J Fund App Life Sci 2: 120-126.

Sharma BD and Vyas MS (1985) Ethnobotanical studies on the ferns and fern-allies of Rajasthan. Bull Botan Surv India 27: 90-91.

Sharma HK, Changte L and Dolui AK (2001) Traditional medicinal plants in Mizoram, India. Fitoterapia 72: 146–161.

Sharma RK, Dhyani SK and Shankar V (1979) Some useful and medicinal plants of the district Dehra Dun and Siwalik. J Sci Res Plant Med 1: 17-43.

Shimpi SN and Dutta RS (2004) Noteworthy Pteridophytes from Apalchand Reserve, Jalpaiguri District, West Bengal. Bulletin Botanical Survey of India 46: 221-225.

Siddiqi SH (1996) BACTEC 460TB System. Product and Procedure Manual, Revision E. Becton Dickinson Diagnostic Instrument Systems. Maryland, USA.

Singh AK and Nath G (1999) Antimicrobial activity of *Elaeocarpus sphaericus*. Phytoth Res 13: 448-450.

Singh DK and Hajra PK (1996) Changing Perspectives of Biodiversity Status in the Himalaya. In: (Eds Gujral GS, Sharma V), British Council Division, New Delhi. pp. 23–38.

Singh DK and Hajra PK (1996) Floristic diversity. In: Biodiversity Status in the Himalaya. British Council, New Delhi. pp. 23-38.

Singh KK, Upreti DK and Kaushal K (2000) Ethnobotanical note on some less known Himalaya lichens. Bulletin of the British Lichen Society 86: 36–37.

Singh M, Rawat AKS and Govindarajan R (2007) Antimicrobial activity of some Indian mosses. Fitoterapia 78: 156-158.

Singh M, Singh N, Khare PB and Rawat AK (2008) Antimicrobial activity of some important *Adiantum* species used traditionally in indigenous systems of medicine. J Ethnopharm 115: 327-329.

Singh U and Lahiri N (2010) Ancient India: New Research. Oxford University Press, India.

Song JH, Kim HS, Kim YG, Son BG, Choi YW and Kang JS (1999) Antimicrobial activity of extract from *Smilax china*. J Agri Tech Develop Insti 3:163-168.

Spencer CF, Koniuszky FR, Rogers EF, Shavel J Jr, Easton NR, Kaczka EA, Kuehl FA, Phillips Jr RFA, Walti K., Malanga FC and Seeler AC (1947) Survey of plants for antimalarial activity. Lloydia 10: 145-174.

Srinivasan D, Nathan S, Suresh T and Perumalsamy PL (2001) Antimicrobial activity of certain Indian medicinal plants used in folkloric medicine. J Ethnopharm 74: 217-220.

Stavri M, Piddock LJV and Gibbons S (2007) Bacterial efflux pump inhibitors from natural sources. J Antimicrob Chemother 59:1247-1260.

Stermitz FR, Lorenz P, Tawara JN, Zenewicz LA, Lewis K (2000) Synergy in a medicinal plant: Antimicrobial action of berberine potentiated by 5*-methoxyhydnocarpin, a multidrug pump inhibitor. Proc Natl Acad Sci 97:1433–1437.

Stoll A, Brack A and Renz J (1950) Die Wirkung von Flechtenstoffen auf Tuberkelbakterien und auf einige andere Mikroorganismen. Schweizer Zeitschrift für Allgemeine. Path Bakteriol 13: 729–751.

Taylor RSL, Manandhar NP, Hudson JB and Tower GHN (1996) Antiviral activities of Nepalese Medicinal Plants. J Ethnopharm 52:157-163.

Tenover FC (2006) Mechanisms of antimicrobial resistance in bacteria. Am J Infect Cont 34:S3-S10.

Thatoi HN, Panda SK, Rath SK and Dutta SK (2008) Antimicrobial Activity and Ethnomedicinal Uses of Some Medicinal Plants from Similipal Biosphere Reserve, Orissa. 7: 260-267.

Turk AO, Yilmaz M, Kivank M and Turk H (2003) Antimicrobial activity of extracts of the lichen *Cetraria aculeata* and its protolichesterinic acid constitient. Verlag der Zeitschrift fiir Naturforschung (C) 58: 850-854.

Turk H, Yilmaz M, Tay T, Turk AO and Kivanc M (2006) Antimicrobial activity of extracts of chemical races of the lichen Pseudevernia furfuracea and their physodic acid, chloroatranorin, atranorin, and olivetoric acid constituents. Z Naturforsch C 61:499-507.

Umachigi SP, Kumar GS. Jayaveera KN, Kumar KDV, Kumar ACK and DhanapalR (2007) Antimicrobial, wound healing and antioxidant activities of *Anthocephalus cadamba*. Afri J Trad Compli Alter Med 4: 481 – 487.

Upadhyay HC, Dwivedi GR, Darokar MP, Chaturvedi V and Srivastava SK (2012) Bioenhancing and Antimycobacterial Agents from *Ammannia multiflora*. Planta Medica 78:79-81.

Upadhyay HC, Thakur JP, Saikia D and Srivastava SK (2013) Anti-tubercular agents from *Ammannia baccifera* (Linn.). Med Chem Res 22: 16–21.

Upreti DK (1996) Studies in Indian ethnolichenology: An overview. Pages 384–387 In: Jain SK (Ed), *Ethnobiology in human welfare*. Deep Publication, New Delhi, India

Van Hoof L, Berghe VDA, Petit E and Vlietinck AJ (1981) Antibacterial and antiviral screening of Bryophyta. Phytoterapia **5:** 223-229.

Victor B, Maridass M, Ramesh U, Prabhu JMA (2003) Antibacterial activity of Essential Oils from the Leaves of *Adiantum capillus -veneris* Linn. Mal J Sci 22: 65-66.

Viveiros M, Jesus A, Brito M, Leandro C, Martins M, Ordway D et al. (2005) Inducement and reversal of tetracycline resistance in *Escherichia coli* K-12 and expression of proton gradient dependent multidrug efflux pump genes. Antimicrob Agents Chemother 49: 3578-82.

Vohora SB, Kumar I, Naqvi SAH and Afaq SH (1972) Pharmacological investigations on *Picrorhiza kurroa* roots (kutki) with special reference to its chloretic and antimicrobial Properties. Ind J Pharm 34: 17-19.

WHO/IUCN/WWF (1993) Guidelines on Conservation of Medicinal Plants, IUCN Gland, Switzerland.

Wu PC (1982) Some uses of mosses in China. Bryological Times 13: 5.

Yeung Him-Che (1985) Handbook of Chinese Herbs and Formulas. Vol. I. Redwing Book Co, USA.

Yilmaz M, Tu¨rk OA, Tay T and Kivanc M (2004) The antimicrobial activity of the lichen *Cladonia foliacea* and its (–)-usnic acid, atranorin, and fumarprotocetraric acid constituents. *Z* Naturforsch C 59: 249–254.

Yoshimura I, Kinoshita Y, Yamamoto Y, Huneck S and Yamada Y (2007) Analysis of secondary metabolites from Lichen by high performance liquid chromatography with a photodiode array detector. Phytochem Analy 5: 197-205.

Zysk KG (1996) Medicine in the Veda: Religious healing in the Veda Motilal Banarsidass, Delhi.

Zysk KG (1998) Asceticism and Healing in Ancient India: Medicine in the Buddhist Monastery. Motilal Banarisidas, Delhi.

Zgurskaya HI and Nikaido H (2000) Multidrug resistance mechanisms: drug efflux across two membranes. Mol Microbiol 37: 219-225.

Web-sites used in this study

http://www.atindia.org/ringal.htm

http://en.wikipedia.org/wiki/Himalaya

http://en.wikipedia.org/wiki/Gmelina_arborea

http://www.stuartxchange.org/Dita.html

http://www.biodiversityhotspots.org/xp/hotspots/himalaya/Pages/default.aspx

http://www.eoearth.org/article/Biological_diversity_in_the_Himalayas%20/

http://www.horticulturedevelopment.com/medicinalplant/introduction.html

http://www.who.int/infectious-disease-report/pages/textonly.html

9

DNA and Protein-based Detection Approaches for Genetically Modified Crops/Produce

Rajesh Kumar, Kamlesh K Mishra and Rajeshwar P. Sinha

Abstract

Despite several controversies surrounding the genetically modified (GM) crops or their produce, the production of such crops is constantly increasing, especially in developing countries. To regulate the GM produce in the market, several countries have set very strict regulatory procedures for the approval to grow, import or utilize these GM crops as food or food ingredients. Hence, for regulatory compliance on GM labeling and identification of genetic traits, the development of reliable, sensitive and precise detection assays have become essential. The detection of GM produce is carried out by detecting either the inserted foreign DNA or the novel transgenic protein specifically expressed in GM crops. This article broadly focuses on some DNA and protein-based detection methods employed for GM produce and significant current advances in such detection technologies.

Keywords: Genetically modified crops; Transgene; PCR; ELISA; Dipstick

Introduction

The world population is growing very fast. Due to mounting this pressure, foods are challenged to meet the current needs and also for the upcoming future. In developing countries, enormous increase in population has resulted in poverty, food insecurity, and poor nutrition among people. The productivity of several crops is tremendously decreasing annually due to damage by insect pests (Estruch et al. 1997). For many years, human beings have opted for crop plants that show resistance to pests and yield sufficient produce. The extensive use of chemical pesticides to protect crops has long-term detrimental effects, leading to environmental degradation (Miller 2004). Several insect and mite species have also developed resistance to one or more chemical insecticides (Rossi and Presciuttini 1996; Wilson and Cain 1997). Therefore, genetically modified (GM) crop has been introduced into the market as an environmentally safe pest control alternative to maintain sustainability of the environment.

A GM crop is a plant into which foreign gene(s) has been artificially inserted instead of the plant acquired them by cross-breeding or natural recombination process. The insertion of foreign genes (transgenes) has often been used in crops to produce novel proteins that confer tolerance against pests. The GM crops have new properties such as tolerance to herbicides, protection from insects, improved nutritional value, stress tolerance, etc (Engel et al. 2002). Since the first commercial introduction of GM crop in 1996, several GM crops have been developed and commercialized in many countries (Agbios 2013). Although no long-term harmful effects of these crops on human health have been detected so far, but it has led to considerable fear regarding their stability and safety in the environment and therefore, is a matter of concern.

With increasing diversity of GM produce into the food chain or the environment, the precise and accurate detection methods are a prerequisite for their reliable control and to discriminate the GM materials which mixed intentionally or unintentionally. The detection and characterization of any genetically modified organism (GMO) is also important from safety point of view. To make sure the GM traces in the sample, various DNA and protein based detection techniques are being employed successfully.

Development of GM crops

A typical transgene construct insert (rDNA) in a GM crop is composed of at least three elements such as promoter element, transgene, and terminator element (Fig. 1). Cauliflower mosaic virus promoter (Pe35S) is frequently being used to drive the expression of transgenes in GM crops while some other promoters like actin-2, ubi-3 and trfA are also being used in GM cotton (Agbios 2013). The inserted transgenes (e.g., *cry1Ac, cry1Ab, cry2Ab, vip3A, cp4epsps, etc.)* encode a specifically selected traits e.g., insect-resistant, herbicide-resistant, stress-tolerant, etc. The T-nos terminator from the Ti plasmid of *Agrobacterium tumefaciens* is the most common terminator and functions as a stop signal when reading the inserted transgene. A GM crop also contains marker gene with its respective promoter and terminator to monitor the plant transformation whether transgene is successfully transferred into recipient cell or not. Two types of marker genes; selectable marker genes (*nptI, nptII, hpt*) and nonselectable marker genes or reporter genes (*uidA*, *luc*), are used to screen GM crops with non-GM crops (Chawla 2002).

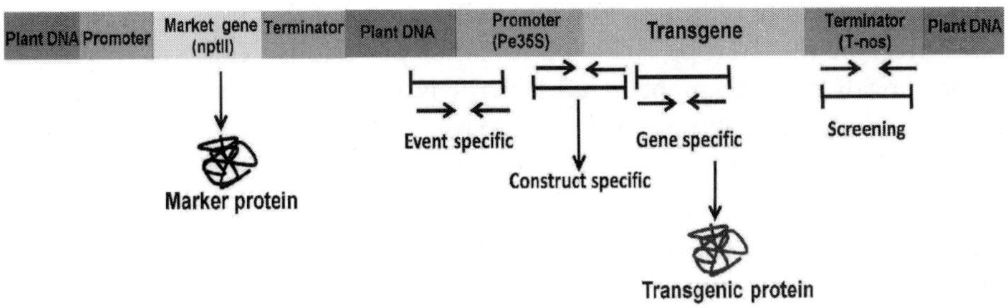

Fig. 1: Schematic of a typical transgene construct and four types of PCR-based assays. The host DNA is genomic DNA of the GM crop; the transgene consists of a promoter, the gene itself, and a terminator. Arrows indicate the primer pairs targeting particular sequences around and within the transgene integration site. Four assay types illustrated targeting screening sequences (promoter P-35S, transgene, terminator T-nos), sequences within the gene of interest; construct-specific junction sequences; and event-specific junction sequences.

Gene gun and *Agrobacterium* infection are two common methods which are being used for transformation of target crops (Allen et al. 2005; Broothaerts et al. 2005; Cho et al. 2004; Gelvin 2005). *Agrobacterium* has ability to infect plants and insert DNA into a plant's genome. This method is mainly used to modify broad leaved plants, such as cotton, brinjal, tobacco, and now, it is also being applied to monocot species, such as maize and rice. The other method, particle bombardment or biolistics employs DNA coated on minute gold particles, used to insert the desired DNA into target plants cells generally monocots such as maize and rice.

Global status of commercialized GM crops

The adoption and cultivation of commercialized GM crops is increasing year by year. In 2012, the global area of GM crops has reached to 170.3 million hectares spread across 28 countries (James 2012). USA continues to be the lead producer of biotech crops globally with 69.5 million hectares (Table 1). Till date, several GM crops have been commercialized and entered into the global market such as soybean, maize, cotton, canola, papaya, alfalfa, squash, sugar beet, tomato, potato, wheat (Agbios 2013). GM soybean was continued to be the principal GM crop in 2012 (James 2012). The increased adoption rate for cultivation of GM crops by farmers across the developing and industrial countries indicates its significant socio-economic, environmental and health benefits.

Labeling regulations of GM crops/produce

Though the extent of production and cultivation of GM crops is increasing, the biosafety of these crops has been a matter of serious debate in recent years. The biosafety assessment of GM crops is required prior to its commercial release in the market. Sometimes, unapproved GM material has been found in the market which was released without proper evaluation of environmental impacts which further led to crop failures and commercial losses (ISIS Report 2005). This led regulatory authorities to implement a set of procedures for the approval as food ingredients worldwide and implemented appropriate mandatory and voluntary labeling thresholds for GM produce (Viljoen 2005; Marmiroli et al. 2008). Table 2 summarizes the state of relevant legislation in various countries and indicates the significant international heterogeneity in the labeling regulations for GM products. Some countries have implemented voluntary labeling guidelines (e.g., USA, Canada) while others implemented mandatory labeling guidelines for GM ingredients. Some countries have implemented both types of labeling for GM food. These labeling regulations also necessitate the availability of certified reference materials (CRM), suitable sampling protocols and detection technologies that allow the monitoring and accurate determination of GM content.

Table 1: Global status of GM crops in 2012.

Rank	Country	Area (million hectares)	Biotech crops
1*	USA*	69.5	Soybean, maize, cotton, canola, squash, papaya, alfalfa, sugarbeet
2*	Brazil*	36.6	Soybean, maize, cotton
3*	Argentina*	23.9	Soybean, maize, cotton
4*	Canada*	11.6	Canola, maize, soybean, sugarbeet
5*	India*	10.8	Cotton
6*	China*	4.0	Cotton, tomato, poplar, papaya, sweet pepper
7*	Paraguay*	3.4	Soybean, Maize, cotton
8*	South Africa*	2.9	Maize, soybean, cotton
9*	Uruguay*	1.4	Soybean, maize
10*	Bolivia*	1.0	Soybean
11*	Philippines*	0.8	Maize
12*	Australia*	0.7	Cotton, canola
13*	Burkina Faso*	0.3	Cotton
14*	Mexico*	0.2	Cotton, soybean

Contd.

Rank	Country	Area (million hectares)	Biotech crops
15*	Spain*	0.1	Maize
16	Chile	<0.1	Maize, soybean, canola
17	Colombia	<0.1	Cotton
18	Honduras	<0.1	Maize
19	Czech Republic	<0.1	Maize
20	Portugal	<0.1	Maize
21	Romania	<0.1	Maize
22	Poland	<0.1	Maize
23	Costa Rica	<0.1	Cotton, soybean
24	Egypt	<0.1	Maize
25	Slovakia	<0.1	Maize

Table 2: Labeling regulations of GM crops/produce in different countries

Country	% Threshold	Type of labeling
European Union	0.9%	Mandatory
USA	N.A.	Voluntary
Australia and New Zealand	1%	Mandatory and voluntary
Japan	5%	Mandatory and voluntary
Korea	3%	Mandatory and voluntary
China	1%	Mandatory
Taiwan	5%	Mandatory and voluntary
Brazil	1%	Mandatory
Saudi Arabia	1%	Mandatory
Canada	5%	Voluntary
Indonesia	5%	Mandatory
Israel	0.9%	Mandatory
Philippines	N.A.	Voluntary
Russia	0.9%	Mandatory
Switzerland	0.9%	Mandatory
Thailand	5%	Mandatory
South Africa	N.A.	Voluntary
India	N.A.	Mandatory[a]

Source: Michelini et al. (2008); Marmiroli et al. (2008); N.A.- Not Applicable, [a]http://www.thehindu.com/news/national/centre-makes-labelling-of-gm-foods-mandatory/article3551679.ece

Detection methods of GM crops

The introduction of GM crops/produce has generated a demand for appropriate analytical methods to trace the GM produce and to fulfill the requirement of regulatory compliance. The detection of GM crops has also become necessary to allow consumers to make an informed- choice. Although legislation governs the release of new GM produce in market through strict approval process, still consumers are reluctant to buy plant-derived GM products.

The detection and quantification of GM materials relies, either on transgenic DNA or the foreign protein(s) produced in GMOs (Fig. 2).

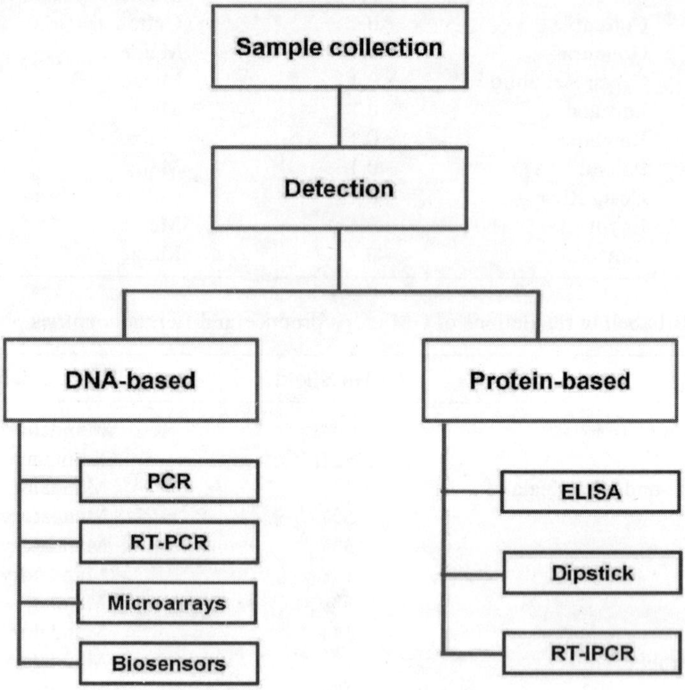

Fig. 2: Outline of GMO testing methods

There are several methods for the detection of transgenes and its expressed proteins in the GM samples such as DNA-based polymerase chain reaction (PCR), southern blot, real-time PCR, and protein-based enzyme linked immunosorbent assay (ELISA), dipstick strip, immuno-PCR, western blot, and insect based bioassay. Other rapidly emerging methods such as chromatography (Byrdwell and Neff 1996), mass spectrophotometry (Ocana et al. 2007), near infrared (NIR) spectroscopy (Xie et al. 2007a; Xie et al. 2007b), microarray and biosensors are being used for detection of GM materials. The advances in molecular techniques for the detection and quantification of GMOs have also been widely discussed in several reports (Ahmed 2002; Anklam et al. 2002; Deisingh and Badrie 2005; Elenis et al. 2008; Marmiroli et al. 2008; Michelini et al. 2008). Several GMO detection methods have been developed till date, and their number is increasing rapidly. Recently, Dong et al. (2008) developed a database "GMO Detection method Database (GMDD)" which collected almost all the previously developed and reported GMOs detection methods.

DNA based detection

As DNA has much higher stability over protein, therefore, it is preferred target for detection of raw ingredients as well as processed materials. The DNA based detection includes PCR, southern blot, microarray, and biosensors. Although these methods are very sensitive, specific and precise, they have some drawbacks, including long analysis time and the need for special equipments. Only PCR method has found broad application in GMO detection as a generally accepted method for regulatory purpose.

Qualitative PCR method

PCR is a highly sensitive and even detection of single molecule of DNA is possible. Therefore, material in very small quantity would be enough to identify the transgene. The PCR based method is a valuable tool for GM crops detection and traceability, since they allow a rapid and relatively low-cost first-level screening procedure. For routine PCR-based screening purposes generic markers representing genetic control elements such as 35S promoter (P-35-S) (Akiyama et al. 2009; Cankar et al. 2005; Cankar et al. 2008; Fernandez et al. 2005; Feinberg et al. 2005; Hardegger et al. 1999; Holden et al. 2010; Tozzini et al. 2000), *npt-II* marker (Su et al. 2003; Randhawa et al. 2009) and T-nos3' terminator (Akiyama et al. 2009; Hardegger et al. 1999) are being used, as these are present in majority of GM samples cultivated commercially.

Several formats of qualitative PCR method such as singlex PCR, multiplex PCR, and nested PCR etc. are available to screen single or multiple transgenes. Generally, the qualitative PCR products are distinguished by size in agarose gel electrophoresis (Fig. 3). Gel allows the separation of amplified products based on fragment size and visualization in the gel matrix using dyes such as ethidium bromide. The singlex PCR that amplify a single target includes construct-specific PCR (junctions between the different components of the construct), event-specific PCR (junction between the transgene and the plant's original DNA) and gene-specific PCR (Fig. 1). The screening of GM produce by singlex PCR has been widely reported for various transgenes e.g., *cry1Ab, cry2Ab, cry1Ac, vip3A, cp4-epsps* (Singh et al. 2007a; Singh et al. 2008; Wu et al. 2008; Yang et al. 2005a; Ballari et al. 2013). The nested PCR method which discriminate the target and non-target amplicons is also suitable method to confirm the presence of transgene in GM samples (Angonesi Brod et al. 2007). Zhang et al. (2007) and Brod et al. (2007) both reported the presence of RR soybean with LOD (Limit of Detection) of 0.5% and 0.01% respectively with the help of nested PCR assay.

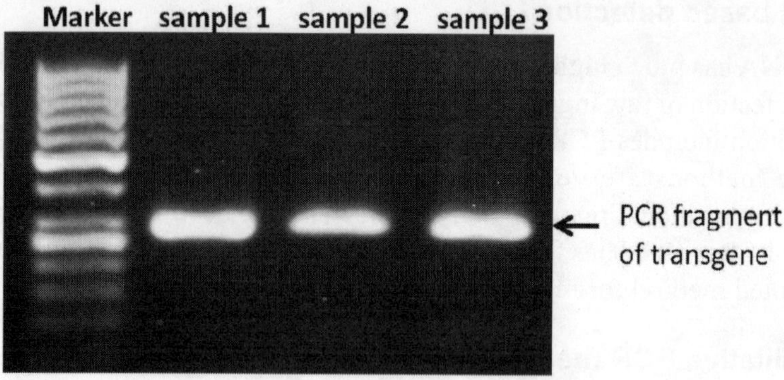

Fig. 3: Agaroge gel electrophoresis of PCR products for qualitative analysis GM samples.

In multiplex PCR, several primer pairs are included to permit the simultaneous detection of multiple target DNA sequences in a single reaction that has advantage over other PCR methods (Mano et al. 2009a; Rudi et al. 2003; Singh et al. 2007b; Randhawa et al. 2009; Ballari et al. 2013). Since the number of authorized transgenic events and their cultivated areas are rapidly increasing, the need arises to accelerate the quick GM detection with less effort. Moreover, As far as we know, Randhawa et al. (2010) and Randhawa et al. (2013) have developed a decaplex PCR assay for simultaneous detection of transgene sequences, specific transgene constructs, and endogenous stearoyl acyl desaturase (Sad1) gene in two events of Bt cotton, i.e., MON531 and MON15985, which contained the most targets reported ever with LOD approximately 0.01%. In another studies, Onishi et al. (2005) and Shrestha et al. (2008) reported respectively nine-target multiplex PCR method to detect eight maize lines as well as the endogenous Zein gene simultaneously in a single reaction tube and LODs were approximately 0.25% in both assays.

The sensitivity of detection and resolution of amplified fragments is limited in agarose gel electrophoresis due to the requirement of amplicons with apparent differences in size and the longer separation time. Therefore, to overcome the above problem, the classical agarose electrophoretic techniques are being replaced by capillary gel electrophoresis (CGE), which allows rapid separation and automation. The combination of PCR and CGE was developed for simultaneous detection of multiple DNA targets with higher resolutions and shorter assay time. (Nadal et al. 2006; Moreano et al. 2006; Nadal et al. 2009; Garcia-Canas et al. 2010). Heide et al. (2008) developed a combination of nineplex PCR and fluorescence CGE method for eight GM maize events and a endogenous *hmga* maize reference gene with LOD approximately 0.1%, which is suitable for screening purposes for food and feed containing the most common

GM maize. Recently, Guo et al. (2011) developed a robust high-throughput analytical approach named multiplex microdroplet PCR implemented capillary gel electrophoresis (MPIC). This assay combines the advantages of bipartite primers, microdroplet PCR and CGE for multiple target DNA analysis, and at least 24 different targets can be simultaneously detected and identified with a relative LOD of 0.1%.

Quantitative PCR method

To fulfill the requirement of mandatory labeling above the threshold value of GM content, the development of quantitative detection system which is required. Earlier, the quantification was based on quantitative competitive PCR (QC-PCR) which principled on co-amplification of specific target gene of unknown amounts with internal control template in known amounts using identical primer pairs in same reaction tube. This technique has been successfully used for quantification of RR soybean (Hubner et al. 1999; Studer et al. 1998; Hardegger et al. 1999; Wurz et al. 1999), and GM maize (Studer et al. 1998; Hupfer et al. 2000). However, its application is limited in the determination of the relative GMO content to a predefined threshold and can't be employed for exact GMO determination. Still QC-PCR qualifies for the detection and quantification of GM samples (Dinelli et al. 2006; Narra and Srivatsava 2012) with little convenient method of choice.

Today, routine quantification of GM ingredients are based on quantitative real-time PCR (qRT-PCR), representing the most powerful and sensitive technique. The qRT PCR allows continuous monitoring of the amplification products using the strength of the fluorescence signal (Fig. 4).

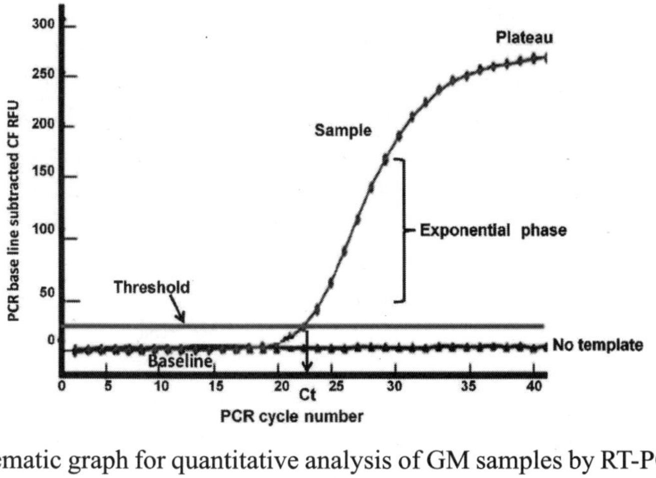

Fig. 4: Schematic graph for quantitative analysis of GM samples by RT-PCR. The GM contents are calculated by threshold cycle (Ct) value.

The non-requirement of post-PCR analysis rendered the qRT- PCR as the most widely used approach in screening the GM produce. Quantification is carried out in the logarithmic phase of PCR, by comparing the number of amplification cycles required to reach a preset threshold fluorescence signal to the number of target DNA copies in the sample. The obtained fluorescence signal in the assay may use either non-specific intercalating dyes (SYBR Green I) or specific probes such as hybridization probes (molecular beacon), hydrolysis probes (TaqMan, CPT, LNA, and MGB) and primer-based technologies (AmpliFluor, Plexor, Lux primers). The specificity of qRT-PCR depends on both the chemistry used to monitor amplification, and the instrumentation used to monitor the signal. To assess the quantity of GM material in a sample, two parallel reactions each containing the same amount of template DNA are performed, one targets the endogenous reference sequence and other, the GM-specific sequence. Relative quantification is achieved by comparison based on the cycle threshold of the two amplified sequences (ÄCt method) and absolute quantification determined by comparison to the standard curve.

Currently, a number of RT-PCR fluorogenic signal reagents have been applied for qualitative and quantitative purposes (Buh Gasparic et al. 2008; Buh Gasparic et al. 2010). The RT-PCR utilizing SYBR Green I fluorescence dye is the simplest and cost-effective technique and has been successfully employed for screening and quantification of wide range of GM samples (Terry et al. 2002; Hernandez et al. 2003; Andersen et al. 2006). Furthermore, the specific probes are currently being used for more specific detection of various transgenes in GM soybean, cotton samples such as TaqMan (Terry et al. 2002; Yang et al. 2005b; Andersen et al. 2006; Yang et al. 2007; Yang et al. 2008; Mano et al. 2009b; Oguchi et al. 2009; Randhawa et al. 2010; Bahrdt et al. 2010), MGB TaqMan probe (Andersen et al. 2006), molecular beacons (Andersen et al. 2006), scorpion (Terry et al. 2002), LNA (locked nucleic acid) probe (Schmidt and Rott 2006; Salvi et al. 2008). Among them, TaqMan probes and SYBR green I are the most widely used RT-PCR chemistries. Furthermore, multiplex TaqMan based RT-PCRs are being targeted to reduce cost and time of the assay (Oguchi et al. 2009; Bahrdt et al. 2010; Dorries et al. 2010; Kim et al. 2011). Recently, Huber et al. (2013) reported the duplex, triplex, and pentaplex real-time PCR screening assays for the detection of the most common screening elements found in commercialized GMOs: P-35S, T-nos, ctp2-cp4-*epsps*, *bar*, and *pat*. These assays are robust as well as cost- and time-effective for GMO screening and allow the coverage of many GMO events globally approved for commercialization. Currently, some commercial companies are also providing the services for RT-PCR based screening of GM samples (Bio-Rad, Applied Biosystems)

Microarrays

Another emerging DNA based analytical method is DNA microarrays or "DNA chips". The arrays consist of oligonucleotide probes immobilized on a glass support (Fig. 5). This approach enables the parallel multi-target detection of several PCR products in a single assay. These probes are specific for screening elements, species reference elements, and control elements. This method relies on direct hybridization of total genomic DNA in microarrays having designed probes with a set of reference sequences. The microarray technique enables the detection of GMOs and also the suspected presence of unidentified varieties by the detection of common GMO screening elements, i.e. Pe35S promoter, T-nos terminator, nptII gene in a sample.

Several microarray-based assays combined with multiplex PCR have been reported for detection of GM maize, canola, cotton and soybean events (Bordoni et al. 2005; Leimanis et al. 2006; Xu et al. 2007; Schmidt et al. 2008; Zhou et al. 2008; Kim et al. 2010) by using fluorescent probes. The primary application of microarrays is screening of authorized or unauthorized GMOs in food chain. DNA-based microarrays have the highest number of multiplexing capabilities and so far maximum simultaneous detection of nineteen GMOs events has been reported (Kim et al. 2010). Recently, a novel multiplex quantitative DNA-based target amplification method, named NASBA implemented microarray analysis (NAIMA) has been developed which is suitable for sensitive, specific and quantitative detection of GMOs on a microarray (Dobnik et al. 2010; Morisset et al. 2008). Furthermore, Feng et al. (2013) developed a microarray based on microfluidics for the detection of GM tomato which is a high-throughput and more efficient screening method.

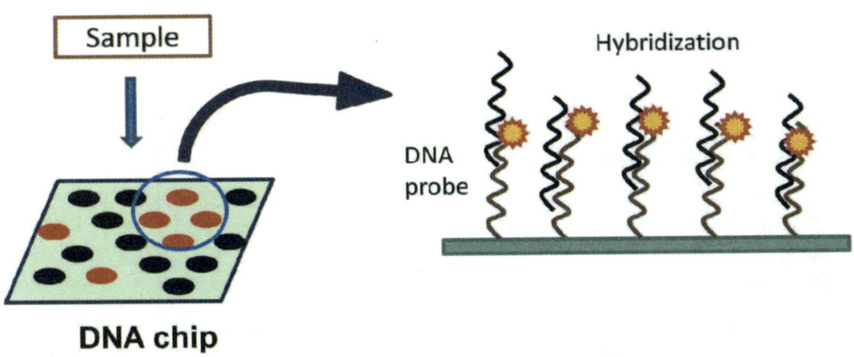

Fig. 5: Schematic illustration of microarray based testing.

The testing laboratories are building up the microarray chip to identify/ quantify GMOs for minimizing the time required and the cost. A microarray chip "DualChip GMO kit" (Eppendorf, Germany) is commercially available which identifies GM events by screening simultaneously multiple genetic elements. Although the microarray analytical approach is relatively expensive, but it is one of the most promising reliable screening method at present owing to its flexibility and automatic and high-throughput capability and proved to be a new method for routine analysis. In spite of producing a large amount of data, the microarrays also have their own limitations such as requiring complex analysis.

Biosensors

Great progress has been achieved in the past few years in the development of DNA sensors allowing for a simple, fast, and reliable DNA testing (Sassolas et al. 2008; Arugula et al. 2014). DNA-biosensors are the leading edge technology that focuses on rapid and inexpensive way of testing and represents an interesting alternative in detection of GMOs (Nica et al. 2004). It is based on hybridization reaction between an immobilized DNA probe attached to the surface of the sensor and a target DNA, consisting of a probe complementary sequence. Recognition between the probe and DNA target generates a signal which is transmitted through a transducers (electrochemical, optical, and piezoelectric) and the signal is detected, measured, and processed to provide both qualitative and quantitative information (Fig. 6). DNA biosensors based on optical (Kalogianni et al. 2006; Bai et al. 2007), electrochemical (Ahmed et al. 2009) and piezoelectric transducers (Passamano and Pighini 2006) have been developed successfully for the detection of amplified GMO-related sequences.

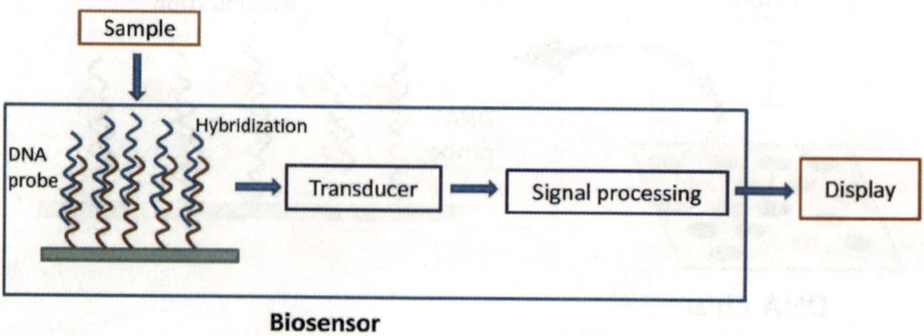

Fig. 6: Schematic illustration of biosensor based testing

The biosensor is suitable for screening of raw ingredients as well as processed food products. It provides the detection of amplified Pe35S promoter and T-nos terminator sequences in 0.1% GMO content in soybean powder within a few minutes without the need of special instrumentation (Kalogianni et al. 2006). Recently, Bai et al. (2010) developed a system for detecting six GM maize lines (Bt11, Bt176, GA21, MON810, NK603, and T25) using optical silicon thin-film biosensor chips which are extremely robust, exhibits high sensitivity and specificity.

Protein based detection

Protein based detection methods are primarily based on immunoassays i.e. specific binding between an antigen and an antibody. The availability of specific antibodies against transgenic protein with the desired affinity is the most important factor for developing immunoassay. Both monoclonal (highly specific) and polyclonal (often more sensitive) antibodies can be used in assays. This technology is ideal for both qualitative and quantitative detection of proteins. Protein based detection methods include enzyme linked immunosorbant assay (ELISA), immuno-PCR, dipstick strip assay, dot immuno-binding assay, western blot and HPLC/Peptide profiling. ELISA and dipstick are typical protein-based tests that have been generally used to detect expressed transgenic protein in plants (Grothaus et al. 2006). The protein based assay is not suitable for detecting expressed transgenic proteins in processed foods because protein gets degraded in the processed foods/produce, although, Romer Labs® Technology (USA) has developed tests that would detect transgenic proteins in highly processed material.

Enzyme-linked immunosorbent assay (ELISA)

ELISA combines the specificity of antibodies with the sensitivity of enzyme assays, by using antibodies or antigens coupled to an easily assayed enzyme. ELISA test is generally performed in a 96 well microtiter plate format (Fig. 7a) and require a suitable laboratory setup. Generally, three types of ELISA are used for protein quantification; Direct, Indirect and Sandwich ELISA (Fig. 7b). The sensitivity of ELISA depends on the colorimetric or fluorometric reaction between substrate and enzyme in final step of assay which can be visualised and measured when the protein and specific antibody bind together. Currently, ELISA test has become important tool for routine analysis of expression levels of transgenic proteins. The assay is very helpful in evaluating the consistency in the expression of protein during the growing season.

Several ELISA tests have been developed for the detection and quantification of different transgenic proteins in various GM samples. Fantozzi et al. (2007) reported ELISA for quantification of Cry1Ab present in MON810 and Bt11 GM maize lines with LOD and LOQ of 0.018 and 0.054%, respectively. Roda et al. (2006) reported ELISA test for Cry 1Ab expressed in GM maize (MON810) with LOD of 0.1% sample. Ermolli et al. (2006) developed ELISA for CP4EPSPS in soy and Cry1Ab in GM maize with LOD of 0.0056% sample. Walschus et al. (2002) developed a sandwich-type ELISA for Cry1Ab with LODs of 0.10% for Bt-11 and 0.23% for MON810. Paul et al. (2008) developed highly specific and sensitive sandwich ELISA for the surveillance of transgenic Cry1Ab protein from Bt-maize in the blood plasma of cows fed on Bt-maize. Several sensitive and specific ELISA assays were also developed for Cry1Ac in GM cotton (Shan et al. 2007; Wang et al. 2007). The ELISA kits are commercially available for detection of several transgenic proteins in GM crops (Agdia; Biogenetic Services; Mahyco; EnviroLogix; Neogen; Romer Labs) and sensitivities of the ELISA assay have been reported upto pg/ml level (EnviroLogix).

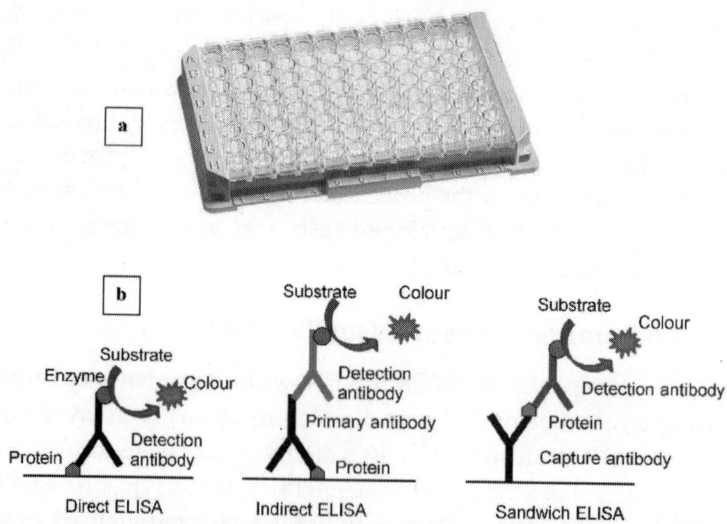

Fig. 7: (a) ELISA plate (Thermo Scientific), **(b)** Formats of ELISA: In direct ELISA one antibody is used. In indirect ELISA two antibodies are used, one antibody specific to protein. In sandwich ELISA two antibodies are used, both specific to protein.

Dipstick or Immunochromatographic (IC) strip assay

The dipstick assay uses a membrane-based detection system. It is a variant of ELISA, using nitrocellulose (NC) strips rather than microtiter wells. The dipstick strip format was initially developed in the 1970s. This assay is also called lateral flow immunoassay or Immunochromatographic (IC) strip assay. The dipstick includes a plastic backing card, NC membrane (5 – 15 μ pore size), sample pad, absorbent pad and conjugate pad (Fig. 8). The NC membrane contains immobilized antibody on test line which is specific for the target protein. In dipstick strip system different label such as latex, silver, colloidal gold, enzyme, were used for making probe which is coupled to antibody specific for protein for visual detection of the results. The colloidal gold has been generally used in most of the assay development. Colloidal gold particles varying from 10-40 nm have been employed by several investigators (Shim et al. 2009; Shyu et al. 2002; Xiulan et al. 2005; Zhao et al. 2008; Zhou et al. 2009) which can suitably detect the antigens. The NC membrane contains two captured zones, one specific for the transgenic protein and another specific for untreated antibodies coupled to the color reagent. The conjugate pad containing immobilised labelled antibody specific for protein was separately placed on the strip to respective position. When strip is placed in a vial containing an extract from plant tissue or seed harbouring transgenic protein, samples moves in upward by capillary action and develops color on the captured zones. The dipstick format gives qualitative results either presence or absence of protein in GM sample within a few minutes.

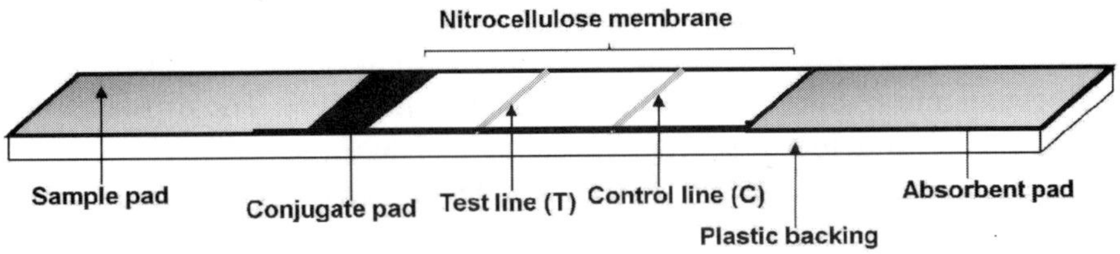

Fig. 8: Schematic representation of typical dipstick strip consists of a sample pad, a conjugate pad, a nitrocellulose membrane, and an absorption pad. The conjugate pad contains gold conjugated antibody. Anti-protein IgG is immobilized at the test line (T) and anti-species IgG is immobilized at the control line (C).

There are two main formats of dipstick strips used for testing; double antibody sandwich and competitive assay. In sandwich format positive results were visualized by formation of two red colored lines on the membrane (Fig. 9a) whereas in competitive format formation of only one red colored line indicates positive results (Fig. 9b). Sandwich assay is most suitable for larger analytes, such as transgenic proteins, bacterial pathogens and viruses whereas competitive dipstick assay for testing small molecules, such as mycotoxins.

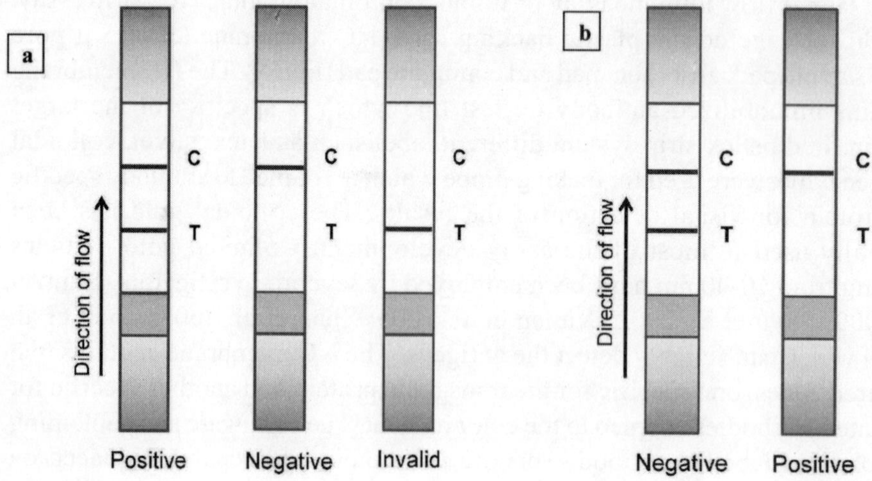

Fig. 9: The illustration of dipstick assay results: **(a)** sandwich format and the positive result is judged by the appearance of red line in test region; **(b)** competitive assay format and the positive result is judged by the non-appearance of red line in test region. The red line in control region should always visible showing valid test assay.

The dipstick is economical, more amenable to point-of-sale application, and suitable as an initial screening method early in the food chain. These test strips are fast, cheap and require minimal training and equipment. The dipstick can suitably be used to detect transgenic proteins in leaves, seeds and grains of different GM crops. Kumar et al. (2010) have developed dipstick for detection of transgenic Vip3A protein in GM crops with LOD of 100 ng/ml protein. Currently, dipsticks are commercially available to detect several transgenic proteins such as Vip, Cry1Ac, Cry1Ab, Cry2Ab, CP4-EPSPS, etc. in different GM crops/produce (Agdia; EnviroLogix; Romer Labs® Technology). The commercial strips for transgenic proteins can detect up to 5 ppb protein in GM samples (EnviroLogix). Furthermore, several commercial strips have been developed for simultaneous detection of two or more transgenic proteins, such as Cry1Ac, CP4EPSPS, and Cry2A/Cry1F to reduce cost and time of assay (Agdia; Mahyco; EnviroLogix; Neogen; Romer Labs Technology). Recently, Kumar (2012a) developed dipsticks for simultaneous detection of Vip3A and Cry1Ab/Cry1Ac transgenic proteins suitable for on-site simultaneous screening of GM crops bearing two proteins.

Immuno-PCR method

The immuno-PCR (IPCR) technique was originally developed by Sano et al. (1992) for the detection of bovine serum albumin (BSA). The IPCR method is very similar to ELISA but here reporter DNA is replaced by enzyme which can further be amplified either by conventional PCR or by real-time PCR (Fig. 10a). It combines the advantages of specificity of immunoassays and sensitivity of PCR. In conventional IPCR the amplified reported DNA can be analysed by agarose gel electrophoresis but sometimes the non-specific results would also be observed. To minimise this non-specific results, RT-IPCR would be employed which involves the RT-PCR technology (Fig. 10b). RT-IPCR assay has been extensively used in field of immunological research and clinical diagnostics to increase the sensitivity of detection where complex biological matrices and less amount of sample available. Due to some limitations such as high cost, time consuming, and requirement of expensive instruments, this technique is not routinely utilized for transgenic detection. Still we find some progress in transgenic detection and quantification such as Cry1Ac protein (Allen et al. 2006), Cry1Ab protein (Kumar 2012b) and Vip3A protein (Kumar 2011) owing to high sensitivity and requirement of less amount of protein over other protein-based methods. Sensitivity of detection in RT-IPCR assay for Cry1Ab been reported upto 0.1 ng/ml protein (Kumar 2012) while in the range of fg/ml for antigens other than transgenic (Chen and Zhuang 2009; Zhuang and Zhou 2009).

Reference materials

Reference samples are an indispensable part of any quantification assay. The percentage of GM material is calculated from certified reference material (CRM) or endogenous reference gene. The selected endogenous reference gene must be taxon-specific and present at a low copy number with highly conserved in nature among cultivars. In general, GM contents are expressed as mass/mass ratio or copy number ratio of GM over non-GM of the assayed material. Determination of percentage of GM contents in the sample utilizing reference gene is normally based on quantification of transgenic DNA by RT-PCR. In this, initial amounts of GM and non-GM DNA templates are quantified by standard curves, and GM contents in terms of percentage can be calculated by the ratios of specific transgene target to endogenous reference gene (Ding et al. 2004; Yang et al. 2005c). To date, many endogenous reference genes and their RT- PCR assays have been developed for commercial GM crops, such as *Sad1* for cotton; *Lectin* for soybean; *Invertase I*, *zein*, *zSSIIb*, *adh* and *hmg-A* for maize; *HMG I/Y, FatA, CruA*, and *BnACCg8* for rapeseed; *PLD*, *SPS, GOS9*, and *ppi-PPF* for rice (Zhang and Guo 2011).

The calculation of GM content can also be carried out in a reliable manner by constructing calibration curves utilizing CRM. A CRM contains different certified mass fractions of GMO. The availability of suitable CRMs is a fundamental requirement for the validation of detection methods and quality control material in PCR amplification (Trapmann et al. 2010). The JRC Institute for Reference Materials and Measurements (IRMM, Belgium) is an institute responsible for the production of appropriate CRMs in Europe. However, current CRMs have some limitations in practical application, such as the inadequate availability of GM crop materials, limited quantitative ranges, and high costs (Li et al. 2009). Considering the problems of availability and high cost of CRM production, reference plasmid or genomic DNA could be used as valid alternative reference material for GMO content analysis. The reference plasmid containing one or more functional transgene fragments has been developed and proved to be a good alternative for CRM in GMO detection (Yang et al. 2007).

GM products: benefits and controversies

GM foods promise to meet the need of booming population in a number of ways, ensuring an adequate food supply. The potential benefits of such GM crops includes- reducing cost of production and pesticide use, high crops yield, enhanced taste and quality of foods, increased nutrients and stress tolerance, improved resistance to disease, pests and herbicides (Untied States Department of Energy Office of Science 2008). On the other hand, the worldwide commercial cultivation of GM crops has also raised concerns about potential adverse effects associated with human health and environment sustainability.

The risks are associated with introduction of allergenic/harmful proteins into the foods, detrimental effects on non-target species and the environment, increase invasiveness and weediness of crop plants, pest resistance and impact on biodiversity (Sakko 2002). A recent report on Monsanto's Bt-cotton also indicated that it kills the soil microbes as well as farmers (ISIS Report 2009).

With the introduction of GM crops and foods in the market, environmental activists, scientific groups/experts and other organizations have raised concerns for potential hazards and undesirable effects on human health and environment and criticizing this type of agribusiness for pursuing profit without proper inspecting the biosafety aspect. Further, controversies surrounding GM foods and crops commonly focus on human and environmental safety, labeling and consumer choice, intellectual property rights, ethics, food security, poverty reduction, and environmental conservation (Untied States Department of Energy Office of Science, 2008).

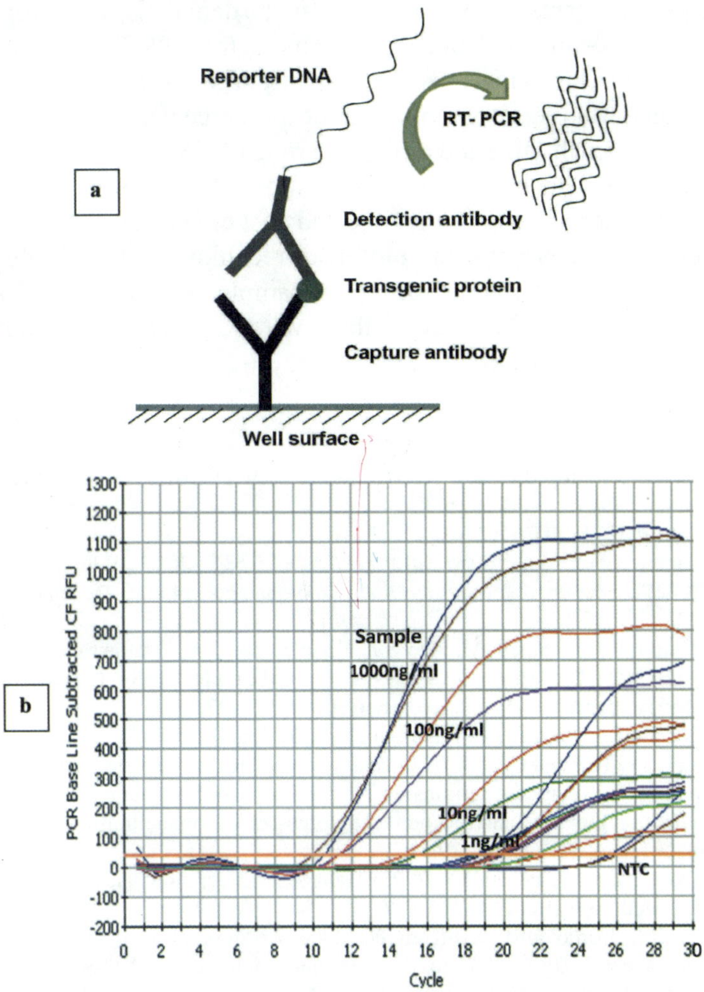

Fig. 10: (a) RT-IPCR assay sandwich type using two combinations of antibodies against protein. Bound reporter DNA was amplified by RT-PCR. **(b)** Amplification curves visualized after RT-IPCR assay using 10-fold dilutions of Vip protein.

Concluding Remarks

In spite of many controversies over GM crops, its adaptation is continuous increasing worldwide. Considering the countries' specific labeling legislations over release of GM crops as well as consumers high demand for the freedom of their choice between GM and non-GM materials, the need is also simultaneously arises for developing specific and sensitive detection methods to discriminate commercially approved GM crops/produce from non-GM materials. Although a diverse detection methods and strategies have been developed and commercially available to fulfil the demanding issues of labelling regulations and screening of their produce. Among them, PCR, RT-PCR, ELISA, and dipstick are being widely used for such purposes. RT-PCR and ELISA provide the quantitative information about the percentage of GMO whereas dipstick gives the qualitative and rapid screening of GMOs. Furthermore, some advances in such detection assays have allowed multiplexing ability and traceability of materials to save the time and cost of analysis. However, recent advancements in microarrays and biosensor techniques have enable the high throughput screening of a large number of samples with high sensitivity and specificity, and play a leading role in the low-cost, rapid and routine advance detection systems for GM produce.

References

Agbios (2013) Biotech crop database, Agbios, Merrickville, Ontario, Canada. Available from http://www.agbios.com/dbase.php.

Agdia Inc. USA (2013) GMO/ Trait Tests. https://orders.agdia.com/inventory.asp?CatId={7B7AFD85-5372-4029-BB76-D26DFC9316F3}

Ahmed FE (2002) Detection of genetically modified organisms in foods. Trends Biotechnol 20: 215–223.

Ahmed MU, Saito M, Hossain MM, Rao SR, Furui S, Hino A, Takamura Y, Takagi M and Tamiya E (2009) Electrochemical genosensor for the rapid detection of GMO using loop-mediated isothermal amplification. Analyst 134: 966-972.

Akiyama H, Nakamura F, Yamada C, Nakamura K, Nakajima O, Kawakami H, Harikai N, Furui S, Kitta K and Teshima R (2009) A screening method for the detection of the 35S promoter and the nopaline synthase terminator in genetically modified organisms in a real-time multiplex polymerase chain reaction using high-resolution melting-curve analysis. Biol Pharmaceut Bull 32: 1824-1829.

Allen GC, Spiken S and Thompson WF (2005) Transgenic integration: use of matrix attachment regions. Meth Mol Biol 286: 313–326.

Allen RC, Rogelj S, Cordova SE and Kieft TL (2006) An immuno-PCR method for detecting *Bacillus thuringiensis* Cry1Ac toxin. J Immunol Meth 308: 109-115.

Anklam E, Gadani F, Heinze P, Pijnenpurg H and Eede GVD (2002) Analytical methods for detection and determination of genetically modified organisms in agricultural crops plant-derived food products. Eur Food Res Technol 214: 3–26.

Andersen CB, Holst-Jensen A, Berdal KG, Thorstensen T and Tengs T (2006) Equal Performance of TaqMan, MGB, Molecular Beacon, and SYBR Green-Based Detection Assays in Detection and Quantification of Roundup Ready Soybean J Agri Food Chem 54: 9658-9663.

Angonesi Brod FC, dos Santos Ferrari C, Valente LL and Arisi ACM (2007) Nested PCR detection of genetically modified soybean in soybean flour, infant formula and soymilk. LWT-Food Sci Tech 40: 748-751.

Applied Biosystems Inc, USA (2013) TaqMan® GMO Detection and Quantitation System. https://products.appliedbiosystems.com/ab/en/US/adirect/ab? cmd=cat Navigate2&cat ID=601230.

Arugula MA, Zhang Y and Simonian AL (2014) Biosensors as 21st Century Technology for Detecting Genetically Modified Organisms in Food and Feed. Anal Chem 86:119-129.

Bahrdt C, Krech AB, Wurz A and Wulff D (2010) Validation of a newly developed hexaplex real-time PCR assay for screening for presence of GMOs in food, feed and seed. Anal Bioanal Chem 396: 2103-2112.

Bai SL, Zhong X, Ma L, Zheng W, Fan LM, Wei N and Deng XW (2007) A simple and reliable assay for detecting specific nucleotide sequences in plants using optical thin-film biosensor chips. Plant J 49: 354-366.

Bai S, Zhang J, Li S, Chen H, Terzaghi W, Zhang X, Chi X, Tian J, Luo H, Huang W, Chen Y and Zhang Y (2010) Detection of six genetically modified maize lines using optical thin-film biosensor chips. J Agric Food Chem 58: 8490-8494.

Ballari RV, Martin A, and Gowda LR (2013) Detection and identification of genetically modified EE-1 brinjal (*Solanum melongena*) by single, multiplex and SYBR® real-time PCR. J Sci Food Agri 93: 340-347.

Biogenetic Services, Inc. USA (2013) http://www.biogeneticservices.com/plant_dnagmo.htm.

Bio-Rad Laboratories, USA (2013) The GMO Investigator kit. http://www.bio-rad.com/en-is/product/gmo-investigator-real-time-pcr-starter-kit.

Bordoni R, Germini A, Mezzelani A, Marchelli R and De Bellis G (2005) A microarray platform for parallel detection of five transgenic events in foods: a combined polymerase chain reaction-ligation detection reaction-universal array method. J Agri Food Chem 53: 912-918.

Brod FCA, Ferrari CS, Valente LL and Arisi ACM (2007) Nested PCR detection of genetically modified soybean in soybean flour, infant formula and soymilk. LWT- Food Sci Technol 40: 748-751.

Broothaerts W, Mitchell HJ, Weir B, Kaines S, Smith LMA, Yang W, Mayer JE, Rodriguez C and Jefferson RA (2005) Gene transfer to plants by diverse species of bacteria. Nature 433: 629–633.

Buh Gasparic M, Cankar K, Zel J and Gruden K (2008) Comparison of different real-time PCR chemistries and their suitability for detection and quantification of genetically modified organisms. BMC Biotechnol 8: 26.

Buh Gasparic M, Tengs T, La Paz JL, Holst-Jensen A, Pla M, Esteve T, Zel J and Gruden K (2010) Comparison of nine different real-time PCR chemistries for qualitative and quantitative applications in GMO detection. Anal Bioanal Chem 396: 2023-2029.

Byrdwell WC and Neff WE (1996) Analysis of Genetically Modified Canola Varieties by Atmospheric Pressure Chemical Ionization Mass Spectrometric and Flame Ionization Detection. J Liq Chrom Relat Tech 19: 2203–2225.

Cankar K, Ravnikar M, Zel J, Gruden K and Toplak N (2005) Real-time polymerase chain reaction detection of cauliflower mosaic virus to complement the 35S screening assay for genetically modified organisms. J AOAC Int 88: 814-822.

Cankar K, Chauvensy-Ancel V, Fortabat MN, Gruden K, Kobilinsky A, Zel J and Bertheau Y (2008) Detection of nonauthorized genetically modified organisms using differential quantitative polymerase chain reaction: application to 35S in maize. Anal Biochem 376: 189-199.

Chawla HS (2002) Introduction to Plant Biotechnology (2nd ed.) Science Publishers Inc, New Hampshire, USA. pp. 360-363.

Chen HY and Zhuang HS (2009) Real-time immuno-PCR assay for detecting PCBs in soil samples. Anal Bioanal Chem 394: 1205-1211.

Cho MJ, Yano H, Okamoto D, Kim HK, Jung HR, Newcomb K, Le VK, Yoo HS, Langham R, Buchanan BB and Lemaux PG (2004) Stable transformation of rice (*Oryzae sativa* L.) via microprojectile bombardment of highly regenerative, green tissues derived from mature seed. Plant Cell Rep 22: 483–489.

Deisingh AK and Badrie N (2005) Detection approaches for genetically modified organisms in foods. Food Res Int 38: 639-649.

Dinelli G, Bonetti A, Marotti I, Minelli M, Navarrete-Casas M, Segura-Carretero A and Fernandez-Gutierrez A (2006) Quantitative-competitive polymerase chain reaction coupled with slab gel and capillary electrophoresis for the detection of roundup ready soybean and maize. Electrophoresis 27: 4029-4038.

Ding J, Jia J, Yang L, Wen H, Zhang C, Liu W and Zhang D (2004) Validation of a rice-specific gene, sucrose-phosphate synthase, used as the endogenous reference gene for qualitative and real-time quantitative PCR detection of transgenes. J Agri Food Chem 52: 3372–3377.

Dobnik D, Morisset D and Gruden K (2010) NAIMA as a solution for future GMO diagnostics challenges. Anal Bioanal Chem 396: 2229-2233.

Dong W, Yang L, Shen K, Kim B, Kleter GA, Marvin HJ, Guo R, Liang W and Zhang D (2008) GMDD: a database of GMO detection methods. BMC Bioinform 9: 260.

Dorries HH, Remus I, Gronewald A, Gronewald C and Berghof-Jager K (2010) Development of a qualitative, multiplex real-time PCR kit for screening of genetically modified organisms (GMOs). Anal Bioanal Chem 396: 2043-2054.

Elenis DS, Kalogianni DP, Glynou K, Ioannou PC and Christopoulos TK (2008) Advances in molecular techniques for the detection and quantification of genetically modified organisms. Anal Bioanal Chem 392: 347-354.

Engel KH, Frenzl T and Miller A (2002) Current and future benefits from the use of GM technology in food production. Toxicol Lett 127: 329–336.

EnviroLogix Inc., USA (2013) GMO Test Kits. http://www.envirologix.com/artman/publish/cat_index_5.shtml, http://www.envirologix.com/artman/publish/cat_index_44.shtml

Ermolli M, Prospero A, Balla B, Querci M, Mazzeo A and Van Den Eede G (2006) Development of an innovative immunoassay for CP4EPSPS and Cry1AB genetically modified protein detection and quantification. Food Addit Contam 23: 876-882.

Estruch JJ, Carozzi NB, Desai N, Duck NB, Warren GW and Koziel MG (1997) Transgenic plants: an emerging approach to pest control. Nat Biotechnol 15: 137-141.

Fantozzi A, Ermolli M, Marini M, Scotti D, Balla B, Querci M, Langrell SR and Van den Eede G (2007) First application of a microsphere-based immunoassay to the detection of genetically modified organisms (GMOs): quantification of Cry1Ab protein in genetically modified maize. J Agric Food Chem 55: 1071–1076.

Feinberg M, Fernandez S, Cassard S and Bertheau Y (2005) Quantitation of 35S promoter in maize DNA extracts from genetically modified organisms using real-time polymerase chain reaction, part 2: Interlaboratory study. J AOAC Int 88: 558-573.

Feng J, Liang Y, Wang F and Chen J (2013) Detection of Genetically Modified Tomato Using PCR Coupled with iParaflo™ Microfluidics Microarrays. J Nanosci Nanotechnol 13: 8266-8274.

Fernandez S, Charles-Delobel C, Geldreich A, Berthier G, Boyer F, Collonnier C, Coue-Philippe G, Diolez A, Duplan MN, Kebdani N, Romaniuk M, Feinberg M and Bertheau Y (2005) Quantification of the 35S promoter in DNA extracts from genetically modified organisms using real-time polymerase chain reaction and specificity assessment on various genetically modified organisms, part I: operating procedure. J AOAC Int 88:547-557.

Garcia-Canas V, Mondello M, Cifuentes A (2010) Simultaneous detection of genetically modified organisms by multiplex ligation-dependent genome amplification and capillary gel electrophoresis with laser-induced fluorescence. Electrophoresis 31: 2249-2259.

Gelvin SB (2005) Gene exchange by design. Nature 433: 583–584.

Grothaus GD, Bandla M, Currier T, Giroux R, Jenkins GR, Lipp M, Shan G, Stave JW and Pantella V (2006) Immunoassay as an analytical tool in agricultural biotechnology. J AOAC Int 89: 913–928.

Guo J, Yang L, Chen L, Morisset D, Li X, Pan L and Zhang D (2011) MPIC: One high-throughput analytical method for multiple DNA targets. Anal Chem 83: 1579–1586.

Hardegger M, Brodmann P and Herrmann A (1999) Quantitative detection of the 35S promoter and the NOS terminator using quantitative competitive PCR. Eur Food Res Technol 209: 83-87.

Heide BR, Heir E and Holck A (2008) Detection of eight GMO maize events by qualitative, multiplex PCR and fluorescence capillary gel electrophoresis. Eur Food Res Technol 227: 527–535.

Hernandez M, Rodr´ýguez-L´azaro D, Esteve T, Prat S and Pla M (2003) Development of melting temperature-based SYBR Green I polymerase chain reaction methods for multiplex genetically modified organism detection. Anal Biochem 323: 164–170.

Holden MJ, Levine M, Scholdberg T, Haynes RJ and Jenkins GR (2010) The use of 35S and Tnos expression elements in the measurement of genetically engineered plant materials. Anal Bioanal Chem 396: 2175-2187.

Hubner P, Studer E and Luthy J (1999) Quantitative competitive PCR for the detection of genetically modified organisms in food. Food Control 10: 353–358.

Huber I, Block A, Sebah D, Debode F, Morisset D, Grohmann L, Berben G, Stebih D, Milavec M, Zel J and Busch U (2013) Development and Validation of Duplex, Triplex, and Pentaplex Real-Time PCR Screening Assays for the Detection of Genetically Modified Organisms in Food and Feed. J Agric Food Chem 61: 10293-10301.

Hupfer C, Hotzel H, Sachse K, Moreano F and Engel KH (2000) PCR-based quantification of genetically modified Bt maize: single-competitive versus dual-competitive approach. Eur Food Res Technol 212: 95-99.

ISIS Report (2005) India's Bt Cotton Fraud. http://www.i-sis.org.uk/IBTCF.php.

ISIS Report (2009) Monsanto's Bt Cotton Kills the Soil as Well as Farmers. www. globalresearch. ca/index.php?context=va&aid=12432.

James C (2012) Global Status of Commercialized Biotech/GM Crops: 2012. ISAAA Brief No. 44. ISAAA: Ithaca, NY.

Kalogianni DP, Koraki T, Christopoulos TK and Ioannou PC (2006) Nanoparticle-based DNA biosensor for visual detection of genetically modified organisms. Biosens Bioelectron 21: 1069-1076.

Kim JH, Kim SY, Lee H, Kim YR and Kim HY (2010) An event-specific DNA microarray to identify genetically modified organisms in processed foods. J Agric Food Chem 58: 6018-6026

Kim JH, Kim JH, Baek MA, Kim HY and Nam YS (2011) Multiplex real-time PCR kit for qualitative detection of genetically modified roundup ready soybean. J Korean Soc Appl Biol Chem 54: 145-148.

Kumar R, Singh CK, Kamle S, Sinha RP, Bhatnagar RK and Kachru DN (2010) Development of nanocolloidal gold based immunochromatographic assay for rapid detection of transgenic vegetative insecticidal protein in genetically modified crops. Food Chem 122: 1298–1303.

Kumar R (2011) A quantitative immuno-PCR method for detection of vegetative insecticidal protein in genetically modified crops. J Agri Food Chem 59: 10448-10453.

Kumar R (2012a) Development of dipsticks for simultaneous detection of vip3A and cry1Ab/cry1Ac transgenic proteins. J AOAC Int 95: 1131-1137.

Kumar R (2012b) A real-time immuno-PCR assay for detection of transgenic Cry1Ab protein. Euro. Food Res Technol 234: 101-108.

Leimanis S, Hernandez M, Fernandez S, Boyer F, Burns M, Bruderer S, Glouden T, Harris N, Kaeppeli O, Philipp P, Pla M, Puigdomenech P, Vaitilingom M, Bertheau Y and Remacle J (2006) A microarray-based detection system for genetically modified (GM) food ingredients. Plant Mol Biol 61: 123-139.

Li X, Yang L, Zhang J, Wang S, Shen K, Pan L and Zhang D (2009) Simplex and duplex polymerase chain reaction analysis of Herculex RW (59122) maize based on one reference molecule including separated fragments of 5' integration site and endogenous gene. J AOAC Int 92: 1472–1483.

Maharashtra Hybrid Seeds Company Limited (Mahyco) (2013), Dawalwadi, India. http://www.mahyco.com/quality_assurance.html.

Mano J, Oguchi T, Akiyama H, Teshima R, Hino A, Furui S and Kitta K (2009a) Simultaneous detection of recombinant DNA segments introduced into genetically modified crops with multiplex ligase chain reaction coupled with multiplex polymerase chain reaction. J Agri Food Chem 57: 2640-2646.

Mano J, Shigemitsu N, Futo S, Akiyama H, Teshima R, Hino A, Furui S and Kitta K (2009b) Real-time PCR array as a universal platform for the detection of genetically modified crops and its application in identifying unapproved genetically modified crops in Japan. J Agri Food Chem 57: 26-37.

Marmiroli N, Maestri E, Gulli M, Malcevschi A, Peano C, Bordoni R, and De Bellis G (2008) Methods for detection of GMOs in food and feed. Anal Bioanal Chem 392: 369-384.

Michelini E, Simoni P, Cevenini L, Mezzanotte L and Roda A (2008) New trends in bioanalytical tools for the detection of genetically modified organisms: an update. Anal Bioanal Chem 392: 355-367.

Miller GT (2004) *Sustaining the Earth* (6th ed.). Thompson Learning, Inc. Pacific Grove, California. pp. 211-216.

Moreano F, Ehlert A, Busch U and Engel KH (2006) Ligation-dependent probe amplification for the simultaneous event-specific detection and relative quantification of DNA from two genetically modified organisms. Eur Food Res Technol 222: 479–485.

Morisset D, Dobnik D, Hamels S, Zel J and Gruden K (2008) NAIMA: target amplification strategy allowing quantitative on-chip detection of GMOs. Nucleic Acids Res 36: e118.

Nadal A, Coll A, La Paz JL, Esteve T and Pla M (2006) A new PCR-CGE (size and color) method for simultaneous detection of genetically modified maize events. Electrophoresis 27: 3879–3888.

Nadal A, Esteve T and Pla M (2009) Multiplex polymerase chain reactioncapillary gel electrophoresis: a promising tool for GMO screeningassay for simultaneous detection of five genetically modified cotton events and species. J AOAC Int 92: 765–772.

Narra D and Srivatsava V (2012) Quantitative competitive PCR for the detection and quantification of genetically modified cotton event MON-531. Int J Basic Appl Sci 1: 92-102.

Neogen Corporation, USA (2013) GMO test kits. http://www.neogen.com/FoodSafety/FS_Product_List.asp?Test_Kit_Cat=205

Nica AG, Mascini M and Ciucu AA (2004) DNA-based biosensor for detection of genetically-modified organisms. Chim Annu 23: 84"94.

Ocana MF, Fraser PD, Patel RK, Halket JM and Bramley PM (2007) Mass spectrometric detection of CP4EPSPS in genetically modified soya and maize. Rapid Commun Mass Spectrom 21: 319-328.

Oguchi T, Onishi M, Minegishi Y, Kurosawa Y, Kasahara M, Akiyama H and Teshima R (2009) Futo S, Furui S, Hino A and Kitta K, Development of quantitative duplex real-time PCR method for screening analysis of genetically modified maize. J Food Hyg Soc Japan 50: 117-125.

Onishi M, Matsuoka T, Kodama T, Kashiwaba K, Futo S, Akiyama H, Maitani T, Furui S, Oguchi T and Hino A (2005) Development of a multiplex polymerase chain reaction method for simultaneous detection of eight events of genetically modified maize. J Agri Food Chem 53: 9713–9721.

Passamano M and Pighini M (2006) QCM DNA-sensor for GMOs detection. Sensors Actuat B: Chem 118: 177–181.

Paul V, Steinke K and Meyer HH (2008) Development and validation of a sensitive enzyme immunoassay for surveillance of Cry1Ab toxin in bovine blood plasma of cows fed Bt-maize (MON810). Anal Chim Acta 607: 106–113.

Randhawa GJ, Chhabra R and Singh M (2009) Multiplex PCR-based simultaneous amplification of selectable marker and reporter genes for the screening of genetically modified crops. J Agric Food Chem 57: 5167-5172.

Randhawa GJ, Chhabra R and Singh M (2010) Decaplex and real-time PCR based detection of MON531 and MON15985 Bt cotton events. J Agric Food Chem 58: 9875-9881.

Randhawa GJ, Singh M and Chhabra R (2013) DNA-based diagnostics for genetically modified cotton: decaplex PCR assay to differentiate MON531 and MON15985 Bt cotton events. Methods Mol Biol 958: 139-151.

Roda A, Mirasoli M, Guardigli M, Michelini E, Simoni P and Magliulo M (2006) Development and validation of a sensitive and fast chemiluminescent enzyme immunoassay for the detection of genetically modified maize. Anal Bioanal Chem 384: 1269–1275.

Romer Labs® Technology Inc. (2013) AgraStrip® GMO TraitChek™. http://www.romerlabs.com/en/products/gmo/.

Rossi E and Presciuttini S (1996) Development of insecticide resistance in *Piophila casei* (Diptera: Piophilidae) strains selected with low doses of deltamethrin. J Econ Entomol 89: 15-20.

Rudi K, Rud I and Holck A (2003) A novel multiplex quantitative DNA array based PCR (MQDA-PCR) for quantification of transgenic maize in food and feed. Nucleic Acids Res 31: e62.

Sakko K (2002), The Debate over Genetically Modified Foods. http://www.actionbioscience.org/biotech/sakko.html#primer.

Salvi S, D'Orso F and Morelli G (2008) Detection and quantification of genetically modified organisms using very short, locked nucleic acid TaqMan probes. J Agric Food Chem 56: 4320–4327.

Sano T, Smith CL and Cantor CR (1992) Immuno-PCR: very sensitive antigen detection by means of specific antibody–DNA conjugates. Science 258: 120–122.

Sassolas A, Leca-Bouvier BD and Blum L (2008) DNA biosensors and microarrays. J Chem Rev 108: 109"139.

Schmidt AM and Rott ME (2006) Real-time polymerase chain reaction (PCR) quantitative detection of *Brassica napus* using a locked nucleic acid TaqMan probe. J Agric Food Chem 54: 1158–1165.

Schmidt AM, Sahota R, Pope DS, Lawrence TS, Belton MP and Rott ME (2008) Detection of genetically modified canola using multiplex PCR coupled with oligonucleotide microarray hybridization. J Agric Food Chem 56: 6791-800.

Shan G, Embrey SK and Schafer BW (2007) A highly specific enzyme-linked immunosorbent assay for the detection of Cry1Ac insecticidal crystal protein in transgenic Wide Strike cotton. J Agri Food Chem 55: 5974-5979.

Shim WB, Kim KY and Chung DH (2009) Development and Validation of a Gold Nanoparticle Immunochromatographic Assay (ICG) for the Detection of Zearalenone. J Agri Food Chem 57: 4035–4041.

Shrestha HK, Hwu KK, Wang SJ, Liu LF and Chang MC (2008) Simultaneous detection of eight genetically modified maize lines using a combination of event- and construct-specific multiplex-PCR technique. J Agri Food Chem 56: 8962-8968.

Shyu RH, Shyu HF, Liu HW and Tang SS (2002) Colloidal gold-based immunochromatographic assay for detection of ricin. Toxicon 40: 255-258.

Singh CK, Ojha A, Kamle S and Kachru DN (2007a) Assessment of cry1Ab transgene cassette in commercial Bt corn Mon810: gene, event, construct & GMO specific concurrent characterization. Nat. Protocols, published online. http://www.natureprotocols.com/2007/10/23/assessment_of_cry1ab_transgene.php.

Singh CK, Ojha A and Kachru DN (2007b) Detection and characterization of cry1Ac transgene construct in Bt cotton: multiple polymerase chain reaction approach. J AOAC Int 90: 1517-1525.

Singh CK, Ojha A, Bhatanagar RK and Kachru DN (2008) Detection and characterization of recombinant DNA expressing vip3A-type insecticidal gene in GMOs—standard single, multiplex and construct-specific PCR assays. Anal Bioanal Chem 390: 377-387.

Studer E, Rhyner C, Luthy J and Hubner P (1998) Quantitative competitive PCR for the detection of genetically modified soybean and maize. Z Lebensm Unters Forsch A. 207: 207-213.

Su W, Song S, Long M and Liu G (2003) Multiplex polymerase chain reaction/membrane hybridization assay for detection of genetically modified organisms. J Biotechnol 105: 227-233.

Trapmann S, Corbisier P, Schimmel H and Emons H (2010) Towards future reference systems for GM analysis. Anal Bioanal Chem 396: 1969–1975.

Terry CF, Shanahan DJ, Ballam LD, Harris N, McDowell DG and Parkes HC (2002) Real-time detection of genetically modified soya using Lightcycler and ABI 7700 platforms with TaqMan, Scorpion, and SYBR Green I chemistries. J AOAC Int 85: 938–944.

Tozzini AC, Martínez MC, Lucca MF, Vázquez Rovere C, Distéfano AJ, Vas M del and Hopp HE (2000) Semi-quantitative detection of genetically modified grains based on CaMV 35S promoter amplification. Electron J Biotechnol 3: 25-34.

Untied States Department of Energy Office of Science (2008) Human Genome Project Information, "Genetically Modified Foods and Organisms". Available from http://www.ornl.gov/hgmis/elsi/gmfood.html.

Viljoen CD (2005) Detection of Living Modified Organisms (LMOs) and the Need for Capacity Building. Asian

Biotechnol Dev Rev 7: 55-69.

Walschus U, Witt S and Wittmann C (2002) Development of Monoclonal Antibodies Against Cry1Ab Protein from Bacillus thuringiensis and Their Application in an ELISA for Detection of Transgenic Bt-Maize. Food Agri Immunol 14: 231-240.

Wang S, Guo AY, Zheng WJ, Zhang Y, Qiao H and Kennedy IR (2007) Development of ELISA for the Determination of Transgenic Bt-Cottons Using Antibodies against Cry1Ac Protein from *Bacillus thuringiensis* HD-73. Eng Life Sci 7: 149–154.

Wilson TG and Cain JW (1997) Resistance to the insecticides lufenuron and propoxur in natural populations of *Drosophila melanogaster* (Diptera: Drosophilidae). J Econ Entomol 90: 1131-1136.

Wu G, Wu Y, Xiao L and Lu C (2008) Event-specific qualitative and quantitative polymerase chain reaction methods for detection of genetically modified rapeseed Ms8xRf3 based on the right border junctions. J AOAC Int 91: 143-151.

Wurz A, Bluth A, Zeltz P, Pfeifer C and Willmund R (1999) Quantitative analysis of genetically modified organisms (GMO) in processed food by PCR-based methods. Food Control 10: 385-389.

Xie LJ, Ying YB, Ying TJ and Yu HY (2007a) Discrimination of transgenic tomatoes based on visible/ near-infrared spectra. Anal Chim Acta 584: 379-384.

Xie LJ, Ying YB and Ying TJ (2007b) Combination and comparison of chemometrics methods for identification of transgenic tomatoes using visible and near-infrared diffuse transmittance technique. J Food Eng 82: 395-401.

Xiulan S, Xiaolian Z, Jian T, Zhou J and Chu FS (2005) Preparation of gold-labeled antibody probe and its use in immunochromatography assay for detection of aflatoxin B1. Int J Food Microbiol 99: 185–194.

Xu J, Zhu S, Miao H, Huang W, Qiu M, Huang Y, Fu X and Li Y (2007) Event-specific detection of seven genetically modified soybean and maizes using multiplex-PCR coupled with oligonucleotide microarray. J Agri Food Chem 55: 5575-5579.

Yang L, Xu S, Pan A, Yin C, Zhang K, Wang Z, Zhou Z and Zhang D (2005a) Event specific qualitative and quantitative polymerase chain reaction detection of genetically modified MON863 maize based on the 5'-transgene integration sequence. J Agri Food Chem 53: 9312-9318.

Yang L, Pan A, Zhang K, Guo J, Yin C, Chen J, Huang C and Zhang D (2005b) Identification and quantification of three genetically modified insect resistant cotton lines using conventional and TaqMan real-time polymerase chain reaction methods. J Agri Food Chem 53: 6222-6229.

Yang L, Chen J, Huang C, Liu Y, Jia S, Pan L and Zhang D (2005c) Validation of a cotton – specific gene, *sad1*, used as an endogenous reference gene in qualitative and real-time quantitative PCR detection of transgenic cottons. Plant Cell Rep 24: 237–245.

Yang L, Guo J, Pan A, Zhang H, Zhang K, Wang Z and Zhang D (2007) Event-Specific Quantitative Detection of Nine Genetically Modified Maizes Using One Novel Standard Reference Molecule. J Agri Food Chem 55: 15–24.

Yang L, Guo J, Zhang H, Liu J and Zhang D (2008) Qualitative and quantitative event-specific PCR detection methods for oxy-235 canola based on the 3' integration flanking sequence. J Agri Food Chem 56: 1804-1809.

Zhang M, Gao X, Yu Y, Ao J, Qin J, Yao Y and Li Q (2007) Detection of Roundup Ready soy in highly processed products by triplex nested PCR. Food Control 18: 1277-1281.

Zhang D and Guo J (2011) The development and standardization of testing methods for genetically modified organisms and their derived products. J Integr Plant Biol 53: 539-551.

Zhao Y, Zhang G, Liu Q, Teng M, Yang J and Wang J (2008) Development of a lateral flow colloidal gold immunoassay strip for the rapid detection of enrofloxacin residues. J Agri Food Chem 56: 12138–12142.

Zhou PP, Zhang JZ, You YH and Wu YN (2008) Detection of genetically modified crops by combination of multiplex PCR and low-density DNA microarray. Biomed Environ Sci 21: 53-62.

Zhou Y, Pan FG, Li YS, Zhang YY, Zhang JH, Lu SY, Ren HL and Liu ZS (2009) Colloidal gold probe-based immunochromatographic assay for the rapid detection of brevetoxins in fishery product samples. Biosens Bioelectron 24: 2744-2747.

Zhuang HS and Zhou C (2009) Determination of anthracene by real-time immuno-polymerase chain reaction assay. Anal Chim Acta 633: 278-282.

10

Diverse Functions and Applications of Novel Cyanobacterial Sheath Pigment, Scytonemin

Jainendra Pathak, Rajneesh, Richa, Vinod K. Kannaujiya, Arun S. Sonker and Rajeshwar P. Sinha

Abstract

Cyanobacteria were first oxygen evolving photosynthetic organisms on Earth, which not only survived under harsh conditions of primitive Earth but are also dominant and efficient photosynthesizers and N_2 fixers of present time which significantly contribute to fertility of paddy fields as natural biofertilizers. Cyanobacteria require solar radiation to carry out photosynthesis and nitrogen fixation. The stratospheric ozone depletion which has resulted in an increase in ultraviolet-B (UV-B; 280 - 315 nm) radiation on the Earth's surface has been reported to inhibit a number of photochemical and photobiological processes in cyanobacteria. However, they have evolved mechanisms such as synthesis of photoprotective compound scytonemin and their derivatives to counteract the damaging effects of UV-B. In addition to Photoprotection, scytonemin has biotechnological potentials and several applications in biology and allied sciences. This chapter deals with the role of scytonemin as photoprotective compound and endeavours pharmacological, biotechnological, ecological and various other potentials of the multipurpose cyanobacterial sheath pigment, scytonemin.

Keywords: Cyanobacteria; Photoprotection; Scytonemin; Ultravioler radiation (UVR).

Introduction

Cyanobacteria are Gram negative photosynthetic prokaryotes inhabiting a wide range of habitats, ranging from oceans to fresh water, hot springs to the Antarctic and Arctic regions, soil to bare rocks and stones (Fig. 1 A, B), deserts to ice shelves and are also found in the form of endosymbionts in plants, lichens and several protists (Quesada and Vincent 1997; Baracaldo et al. 2005; Thajuddin and Subramanian 2005). These photolysis mediated oxygen evolving, cosmopolitan prokaryotes have survived and flourished on Earth for over two billion years (Sergeev et al. 2002). Nitrogen fixation ability of cyanobacteria and their role in maintaining fertility of rice fields is well documented (Sinha and Häder 1996; Vaishampayan et al. 1998, 2001, Tirkey and Adhikary 2005). Absorption of solar radiation to perform photosynthesis and nitrogen fixation exposes cyanobacteria to lethal ultraviolet-B (UV-B; 280–315 nm) radiation, due to depletion of stratospheric ozone layer, owing to atmospheric pollutants such as chloroflorocarbons, chlorocarbons and organobromides which are released anthropogenically (Crutzen 1992; Kerr and McElroy 1993; Lubin and Jensen 1995; Weatherhead and Andersen 2006). Depletion in ozone layer has been reported in Antarctic (Hodgson et al. 2005) as well as Arctic regions, where more than 70% decrease in ozone layer has been reported during late winter and early spring in the polar vortex (Smith et al. 1992; Von der Gathen et al. 1995). Aquatic ecosystems are also affected by harmful doses of UV-B and especially UV-A radiation which penetrate deep into the water column (up to several hundred meters) (Häder et al. 2007).

Ultraviolet radiation (UVR) adversely affects living systems such as bacteria (Peak et al. 1984), cyanobacteria, phytoplankton, macroalgae (Sinha et al. 2001), plants (Quaite et al. 1992), animals (Kripke et al. 1992) and humans (Stein et al. 1989). The highly energetic UV-B affects DNA and proteins directly as well as indirectly (Björn 2007; Sinha and Häder 2002; Häder and Sinha 2005; Castenholz and Garcia-Pichel 2000; Castenholz 2004; Cadoret et al. 2005; Mazur-Marzec 2006; Rath and Adhikary 2007; Solheim et al. 2006). Direct effect involves the denaturation of DNA and RNA whereas indirect effects include production of reactive oxygen species (Häder and Sinha 2005). UV-A radiation is not absorbed directly by DNA, but induces DNA damage either by producing a secondary photoreaction of existing DNA photoproducts or *via* indirect photosensitizing reactions (Hargreaves et al. 2007). UV-B and UV-A radiation affects a number of physiological and biochemical processes in cyanobacteria, such as survival, growth, pigmentation, photosynthetic oxygen production, motility, N_2 uptake, phycobiliprotein composition and CO_2 uptake (Bhandari and Sharma 2006; Cheng et al. 2007; Helbling 2006; Wright et al. 2005; Wulff et al. 2007; Sinha et al. 2005; Xue et al. 2005; Vincent et al.

1993). However, certain autotrophs including cyanobacteria have evolved mitigation strategies to counteract the toxicity of ultraviolet radiation such as light dependent repair of UV-induced damage of DNA (Britt 1995; Kim and Sancar 1995; Pakker et al. 2000; Sinha et al. 2002; Häder and Sinha 2005), accumulation of carotenoids and detoxifying enzymes or radical quenchers and antioxidants (Rastogi et al. 2013), and synthesis of photoprotective compounds such as mycosporine-like amino acids (MAAs) (Singh et al. 2008) and scytonemin (Sinha et al. 1998; Sinha et al. 1999; Richter et al. 2006; Sinha and Häder 2008; Sinha et al. 2008; Rastogi and Sinha 2009; Rastogi et al. 2013) to counteract the damaging effects of UV radiation. These photoprotective metabolites i.e. scytonemin and mycosporine-like amino acids (MAAs), which are produced by several species of cyanobacteria, play significant role in screening of ultraviolet radiation (Sinha et al. 1998; Sinha and Häder 2008; Rastogi et al. 2013).

Structure of scytonemin

Nägeli (1849) first reported this pigment in some terrestrial cyanobacteria and later termed it as scytonemin (Nägeli and Schwenderer 1877). It is highly stable, lipid soluble, yellow-brown hydrophobic pigment, reported from extracellular polysaccharide sheaths (Fig. 1C, D) of about 300 cyanobacteria. It is a dimer of indole and phenolic subunits and has a molecular mass of 544 Da.

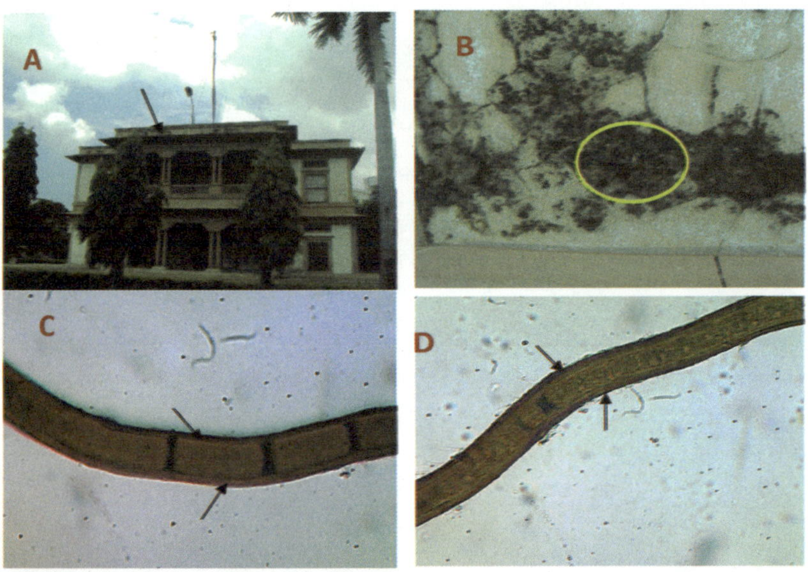

Fig. 1: (A), (B) Cyanobacteria harbouring the stones of the monument, Bharatmata mandir, Varanasi, India, as a black-brown leathery mats. **(C), (D)** Photographs showing the filaments of *Scytonema* sp. with sheaths typically colored by the presence of yellow-brown UV-protective pigment, scytonemin.

The two subunits in scytonemin are linked through an olefinic carbon atom which is unique among natural products. This linkage provides scytonemin a new ring system and has been termed as 'the scytoneman skeleton' by Proteau et al. (1993). Scytonemin exists in two forms ie oxidized (green) and reduced (red) (Garcia-Pichel and Castenholz 1991) were named as fuscochlorin and fuscorhodin, respectively (Kylim 1927, 1937). During the process of extraction, redox and acid-base conditions, determine the existence of the two forms. On the basis of ^{1}H and ^{13}C NMR and MS experiments, three new pigments i.e. dimethoxyscytonemin, tetramethoxyscytonemin and scytonin has been reported (Bultel-Poncé et al. 2004) from the organic extracts of *Scytonema* sp., and all these forms are derived from the 'scytoneman skeleton' of the scytonemin (Fig. 2).

Recently Grant and Louda (2013) have reported a new member of 'scytoneman skeleton', scytonemin-3a-imine (Fig. 2) from cyanobacterium, *Scytonema hoffmani*, growing under intense light conditions. Its structure has been characterized on the basis of UV/Vis spectrophotometry, Fourier transform infrared (FTIR) spectroscopy, matrix assisted laser desorption ionization (MALDI) and electrospray ionization (ESI) mass spectrometry (MS and MS^n), and both 1- and 2-D ^{1}H nuclear magnetic resonance (NMR) spectrometry. Structural features were confirmed utilizing derivatization ($NaBH_4$ reduction, acetylation, deuterium exchange).

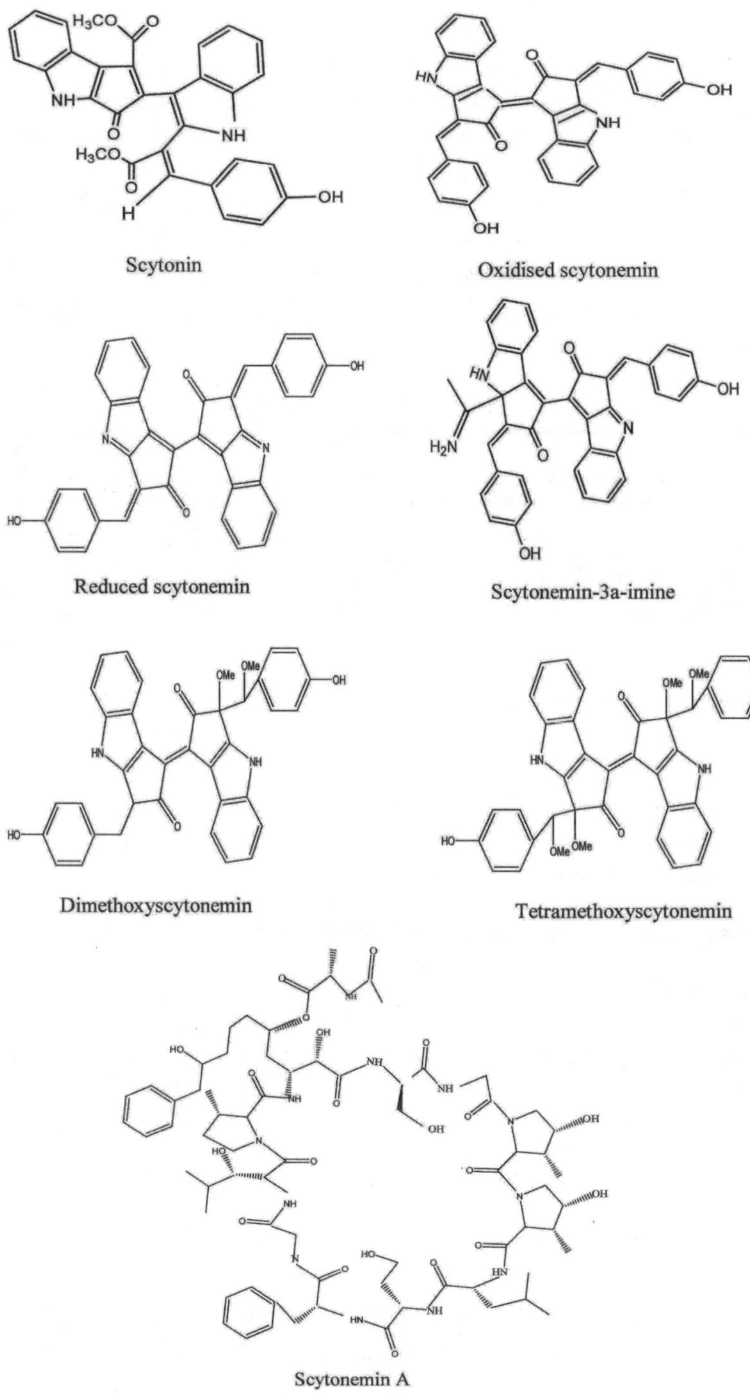

Fig. 2: Chemical structures of different forms of scytonemin

Scytonemin gene cluster

Gene cluster associated with scytonemin biosynthesis has been studied in *N. punctiforme* by Soule et al. (2007). The genomic region flanking the mutation revealed an 18-gene clusters (NpR1276 to NpR1259) (Fig. 3) among which the gene clusters NpR1274-NpR1271 plays central role in scytonemin biosynthesis. NpR1273 was found to be directly involved in scytonemin biosynthesis in *N. punctiforme.* All 18 genes gets induced by UV-A irradiation and relative transcription levels generally peak after 48 h of continuous UV-A exposure. All of the genes in the 18-gene region get co-transcribed as part of a single transcriptional unit

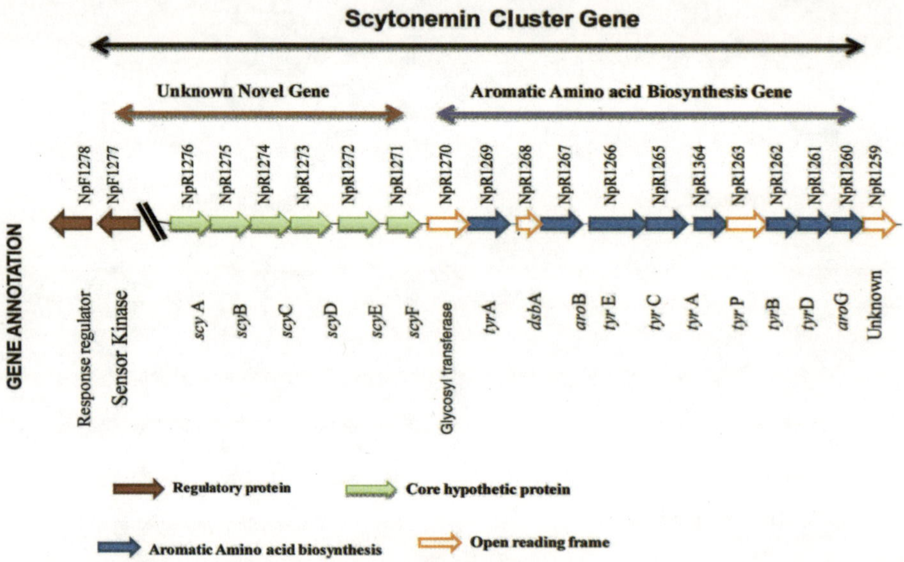

Fig. 3: Structure of genome associated with biosynthesis of scytonemin in *N. punctiforme.* The ORFs from NpR1276 to NpR1259 are labelled with particular proteins. The ORFs in the downstream region of these clusters (NpR1269 to NpR1260) are predicted to encode a set of enzymes involved in the shikimic acid and aromatic amino acid biosynthesis pathways. Most of the upstream ORFs within these clusters (NpR1276 to NpR1270) encode products annotated as hypothetical proteins. The hatch marks indicate a break in the distance scale. (Modified from Soule et al. 2007).

(Soule et al. 2009a). This 18-gene cluster consists of two major cluster architectures which are thought to be evolved from rearrangements of large sections, such as genes responsible for aromatic amino acid biosynthesis and through the insertion of genes that potentially confer additional biosynthetic capabilities. Differential transcriptional expression analysis demonstrated increased transcription of entire gene cluster after exposure to UV radiation.

The knowledge from an evolutionary phylogenetic analysis in combination with the fact that the scytonemin gene cluster is distributed across several cyanobacterial lineages led to conclusion that the distribution of this gene cluster is best explained through an ancient evolutionary origin (Sorrels et al. 2009). Balskus and Walsh (2008) have proposed possible biosynthetic route (Fig. 4) for the scytonemin biosynthesis and the acyloin reaction was found to be the key step in constructing the carbon framework of this ecologically and evolutionary important pigment. Two enzymes encoded by ORF NpR1275 and NpR1276 from the gene cluster has been functionally characterized and are involved in the initial step of scytonemin biosynthesis (Singh et al. 2009). However, the products of NpR1263 and NpR1269 ORFs are still to be functionally characterized. On the basis of conservation and location two additional conserved clusters, NpF5232 to NpF5236 and a putative two-component regulatory system (NpF1277 and NpF1278), are likely involved in scytonemin biosynthesis and regulation, respectively (Soule et al. 2009b). The production of the oxidized scytonemin from *N.commune* has been investigated in different culture conditions. High culture temperature, strong illumination intensity, and light–dark cycle (12:12 h) were found to be good for producing the oxidized scytonemin by *N. commune*.

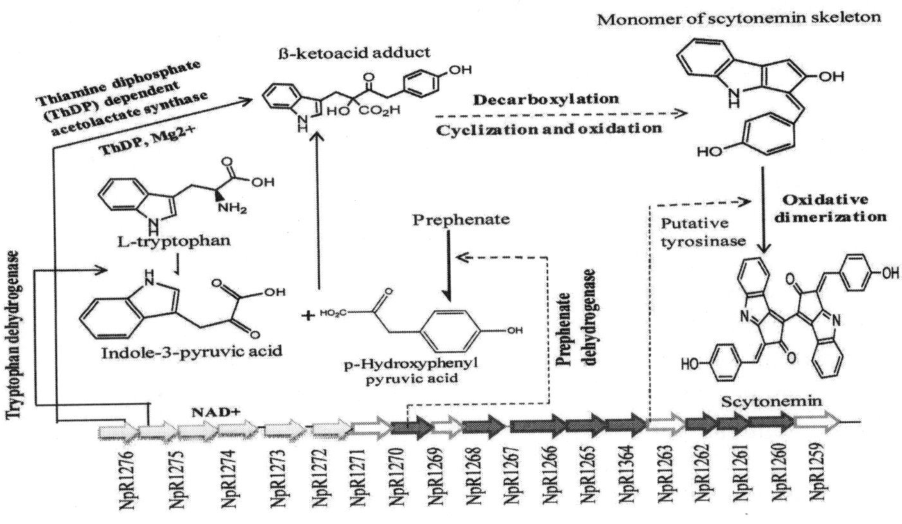

Fig. 4: Biosynthetic route for the scytonemin and corresponding gene products involved in each step. Continuous arrow represents functionally characterized gene product while gene product indicated by broken arrow are still to be functionally characterized for their involvement in corresponding step (Modified from Singh et al. 2009).

Synthesis of scytonemin

Metabolites of aromatic amino acid biosynthesis, are thought to be precursors of scytonemin biosynthesis, which is induced under high photon fluence rate (Garcia-Pichel and Castenholz 1991). Formaldehyde, chloramphenicol and low temperature (4°C) counteracts the effect of bright light in enhancing scytonemin production. Scytonemin synthesis involves protein synthesis pathway. UV-A treatment is very efficient in inducing the synthesis of scytonemin (Garcia-Pichel and Castenholz 1991), whereas blue, green or red light at the same fluence rates do not cause any significant increase in scytonemin. Biosynthesis of scytonemin is also greatly affected under different abiotic stresses such as heat, osmotic and oxidative stress (Dillon et al. 2002). Elevation in temperature and oxidative stress, in combination with UV-A, have a synergistic effect on increased production of scytonemin whereas osmotic stress causes decrease in scytonemin synthesis in the presence as well as in absence of scytonemin-inducing irradiance. As multiple environmental signal influence scytonemin synthesis, it can be said that scytonemin induction regulation is a part of complex stress response pathway. Balskus et al. (2011) examined biosynthesis and accumulation of cyanobacterial sunscreening pigment scytonemin within intertidal microbial mat communities, using a combination of chemical, molecular, and phylogenetic approaches. Both laminated (layered) and non-laminated mats contained scytonemin, with morphologically distinct mats having different cyanobacterial community compositions. Within laminated microbial mats, regions with and without scytonemin had different dominant oxygenic phototrophs, with scytonemin-producing areas consisting primarily of *Lyngbya aestuarii* and scytonemin-deficient areas dominated by a eukaryotic alga. The non-laminated mat was populated by a diverse group of cyanobacteria and did not contain algae. The amplification and phylogenetic assignment of scytonemin biosynthetic gene *scyC* from laminated mat samples confirmed that the dominant cyanobacterium in these areas, *L. aestuarii*, is likely responsible for sunscreen production. Hence, understanding of the molecular basis of scytonemin assembly could be utilised to explore its synthesis and function within natural microbial communities.

Stability of Scytonemin

The stability of UV-absorbing compounds may be critical for their ultimate utility as photoprotective function. Stability of scytonemin was investigated under three different abiotic stress conditions such as UV-B radiation, heat and strong oxidizing agent H_2O_2 (0.25%) (Rastogi et al. 2013). The partially purified MAAs and scytonemin were treated with the above mentioned different stress conditions for an hour after which their remaining contents were analyzed in the HPLC system. The effect of heat was assayed by incubating scytonemin in an incubator shaker at 60 °C. The content of scytonemin was lower under 0.25 % H_2O_2 stress than that under UV and heat stress. High stability of scytonemin can be evident by the fact that about 84 % of the scytonemin screen in desiccated *Nostoc punctiforme* cells remained intact even after almost 2months of constant exposure to UV-A radiation (Fleming and Castenholz 2007).

Role of scytonemin in photoprotection

Cyanobacteria are the first photosynthetic oxygen evolving organisms that are thought to appear in Precambrian era. The presence of a UV-absorbing compound like scytonemin most probably helped them to survive and protect them from lethal effects of UV radiation when there was no stratospheric ozone layer. This assumption is supported by the fact that scytonemin has an *in vivo* absorption maximum at 370 nm whereas purified scytonemin shows maximum absorption at 386 nm, but it also absorb significantly at 252, 278 and 300 nm (Fig. 5). HPLC chromatograms as well as absorption spectra of *Rivularia* sp. extracted with 1:1 (v/v) methanol:ethyl acetate (Rastogi et al. 2013) revealed the presence of a single peak of scytonemin with a retention time of 4.5 min. The presence of scytonemin at RT-1.8 was observed in case of *Scytonema* sp. inhabitaing the monuments of Varanasi (Fig. 6). The evidence for the photoprotective role of scytonemin has been shown in a number of cyanobacteria from various harsh habitats (Sinha et al. 1999; Garcia-Pichel and Castenholz 1991; Hunsucker et al. 2001; Garcia- Pichel et al. 1992; Gröniger et al. 2000). The relevance of UV sunscreens such as scytonemin for the protection have also been reported from cyanobacterial lichens such as *Collema, Gonohymenia* and *Petulla*, growing in high light intensity habitats and it is shown that scytonemin is located extracellularly in the sheath of the outer thallus part (Büdel et al. 1997). Solar radiation is not always required for the production of scytonemin because this pigment has been reported to be synthesized in cyanobacterium *Calothrix* deficient in Fe or Mg and grown under low illumination (Sinclair and Whitton 1977). The UV-sunscreen role of scytonemin has been demonstrated in terrestrial cyanobacterium *Chlorogloeopsis* sp. (Garcia-Pichel et al. 1992). In cyanobacterial cultures, 5% of the cellular dry weight is contributed by the scytonemin while it may be higher in naturally occurring cyanobacteria (Castenholz 1997). Fleming and Castenholz (2007) showed that when the cells were hydrated for two days, in between desiccation periods had high scytonemin synthesis compared to the cells which were hydrated for one day in *Nostoc punctiforme*, but in *Chroococcidiopsis* periodic desiccation inhibits scytonemin synthesis. Fleming and Castenholz (2008) suggested that the greater the restriction in nitrogen accessibility, the greater is the production of scytonemin.

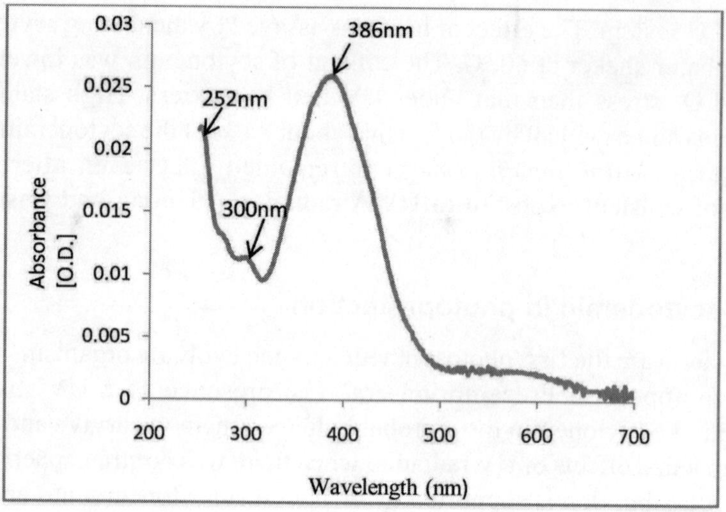

Fig. 5: Absorption spectrum of scytonemin extracted from *Scytonema* sp. having peaks at 252, 300 and 386 nm.

Pentecost (1993) has examined a correlation between UV flux and scytonemin content in population of *Scytonema* and *Rivularia* sp. A variable and even negative correlation was found between UV and scytonemin in *Scytonema*, but a positive correlation was found in *Rivularia*. The high amount of scytonemin is required for uninhibited photosynthesis under high UV flux in a monospecific population of *Calothrix* sp. (Brenowitz and Castenholz 1997). The absorption spectra of methanolic extracts of the terrestrial cyanobacterium, *Tolypothrix byssoidea* showed a prominent absorption at 260 and 384 nm, corresponding to the presence of scytonemin (Adhikary and Sahu 1998).

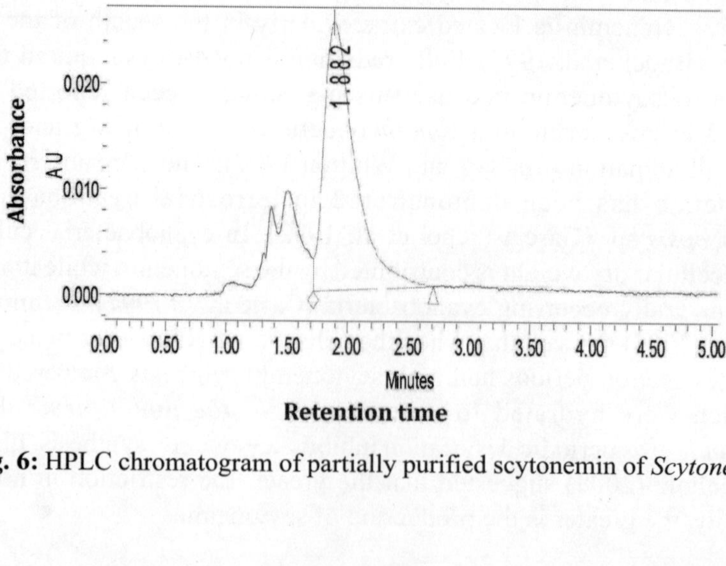

Fig. 6: HPLC chromatogram of partially purified scytonemin of *Scytonema* sp.

The presence of scytonemin in cyanobacterial sheath has been reported to reduce the entry of UV-A radiation in the cell by 90% (Garcia-Pichel et al. 1992; Proteau et al. 1993). The scytonemin is highly stable and perform its screening activity without any further metabolic investment even under prolonged physiological inactivity (e.g. desiccation) when other ultraviolet protective mechanisms such as active repair of biosynthesis damaged cellular component would be ineffective (Sinha et al. 1999; Brenowitz and Castenholz 1997; Ehling-Schulz et al. 1997).

Radical scavenging activity of scytonemin

The photoprotective function of scytonemin has been studied in terms of its radical scavenging capacity (Rastogi et al. 2013). The total antioxidant activities of different concentrations of scytonemin have been determined by the decolorization of D-PPH radicals. Decolorization of D-PPH was monitored visually as well as with a spectrophotometer. The activity of different concentrations of scytonemin was compared with that of ascorbic acid used as a standard. Similar to MAAs, the sun screening pigment scytonemin also exhibited the radical scavenging activity. The dose-dependent antioxidant activity of scytonemin was 12 %, 33 % and 57 % at concentrations of 0.5, 1.0 and 2.0 mg mL^{-1}, respectively. Antioxidant activity of the cyanobacterial pigment scytonemin was found to be relatively moderate in the D-PPH assay. Hence it can be inferred that presence of MAAs and scytonemin might help *Lyngbya* sp. to cope with oxidative stress under intense solar radiation. Scytonemin purified from the terrestrial cyanobacterium *Nostoc commune* quenched an organic radical in vitro and accounted for up to 10% of the total activity of an ethanol extract of *N. commune* (Matsui et al. 2012). Hence it can be concluded that the extracellular UV-absorbing pigment scytonemin has multiple roles, functioning as a UV sunscreen and an antioxidant relevant to anhydrobiosis in *N. commune.*

Scytonemin and salinity stress

Much study has been done on the effect of light quality on scytonemin biosynthesis (Gröniger et al. 2007). A correlation between scytonemin content and photosynthetic resistance to UV exposure has been demonstrated in *Calothrix* sp. (Brenowitz and Castenholz 1997). Scytonemin production constitutes an adaptive strategy of photoprotection against UV irradiation (Garcia-Pichel and Castenholz 1991). Impact of other stresses on scytonemin synthesis is not well studied. Rath et al. (2012) have studied effect of salinity in combination with high solar radiation, UV and desiccation in *Lyngbya aestuarii* growing throughout the Chilika lagoon, India, in different seasons.

L. aestuarii mats were exposed to higher salinity. It was observed that filaments were healthy up to 56 ppt salinity and as the salinity decreased from 28 ppt, an almost proportional decrease in the scytonemin content resulting from sheath degradation was observed. Thus, it can be concluded that scytonemin synthesis is an adaptive strategy for survival against salinity stress in *L. aestuarii*.

Pharmacological potentials of scytonemin

Scytonemin may have application in sunscreens as it has been found to act as ultraviolet sunscreen, with the greatest absorption in the spectral range of UV-A (Proteau et al. 1993, Rastogi and Sinha 2009). Scytonemin production in certain cyanobacteria is believed to be the earliest mechanism evolved for ultraviolet protection, more ancient than the flavonoids or melanins (Garcia-Pichel 1998). Its ring structure, the "scytoneman skeleton", is unique among natural products and is believed to be a condensation product of tryptophan- and phenylpropanoid- derived subunits and is also closely related to nostodione A (Proteau et al. 1993). Lack of chirality, multiple dissection points and phenolic groups that could be easily modified and its relation to other antiproliferative agents make scytonemin a prime candidate having potential to be used as a pharmacophore which can be used for developing new therapies targeting hyperproliferative disorders.

Scytonemin A (Fig. 2), a cyclic peptide characterised from a *Scytonema* sp. has also been shown to be a strong calcium agonist (Helms et al. 1988). Scytonemin inhibits human *polo*-like kinase, a serine/ threonine kinase that play an integral role in regulating the G2/M transition in the cell cycle. Scytonemin was shown to be active, with an IC_{50} of 2 mM (Stevenson et al. 2002a). It also inhibits other cell cycle kinases with similar potency. Scytonemin inhibited cell proliferation (IC_{50} =7.8 mM) in human T-cell leukemia Jurkat cells and induced apoptosis in 24% of the cells (at 3mM). Scytonemin inhibited *polo*-like kinase 1 activity in a concentration-dependent manner with an IC_{50} of 2 mM against the recombinant enzyme. Biochemical analysis showed that scytonemin reduced GST-*polo*-like kinase 1 activity in a time-independent fashion, suggesting reversibility and with a mixed-competition mechanism with respect to ATP.

Although scytonemin was less potent against protein kinase A and Tie2, a tyrosine kinase, it did inhibit other cell cycle-regulatory kinases like Myt1, checkpoint kinase 1, cyclin-dependent kinase 1/cyclin B and protein kinase C-2 with IC_{50} values similar to that seen for *polo*-like kinase 1. Consistent with these effects, scytonemin effectively attenuated, without chemical toxicity, the growth factor- or mitogen-induced proliferation of three cell types commonly implicated in inflamematory hyperproliferation. Similarly, scytonemin (up to

10mM) was not cytotoxic to nonproliferating endotoxin stimulated human monocytes. In addition, Jurkat T cells treated with scytonemin were induced to undergo apoptosis in a non-cell cycle-dependent manner consistent with its activities on multiple kinases. Scytonemin possesses both anti-inflammatory and anti-proliferative properties. The dual kinase inhibitory activity may be of value therapeutically in acute and possibly chronic disorders where both inflammation and proliferation are prevalent. Limiting both neovascularization and the presence of inflammatory mediators at the affected site offers a broader spectrum of activity, which could hypothetically be more efficacious in the management of complex inflammatory disorders. Whether or not the ability of scytonemin to inhibit both inflammation and proliferation is due solely to its effects on these two kinases, scytonemin offers a novel pharmacophore, which may serve as a template for synthesizing more potent and selective inhibitors (Stevenson et al. 2002b).

Role of scytonemin in apoptosis

Itoh et al. (2013) recently reported induction of autophagic cell death in Human T-lymphoid cell line Jurkat cells by reduced scytonemin isolated from *Nostoc commune*. Reduced-scytonemin (R-scy), isolated from *N. commune* Vaucher, has been shown to suppress the human T-lymphoid Jurkat cell growth. To reveal the mechanisms underlying the R-scy-mediated inhibition of Jurkat cell growth, cell morphology, DNA fragmentation, and microtubule-associated-protein light chain 3 (LC3) modifications were examined in these cells. Multiple vacuoles as well as the conversion of LC3-I to LC3-II in R-scy-treated cells were observed suggesting that the R-scy induced Jurkat cell growth inhibition is attributable to the induction of type II programmed cell death (PCD II; autophagic cell death or autophagy). The cells treated with R-scy produced large amounts of reactive oxygen species (ROS), leading to the induction of mitochondrial dysfunction. However, the elimination of R-scy-induced ROS by treatment with N-acetyl-L-cysteine (NAC) markedly opposed R-scy-induced PCDII. Based on these results, it can be conclude that ROS formation plays a critical role in R-scy-induced PCDII.

Role of scytonemin in tracing evolution

Scytonemin can be used as biomarker for tracing the evolution and expansion of cyanobacterial populations, especially in association with elevated UV stress. It is exclusive to cyanobacteria and therefore can serve as a diagnostic biomarker, particularly for UV-exposure growth conditions (Fulton et al. 2012). Scytonemin was found to be preserved in abundance in mid-Holocene sedimentary intervals in the Black Sea, demonstrating that scytonemin is resistant to degradation during erosion and transport. C and N isotopic compositions support the interpretation that scytonemin was derived from cyanobacteria in cryptobiotic desert soil, suggestive of expanding aridity during the subboreal Phase in the Black Sea region. Scytonemin has potential for preservation in black shales, where it may serve as an important biomarker for tracing the evolution and expansion of cyanobacterial

Role of scytonemin in paleobotanical studies

Because of the long term stability of scytonemin, it can be used for understanding the evolutionary history of life in palaeobotanical studies. Attempts are being made to reconstruct the historical ozone and UV-B amount for periods prior to the modern instrumental records and prior to human impacts by analyzing the UV-absorbing pigments in herbarium specimens over a period of upto ~100 years (Huttunen et al. 2005). This attempt is still in a developmental stage, but show promise for the future.

Role of scytonemin in paleoaquatic UV inferences

Spectral absorbance of benthic cladoceran carapaces can be used for inferring past UV exposure of aquatic biota. Aquatic microcrustaceans, such as cladocerans (Branchiopoda), are negatively affected by high intensities of UV (Williamson et al. 2001b; Rautio and Korhola 2002a). However, these animals can also be well adapted to high UV, because they can synthesize and/or accumulate photoprotective compounds such as melanin, carotenoid, scytonemin and mycosporine-like amino acids, repeatedly during their life (Hessen 1996; Rautio et al. 2009). These compounds increase survival under intensive UV (Hairston 1976; Hebert and Emery 1990; Hessen et al. 1999), being adaptive responses to the underwater UV environment (Rhode et al. 2001). Pigment synthesis is considered to be energetically costly and there occurs a trade off between damage from UV exposure and the costs of pigmentation as well as increased predation from fish hunting the most visible prey, i.e. pigmented individuals (Hansson, 2000). Nevalainen and Rautio (2014) have developed a method for measuring fossil cladoceran (Branchiopoda) carapace absorbance to infer past ultraviolet radiation (UV) exposure in lakes. They measured the absorbance of extracted carapaces with a spectrophotometer under visible light and UV wavelengths (700-280 nm range) using an adapter, which was designed to hold the microfossils. Photoprotective pigments present in fossil like melanin, carotenoid, scytonemin and mycosporine-like amino could provide a good estimation of past underwater UV exposure in lakes with large cladoceran carapaces preserved in sediments.

Role of scytonemin containing cyanobacteria in maintaining the fertility of desert soils

Cyanobacteria, algae, bacteria, mosses, lichens, fungi growing on a matrix of extracellular polymeric substances and soil particles forms Biological soil crusts (BSCs) (Johansen 1993). Biological soil crusts (BSCs) are usually found in arid and semiarid areas of various geographical regions (Belnap et al. 2001). Desert microalgae play several important roles such as:

a) Primary colonization and stabilization of desert soil surface (Painter 1993; Mazor et al. 1996; Hu et al. 2002).

b) Contribute to water holding properties of the soils, favouring the growth of the microbial community residing in BSCs (Rossi et al. 2012).

c) In accelerating N and C cycles by fixing nitrogen and increasing the microbial biomass in desert soils (Acea et al. 2003; Hawkes 2003) and

d) Most importantly in the protection of the phototrophic microorganisms residing in the lower soil levels (Bowker et al. 2002; Quesada et al. 1999).

The highest amount of microalgal species with high concentration of UV-shielding pigments have been found in the upper BSC layers, suggesting a role of these microalgae. Terrestrial microalgae growing in desert regions are often subject to intense UV-B irradiance (Chen et al. 2009). The biologically effective doses of UV-B radiation can penetrate deep into the surface soil and cause reactive oxygen species (ROS) generation in autotrophic organisms and thus may affect species composition and vertical distribution in BSCs (Bowker et al. 2002; Hu et al. 2003). These increased ROS can easily destroy proteins, DNA, and other biological molecules and consequently may affect growth and reproduction, survival, photosynthetic energy harvesting enzymes, and the content of photosynthetic pigments (He and Häder 2002; Apel and Hirt 2004; Holzinger and Lütz 2006).

Desertification is a serious environmental problem all over the world, since cyanobacterial inoculation might be a suitable and effective technique to recover BSCs and restore the desert ecological system in desertified areas, more and more attentions have been given to study the use of this technique (Chen et al. 2006; Navarro-González and Rainey 2003; Wang et al. 2009). However, one of the main problems in the use of this technique is the low survival rates of the microalgae utilized as inocula onto the unconsolidated sand (Chen et al. 2006).

Chen et al. (2012) aimed at selecting a good candidate for being used in the inoculation of desert soils by investigating the different responses to UV-B radiation of three cyanobacteria, *Microcoleus vaginatus*, *Nostoc* sp., *Scytonema javanicum*, and one green alga, *Chlorellavulgaris* isolated from naturally occurring BSCs. In this study, *Nostoc*, *S. javanicum*, and *M. vaginatus* showed a high absorption in the UV-B region, possibly due to the synthesis of UV protectants (Soule et al. 2007). *S. javanicum*, *Nostoc* sp., the motile *M.vaginatus*, and *C. vulgaris* showed different capability to survive the intense UV radiation in desert regions due to different protection and repair mechanisms. *Nostoc* sp. was found to be the most suitable species utilized in the inoculation of desert sandy soils initiating their primary colonization. Therfore, microalgae containing photoprotective compounds like scytonemin and MAAs against UV-B radiation can be utilized for inoculating desert soils (Chen et al. 2013). UV-tolerant cyanobacteria in desert environments, are principally composed of *Scytonema* and *Nostoc* species (Powell et al. 2013). These cyanobacteria live on the soil surface and possess direct adaptive mechanisms to the desert environment, particularly to temperature fluctuations and wet/dry cycles. They have developed the ability to reversibly activate their metabolism, limiting photosynthesis and growth to wet periods when the cells are rehydrated. During hot, dry periods, the cells enter into a quiescent state. Although photosystems I and II (PSI, PSII) are damaged during high light and dryness, in vitro studies have shown that PS I and PSII are self-repaired and operational within minutes of the cell's rehydration (Harel et al. 2004). This ability of the cyanobacterial cell to recover from photoinhibition almost immediately after rehydration is paramount to their survival. The cyanobacteria on desert surfaces also protect their photosynthetic mechanisms and genetic material with UV-protective pigments. Cyanobacteria developed protective pigments during a time when UV exposure was far greater than present. In terms of ozone protection, ancient taxa of cyanobacteria survived early environmental times when atmospheric oxygen was 10^{-5} that of its present concentration (Berkner and Marshall 1965). Cyanobacteria evolved pigments, scytonemin and mycosporine-like amino acids, which absorb potentially lethal doses of UV radiation (Proteau et al. 1993; Soule et al. 2009). Scytonemin and its derivatives occur at varying frequency in cyanobacterial species. In stratiform, epithilic crust mats, the heavily pigmented cyanobacteria on top protect deeper microorganisms (Powell et al. 2013). Despite the cyanobacteria's impressive adaptations, they remain vulnerable to anthropogenic disturbance. Recovery times for cyanobacteria disrupted by human activities are considered to be measured in centuries (Belnap 2003)

Role of scytonemin in chemosystematic evaluation of cyanobacteria

Scytonemin can be used as biomarker which remains conserved, even during long-term cultivation, suggesting that it can be used as chemotaxonomic marker in taxonomic and phylogenetic studies of *Scytonema* (Asencio and Hoffmann

2013). Forty unicyanobacterial cultures of the genus *Scytonema*, were isolated from terrestrial habitats from different parts of world (Europe, America, Southeast Asia, Australia and Africa). The pigments detected were four phycobiliproteins (allophycocyanin, phycocyanin, phycoerythrocyanin and phycoerythrin), scytonemin, six carotenoids (β-carotene, canthaxanthin, echinenone, isozeaxanthin, myxoxanthophyll and zeaxanthin) and 14 different mycosporines, including shinorine, stanierine and asterina-330. Cluster analyses indicated three groups among the isolates which were not related to their biogeographical origins. The first was composed of isolates with allophycocyanin and phycocyanin; β-carotene, canthaxanthin, echinenone and zeaxanthin and scytonemin but no MAAs. The second group comprised isolates containing allophycocyanin, phycocyanin and phycoerythrin; β-carotene, canthaxanthin, echinenone, isozeaxanthin and myxoxanthophyll and also both MAAs and scytonemin. The third group comprised isolates with allophycocyanin, phycocyanin and phycoerythrocyanin, β-carotene, canthaxanthin and echinenone, shinorine-type MAA, but no scytonemin and the organisms grew in dimly lit places. Multivariate statistical analyses were performed to identify compounds with especial value for taxonomy. Allophycocyanin, phycocyanin, phycoerythrin and phycoerythrocyanin; scytonemin; isozeaxanthin, myxoxanthophyll andzeaxanthin; and shinorine, asterina-330 and mycosporine 13, showed statistically significant variations among the three groups of *Scytonema* unicyanobacterial cultures. These results along with the fact that the biomarkers studied were conserved, even during long term cultivation, suggest that these compounds can be used as chemotaxonomic markers in taxonomic and phylogenetic studies of *Scytonema*.

Role of scytonemin in space research and geo sciences

Scytonemin can be treated as biosignature of life in the planets such as Mars. Presence of jarosite and related sulphates, which have recently been identified by the NASA Mars exploration rover "Opportunity" in the El Capitan region of Meridiani Planum on Mars, makes the Rio Tinto site as a terrestrial Mars analogue. Acidophilic microbial action makes the colour of the water deep blood-red in Rio Tinto, where the pH varies from 1.5 to 3.0 and the water is rich in iron and sulfur.

With the help of recent Raman spectroscopic characterisation of the mineral phases of the Rio Tinto system, the biological components found in several specimens of deposited minerals and near the waterside , were studied (Edwards et al. 2006). Carotenoids, scytonemin and mycosporine-like aminoacids, were the key biosignatures found, which are indicative of the biological colonisation of exposed mineral substrates. This information would help in targeting Martian sites using a miniaturized Raman instrument, where the biosignatures of relict or extant life could remain in the geological record. Hence presence of compounds such as carotenoids, scytonemin and mycosporine-like amino acids in geological records can be treated as indicators of life in space. The recognition that scytonemin, the radiation protectant pigment produced by extremophilic cyanobacterial colonies in stressed terrestrial environments, is a key biomarker for extinct or extant life preserved in geological scenarios is critically important for the detection of life signatures by remote analytical instrumentation on planetary surfaces and subsurfaces. Scytonemin and its reduced form have been both isolated experimentally and the Raman spectrum for scytonemin has been assigned and characterised experimentally both in extracts and in living extremophilic cyanobacterial colonies (Varnali and Edwards 2013). Scytonemin is recognised as a key biomarker molecule for terrestrial organisms in stressed environments. DFT calculations for scytonemin, reduced scytonemin and the new structure modelled and proposed for oxidised scytonemin are reported along with their Raman spectroscopic data and K_{max} UV-absorption data obtained theoretically. Comparison of the vibrational spectroscopic assignments allows the three forms of scytonemin to be detected and identified and assist not only in the clarification of the major features in the experimentally observed Raman spectral data for the parent scytonemin but also support a protocol proposed for their analytical discrimination. This study provide a basis for the search for molecules of this type in future astrobiological missions of exploration and the search for extinct and extant life terrestrially. The ExoMars mission to seek life signatures on Mars is just one experiment that will rely upon the detection of molecules such as scytonemin in the Martian regolith. Following a detailed structural analysis of the parent scytonemin, first time a similar analysis of several of its methoxy derivatives that have recently been extracted from stressed cyanobacteria was done (Varnali and Edwards 2013). *Ab initio* calculations were carried out to determine the most stable molecular configurations, and the implications of the structural changes imposed by the methoxy group additions on the spectral characteristics of the parent molecule were discussed. The calculated electronic absorption bands of the derivative molecules revealed that their capability of removing UVA wavelengths is removed while preserving the ability to absorb the shorter wavelength UVB and UVC radiation, in contrast to scytonemin itself. This is indicative of a special role for these molecules in the protective strategy of the cyanobacterial extremophiles. Iron (III) oxides, which offer an additional UV-protecting facility to subsurface biological colonization and banded iron formations with zones of iron depletion in rock matrices, have attracted attention with special interest in the mobilisation and transportation of iron compounds through the rock. An iron-scytonemin complex could facilitate the movement of iron through the subsurface rock as part of extremophilic survival strategy (Varnali and Edwards 2013). The predicted Raman wave numbers for the proposed scytonemin complex of iron (III) are derived computationally using DFT calculations. Comparison of the experimentally observed Raman spectra of scytonemin with the theoretically predicted Raman spectra of the iron-scytonemin complex showed that the latter may be discriminated and the expected characteristic bands are reported in relation to structural changes that are effected upon complexation (Varnali and Edwards 2013).

This information will inform the future search for experimental evidence for an iron-scytonemin complex, which has not been recognised hitherto and which could provide a novel biosignature for the extremophilic colonization of terrestrial iron-rich geological matrices. Such a terrestrial scenario would be potentially of significance for the remote robotic analytical exploration of the iron-rich surface and immediate subsurface of Mars.

Conclusion

Scytonemin is not only a very efficient photoprotective compound synthesized by cyanobacteria for providing protection against lethal UV radiation but is also a multipurpose biological compound with numerous applications (Fig. 7). In addition to its important role in cyanobacterial adaptation, this pigment certainly play a vital role in survival of microbial communities exposed to high solar radiation (Proteau et al. 1993). The high concentration of scytonemin in many cyanobacterial sheaths might be providing significant protection to other microorganisms living within and beneath the upper layer of sheathed cyanobacteria (Powell et al. 2013).

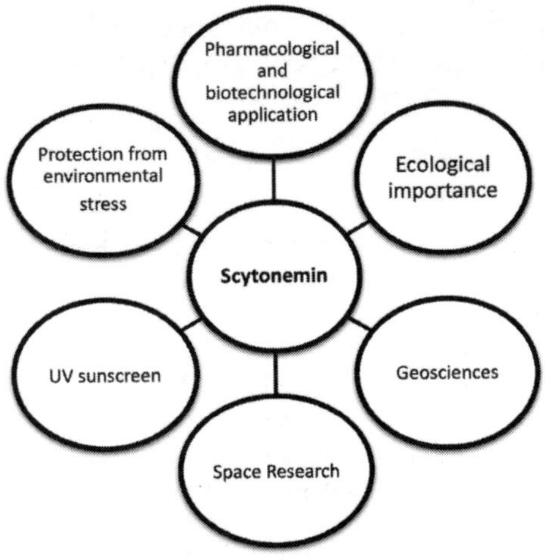

Fig. 7: Various applications of scytonemin

Scytonemin has also been identified and characterized as an antiproliferative pharmacophore that inhibits cell cycle kinases (Stevenson et al. 2002a). Reduced scytonemin isolated from *Nostoc commune* plays role in induction of autophagic cell death in Human T-lymphoid cell line Jurkat cells (Itoh et al. 2013). Because of the long term stability (Garcia-Pichel et al. 1992) of scytonemin, this compound can be used for understanding the evolutionary history of life in palaeobotanical studies in terrestrial as well as aquatic ecosystems. Attempts are being made to reconstruct the historical ozone and UV-B amount for periods prior to the modern instrumental records and prior to human impacts by analyzing the UV-absorbing pigments in herbarium specimens over a period of upto ~100 years (Huttunen et al. 2005). Cyanobacteria containing photoprotective compounds such as scytonemin can be used for inoculating desert soils. Scytonemin can be used as biosignature of life in space especially on planets like Mars and is an important biomarker to be used in geosciences and space research. Being a highly conserved and stable biological compound, scytonemin can be used in chemosystematic evaluation of cyanobacteria. Extensive work is going on to explore the ecological, industrial and pharmaceutical importance of scytonemin which shows potent anti-inflammatory and anti-proliferative properties, reliable biosignature of life in space, combined with UV protection properties, but still much more work is needed to explore potentials of scytonemin as a multipurpose pharmacophore.

Acknowledgements

Jainendra Pathak and Vinod K. Kannaujiya are thankful to CSIR, New Delhi, India for the financial support in the form of fellowships. The work was also partially supported by Department of Science and Technology sponsored project (No. SR/WOS-A/LS-140/2011) sanctioned to Richa. Arun S. Sonker and Rajneesh are thankful to UGC-CSIR and DBT respectively for the research fellowships.

References

Acea MJ, Prieto-Fernandez A and Diz-Cid N (2003) Cyanobacterial inoculation of heated soils: effect on microorganisms of C and N cycles and on chemical composition in soil surface. Soil Biol Biochem 35: 513–524.

Adhikary SP and Sahu JK (1998) UV protecting pigments of the terrestrial cyanobacterium *Tolypothrix byssoidea*. J Plant Physiol 153: 770-773.

Asencioa AD and Hoffmann L (2013) Chemosystematic evaluation of the genus *Scytonema* (Cyanobacteria) based on occurrence of phycobiliproteins, scytonemin, carotenoids and mycosporine-like amino acid compounds. Eur J Phycol 48: 331–344.

Balskus EP, Case RJ and Walsh CT (2011) The biosynthesis of cyanobacterial sunscreen scytonemin in intertidal microbial mat communities. FEMS Microbiol Ecol 77: 322–332.

Balskus EP, Walsh CT (2008) Investigating the initial steps in the biosynthesis of cyanobacterial sunscreen scytonemin. J Am Chem Soc 130: 15260-15261.

Baracaldo PS, Hayes PK and Blank CE (2005) Morphological and habitat evolution in the cyanobacteria using a compartmentalization approach. Geobiology 3: 145–65.

Belnap J, Kaltenecker JH, Rosentreter R and Williams J (2001) Biological soil crusts: ecology and management. United States Department of the Interior Bureau of Land Management, Technical Reference: 1730–1732

Berkner LV, Marshall LC (1965) On the origin and rise of oxygen concentration in the Earth's atmosphere. J Atmos Sci 22: 225-261.

Bhandari R and Sharma PK (2006) High-light-induced changes on photosynthesis, pigments, sugars, lipids and antioxidant enzymes in freshwater (*Nostoc spongiaeforme*) and marine (*Phormidium corium*) cyanobacteria. Photochem Photobiol 82: 702–710.

Björn LO (2007) Stratospheric ozone, ultraviolet radiation and cryptogams. Biol Conservat 135: 326–333.

Bowker MA, Reed SC, Belnap J and Phillips SL (2002) Temporal variation in community composition, pigmentation and F_v/F_m of desert cyanobacterial soil crusts. Microb Ecol 43: 13–25.

Brenowitz S and Castenholz RW (1997) Long-term effects of UV and visible irradiance on natural populations of a scytonemin-containing cyanobacterium (*Calothrix* sp.). FEMS Microbiol Ecol 24: 343–352.

Britt AB (1995) Repair of DNA damage induced by ultraviolet radiation. Plant Physiol 108: 891-896.

Büdel B, Karsten U and Garcia-Pichel F (1997) Ultraviolet-absorbing scytonemin and mycosporine-like amino acid derivatives in exposed, rock-inhabiting cyanobacterial lichens. Oecologia 112: 165-172.

Bultel-Poncé V, Felix-Theodose F, Sarthou C, Ponge JF and Bodo B (2004) new pigment from the terrestrial cyanobacterium *Scytonema* sp. collected on the MitrakaInselberg, French Guyana. J Nat Prod 67: 678-681.

Cadoret JC, Rousseau B, Perewoska I, Sicora C, Cheregi O, Vass I and Houmard J (2005) Cyclic nucleotides, the photosynthetic apparatus and response to a UV-B stress in the cyanobacterium *Synechocystis* sp. PCC 6803. J Biol Chem 280: 33935–33944.

Castenholz RW (1997) Multiple strategies for UV tolerance in cyanobacteria. Spectrum 10: 10-16

Castenholz RW (2004) Phototrophic bacteria under UV stress, in: J. Seckbach (Ed.), Origins: Genesis, Evolution and Diversity of Life, Kluwer Academy Publisher, Dordrecht, pp. 445–461.

Castenholz RW and Garcia-Pichel F (2000) Cyanobacterial responses to UV radiation, in: B.A. Whitton, M. Potts (Eds.), Ecology of Cyanobacteria: Their Diversity in Time and Space, Kluwer Academy Publisher. Dordrecht, pp. 591–611.

Chen J, Zhao L, Xu J, Yang R, He S and Yan J X (2013) Determination of oxidized scytonemin in *Nostoc commune* Vauch cultured on different conditions by high performance liquid chromatography coupled with triple quadrupole mass spectrometry. Appl Phycol 25: 1001–1007.

Chen L, Deng S, Philippis RD, Tian W, Wu H and Wang J (2013) UV-B resistance as a criterion for the selection of desert microalgae to be utilized for inoculating desert soils. J Appl Phycol 25: 1009–1015.

Chen LZ, Wang GH, Hong S, Liu A, Li C and Liu YD (2009) UV-B induced oxidative damage and protective role of exopolysaccharides in desert cyanobacterium *Microcoleus vaginatus*. J Integr Plant Biol 51: 194–200.

Cheng K, Zhao Y, Du X, Zhang Y, Lan S and Shi Z (2007) Solar radiation driven decay of cyanophage infectivity and photoreactivation of the cyanophage by host cyanobacteria. Aq Microb Ecol 48: 13–18.

Cidya S. Grant and J.W. Louda (2013) Scytonemin-imine, a mahogany-colored UV/Vis sunscreen of cyanobacteria exposed to intense solar radiation. Org Geochem 65: 29–36.

Crutzen PJ (1992) Ultraviolet on the increase. Nature 356: 104–105.

Dillon JG, Tatsumi CM and Tandingum PG and Castenholz RW (2002) Effect of environmental factors on the synthesis of scytonemin, a UV screening pigment, in a cyanobacterium (*Chroococcidiopsis* sp.). Arch Microbiol 177: 322-331.

Edwards HGM, Vandenabeele P, Jorge-Villar SE, Carter EA and Perez FR Hargreaves MD (2006) The Rio Tinto Mars Analogue site: An extremophilic Raman spectroscopic study. Spectrochim Acta A 68: 1133–1137

Edwards HGM, Vandenabeele P, Jorge-Villar SE, Carter EA, Perez FR, Hargreaves MD (2007) The Rio Tinto Mars Analogue site: An extremophilic Raman spectroscopic study. Spectrochim Acta A 68: 1133–1137.

Ehling-Schulz M, Bilger W and Scherer S (1997) UV-B induced synthesis of photoprotective pigments and extracellular polysaccharides in the terrestrial cyanobacterium *Nostoc commune*. J Bacteriol 179: 1940-1945.

Fleming ED and Castenholz RW (2007) Effects of periodic desiccation on the synthesis of the UV-screening compound, scytonemin, in cyanobacteria. Environ Microbiol 9: 1448- 1455.

Fleming ED and Castenholz RW (2008) Effects of nitrogen source on the synthesis of the UV-screening compound, scytonemin, in the cyanobacterium *Nostoc punctiforme* PCC 73102. FEMS Microbiol Ecol 63: 301-308.

Fulton JM, Arthur MA and Freeman KH (2012) Subboreal aridity and scytonemin in the Holocene Black Sea. Org Geochem 49: 47–55.

Garcia-Pichel F and Castenholz RW (1991) Characterization and biological implication of scytonemin, a cyanobacterial sheath pigment. J Phycol 27: 395-409.

Garcia-Pichel F, Sherry ND and Castenholz RW (1992) Evidence for a UV sunscreen role of the extracellular pigment scytonemin in the terrestrial cyanobacterium *Chlorogloeopsis* sp. Photochem Photobiol 56: 17-23.

Gröniger A, Sinha RP, Klisch M and Häder D-P (2000) Photoprotective compounds in cyanobacteria, phytoplankton and macroalgae-a database. J Photochem Photobiol B: Biol 58: 115–122.

Häder D-P, Kumar HD, Smith RC and Worrest RC (2007) Effects of solar UV radiation on aquatic ecosystems and interactions with climate change. Photochem Photobiol Sci 6: 267–285.

Häder D-P and Sinha RP (2005) Solar ultraviolet radiation-induced DNA damage in aquatic organisms: potential environmental impact. Mut Res 571: 221-233.

Hairston NG (1976) Photoprotection by carotenoid pigments in the copepod *Diaptomus nevadensis*. Proc Natl Acad Sci U.S.A 73: 971-974.

Hansson LA (2000) Induced pigmentation in zooplankton: a trade-off between threats from predation and ultraviolet radiation. Proc R Soc Lond B 267: 2327-2331.

Harel Y, Ohad I and Kaplan A (2004) Activation of photosynthesis and resistance to photoinhibition in cyanobacteria. Plant Physiol 136: 307–309.

Hargreaves A, Taiwo FA, Duggan O, Kirk SH, Ahmad SI (2007) Near ultraviolet photolysis of b-phenylpyruvic acid generates free radicals and results in DNA damage. J Photochem Photobiol B: Biol 89: 110-116.

Hawkes CV (2003) Nitrogen cycling mediated by biological soil crust and arbuscular mycorrhizal fungi. Ecology 84: 1553–1562.

He YY and Häder D-P (2002) UV-B-induced formation of reactive oxygen species and oxidative damage of the cyanobacterium *Anabaena* sp.: protective effects of ascorbic acid and N-acetyl-L-cysteine. J Photochem Photobiol B: Biol 66: 115–124.

Helbling EW, Gao K, Ai H, Ma Z and Villafan˜e VE (2006) Differential responses of *Nostoc sphaeroides* and *Arthrospira platensis* to solar ultraviolet radiation exposure. J Appl Phycol 18: 57–66.

Helms GL, Moore RE, Niemczura WP, and Patterson GML, Tomer KB and Gross ML (1988) Scytonemin A, a novel calcium antagonist from a blue - green alga. J Org Chem 53: 1298-1307.

Hessen DO (1996) Competitive trade-off strategies in Arctic Daphnia linked to melanism and UV-B stress. Polar Biol 16: 573-576.

Hessen DO, Borgeraas J, Kessler K, Refseth UH (1999) UV-B susceptibility and photoprotection of Arctic Daphnia morphotypes. Polar Res 18: 345-352.

Hodgson DA, Vyverman W, Verleyen E, Leavitt PR, Sabbe K, Squier AH and Keely BJ (2005) Late Pleistocene record of elevated UV radiation in an Antarctic lake, Earth Planet. Sci Lett 236: 765–772.

Holzinger A and Lütz C (2006) Algae and UV irradiation: effects on ultrastructure and related metabolic functions. Micron 37: 190–207.

Hu CX, Liu Y, Song L and Zhang D (2002) Effect of desert soil algae on the stabilization of fine sands. J Appl Phycol 14: 281–292.

Hu CX, Zhang D, Huang Z and Liu Y (2003) The vertical micro distribution of cyanobacteria and green algae within desert crusts and the development of the algal crusts. Plant Soil 257: 97–111.

Hunsucker SW, Tissue BM, Potts M and Helm RF (2001) Screening protocol for the Ultraviolet-protective pigment scytonemin. Anal Biochem 288: 227-230.

Huttunen S, Lappalainen NM, Turunen J (2005) UV-absorbing compounds in sub-arctic herbarium bryophytes. Environ Pollut 133: 303-314.

Itoh T , Tsuzuki R , Tanaka T , Ninomiya M , Yamaguchi Y , Takenaka H , Ando M , Tsukamasa Y and Koketsu M (2013) Reduced scytonemin isolated from *Nostoc commune* induces autophagic cell death in human T-lymphoid cell line Jurkat cells. Food Chem Toxicol 60:76–82.

Johansen JR (1993) Crypogamic crusts of semiarid and arid lands of North America. J Phycol 29: 140–147.

Kerr JB and McElroy CT (1993) Evidence for large upward trends of ultraviolet- B radiation linked to ozone depletion. Science 262: 1032–1034.

Kim ST, Sancar A (1995) Photorepair of non adjacent pyrimidine dimers by DNA photolyase. Photochem Photobiol 61: 171-174.

Kripke ML, Cox PA, Alas LG and Yarosh DB (1992) Pyrimidine dimers in DNA initiate systemic immune suppression in UV-irradiated mice. Proc Natl Acad Sci USA 89: 7516–20.

Kylim H (1927) Über die Karotinoide Farbstoffe der Algen. Hoppe-Scyler'szeitschr. Physiol Chem 166: 33-77.

Kylim H (1937) Über die Farbstoffe und die Farbe der cyanophyceen. Fysiogr Sällsk Förhandl 7: 131-158.

Lubin D and Jensen EH (1995) Effects of clouds and stratospheric ozone depletion on ultaviolet radiation trends, Nature 377: 710–713.

Matsui K, Nazifi E, Hirai Y, Wada N, Matsugo S, and T Sakamoto (2012). The cyanobacterial UV-absorbing pigment scytonemin displays radicals scavenging Activity. J Gen Appl Microbiol 58: 137 144

Mazur-Marzec H, Meriluoto J and M. Plin´ski (2006) The degradation of the cyanobacterial hepatotoxin nodularin (NOD) by UV radiation, Chemosphere 65: 1388–1395.

Nägeli C (1849) Gattungeneinzelliger Algen, physiologisch und Systematisch bearbeitet Neue Denkschrift. Allg Schweiz Nat Ges 10: 1-138.

Nägeli C and Schwenderer S (1877). In: Das Mikroskop, Leipzig: Willhelm Engelmann. 2nd edi.

Nevalainen L and Rautio M (2014) Spectral absorbance of benthic cladoceran carapaces as a new method for inferring past UV exposure of aquatic biota. Quaternary Sci Rev 84: 109-115.

Painter TJ (1993) Carbohydrate polymers in desert reclamation: the potential of microalgal biofertilizers. Carbohyd Polym 20: 77–86.

Pakker H, Beekman CAC and Breeman AM (2000) Efficient photoreactivation of UVBR-induced DNA damage in the sublittoral macroalga *Rhodymenia pseudopalmata* (Rhodophyta). Eur J Phycol 35: 109-114.

Peak MJ, Peak JG, Moehring P and Webb RB (1984) Ultraviolet action spectra for DNA dimer induction, lethality and mutagenesis in *Escherichia coli* with emphasis on the UV-B region. Photochem Photobiol 40: 613–20.

Pentecost A (1993) Field relationships between scytonemin density, growth and irradiance in cyanobacteria occurring in low illumination regimes. Micro Ecol 26: 101-110.

Powell JT, Chatziefthimiou AD, Banack SA, Cox PA, Metcalf JS (2015) Desert crust microorganisms, their environment and human health. J Arid Environ 112: 127-133.

Proteau PJ, Gerwick WH, Garcia-Pichel F and Castenholz RW (1993) The structure of scytonemin, an ultraviolet sunscreen pigment from the sheath of cyanobacteria. Experimentia 49: 825- 829.

Quaite FE, Sutherland BM and Sutherland JC (1992) Action spectrum for DNA damage in alfalfa lowers predicted impact of ozone depletion. Nature 358: 577–8.

Quesada A, Vincent WF, Lean DRS (1999) Community and pigment structure of Arctic cyanobacterial assemblages: the occurrence and distribution of UV-absorbing compounds. FEMS Microbiol Ecol 28: 315–323.

Quesada A, Vincent WF. (1997) Strategies of adaptation by Antarctic cyanobacteria to ultraviolet radiation. Eur J Phycol 32: 335–42.

Rastogi RP and Incharoensakdi A (2014) Characterization of UV-screening compounds, mycosporine-like amino acids and scytonemin in the cyanobacterium *Lyngbya* sp. CU2555. FEMS Microbiol Ecol 87: 244-256.

Rastogi RP and Sinha RP (2009) Biotechnological and industrial significance of cyanobacterial secondary metabolites. Biotechnol Adv 27: 521-539.

Rastogi RP, Sinha RP, Incharoensakdi A (2013) Partial characterization, UV-induction and photoprotective function of sunscreen pigment, scytonemin from *Rivularia* sp. HKAR-4. Chemosphere 93: 1874–1878.

Rath J and Adhikary SP (2007) Response of the estuarine cyanobacterium *Lyngbya aestuarii* to UV-B radiation. J Appl Phycol 19: 529–536.

Rath J, Mandal S and Adhikary SP (2012) Salinity induced synthesis of UV-screening compound scytonemin in the cyanobacterium *Lyngbya aestuarii*. J Photochem Photobiol B: Biol 115: 5-8.

Rautio M and Korhola A (2002a) Effects of ultraviolet radiation and dissolved organic carbon on the survival of subarctic zooplankton. Polar Biol 25: 460-468.

Rautio M, Bonilla S, Vincent WF (2009) UV photoprotectants in arctic zooplankton. Aquat Biol 7: 93-105.

Rhode SC, Pawlowski M, Tollrian R (2001) The impact of ultraviolet radiation on the vertical distribution of zooplankton of the genus *Daphnia*. Nature 412: 69-72.

Rossi F, Potrafka RM, Garcia-Pichel F and De Philippis R (2012) The role of the exopolysaccharides in enhancing hydraulic conductivity of biological soil crusts. Soil Biol Biochem 46: 33–40.

Sergeev VN, Gerasimenko LM and Zavarzin GA (2002) The proterozoic history and present state of cyanobacteria. Microbiology 71: 623-37.

Sinclair C and Whitton BA (1977) Influence of nutrient deficiency on hair formation in the Rivulariaceae. Br Phycol J 12: 297-313.

Singh SP, Häder D-P and Sinha RP (2010) Cyanobacteria and ultraviolet radiation (UVR) stress: mitigation strategies. Ageing Res Rev 9: 79-90.

Sinha RP and Häder D-P (1996) Photobiology and ecophysiology of rice field cyanobacteria. Photochem Photobiol 64: 887-896.

Sinha RP and Häder D-P (2002) UV-induced DNA damage and repair: a review. Photochem Photobiol Sci 1: 225-236.

Sinha RP and Häder D-P (2008) UV protectants in cyanobacteria. Plant Sci 174: 278–89

Sinha RP, Dautz M and Häder D-P (2001) A simple and efficient method for the quantitative analysis of thymine dimers in cyanobacteria. phytoplankton and macroalgae. Acta Protozool 40: 187-95.

Sinha RP, Klisch M, Gröniger A and Häder D-P (1998) Ultraviolet absorbing/screening substances in cyanobacteria, phytoplankton and macroalgae. Photochem Photobiol 47: 83-94.

Sinha RP, Klisch M, Vaishampayan A and Häder D-P (1999) Biochemical and spectroscopic characterization of the cyanobacterium *Lyngbya* sp. inhabiting Mango (*Mangifera indica*) trees: Presence of an ultraviolet-absorbing pigment, Scytonemin. Acta Protozool 38: 291-298.

Sinha RP, Kumar A, Tyagi MB and Häder D-P (2005) Ultraviolet-B-induced destruction of phycobiliproteins in cyanobacteria. Physiol Mol Biol Plants 11: 313-319.

Smith RC, Pre´zelin BB, Baker KS, Bidigare RR, BoucherNP, Coley T, Karentz D, MacIntyre S, Matlick HA, Menzies D, Ondrusek M, Wan Z and Waters KJ (1992) Ozone depletion: ultraviolet radiation and phytoplankton biology in Antarctic waters. Science 255: 952-959.

Solheim B, Zielke M, Bjerke JW and Rozema J (2006) Effects of enhanced UVB radiation on nitrogen fixation in arctic ecosystems. Plant Ecol 182: 109-118.

Sorrels CM, Proteau PJ and Gerwick WH (2009) Organization, evolution, and expression analysis of the biosynthetic gene cluster for scytonemin, a cyanobacterial UV-absorbing pigment. Appl Environ Microbiol 75: 4861-4869.

Soule T, Garcia-Pichel F and Stout V (2009a) Gene expression patterns associated with the biosynthesis of the sunscreen scytonemin in *Nostoc punctiforme* ATCC 29133 in response to UVA radiation. J. Bacteriol. 191: 4639-4646.

Soule T, Palmer K, Gao Q, Potrafka RM, Stout V and Garcia-Pichel F (2009b) A comparative genomics approach to understanding the biosynthesis of the sunscreen scytonemin in cyanobacteria. BMC Genomics 10: 336-345.

Soule T, Stout V, Swingley WD, Meeks JC and Garcia-Pichel F (2007) Molecular genetics and genomic analysis of scytonemin biosynthesis In *Nostoc punctiforme* ATCC 29133. J Bacteriol 189: 4465-4472.

Stein B, Rahmsdorf HJ, Steffen A, Litfin M and Herrlich P (1989) UV-induced DNA damage is an intermediate step in UV-induced expression of human immunodeficiency virus type 1 collagenase C-fos and metallothionein. Mol Cell Biol 9: 5169–81.

Stevenson CS, Capper EA, Roshak AK, Marquez B, Eichman C,Jackson JR, Mattern M, Gerwick WH, Jacobs RS and Marshall LA (2002a) The identification and characterization of the marine natural product scytonemin as a novel antiproliferative pharmacophore. J Pharmacol Exp Ther 303: 858-866.

Stevenson CS, Capper EA, Roshak AK, Marquez B, Grace K, Gerwick WH, Jacobs RS and Marshall LA (2002b) Scytonemin- a marine natural product inhibitor of kinases key in hyperproliferative inflammatory diseases. Inflamm Res 51: 112-114.

Thajuddin N, Subramanian G (2005) Cyanobacterial biodiversity and potential applications in biotechnology. Curr Sci 89: 47–57.

Tirkey J and Adhikary SP (2005) Cyanobacteria in biological soil crusts of India. Curr Sci 89: 515–521.

Vaishampayan A, Sinha RP and Häder D-P (1998) Use of genetically improved nitrogen-fixing cyanobacteria in rice paddy fields: prospects as a source material for engineering herbicide sensitivity and resistance in plants. Bot Acta 111: 176–190.

Vaishampayan A, Sinha RP, Häder D-P, Dey T, Gupta AK., Bhan U and Rao AL (2001) Cyanobacterial biofertilizers in rice agriculture. Bot Rev 67: 453–516.

Varnali T and Edwards HGM (2014) Reduced and oxidised scytonemin: Theoretical protocol for Raman spectroscopic identification of potential key biomolecules for astrobiology. Astrobiology 117: 72-77.

Varnalia T and Edwards HGM (2013) A potential new biosignature of life in iron-rich extreme environments: An iron (III) complex of scytonemin and proposal for its identification using Raman spectroscopy. Planet Space Sci 82–83:128–133.

Vincent WF and Quesada A (1994) Ultraviolet radiation in Antarctica: measurements and biological effects. In: Weiler CS, Penhale PA, editors. Antarctic research series, vol. 63. Washington, DC: American Geophysical Union. p. 111.

Vincent WF and Roy S (1993) Solar ultraviolet-B radiation and aquatic primary production: damage, protection, and recovery. Environ Rev 1: 1–12.

Von der Gathen P, Rex M, Harris NRP, Lucic D, Knudsen BM, Braathen GO H, DeBacker, Fabian R, Fast H, Gil M, Kyra¨E, Mikkelsen IS, Rummukainen M, Sta¨helin J and Varotsos C (1995) Observational evidence for chemical ozone depletion over the Arctic in winter 1991– 1992. Nature. 375: 131–134.

Wang W, Liu Y, Li D, Hu C and Rao B (2009) Feasibility of cyanobacterial inoculation for biological soil crusts formation in desert area. Soil Biol Biochem 41: 926–929.

Weatherhead EC and Andersen SB (2006) The search for signs of recovery of the ozone layer. Nature 441: 39–45.

Williamson CE, Olson OG, Lott SE, Walker ND, Engstrom DR and Hargreaves BR (2001a) Ultraviolet radiation and zooplankton community structure following deglaciation in Glacier Bay, Alaska. Ecology 82: 1748-1760.

Wright DJ, Smith SC, Joardar V, Scherer S, Jervis, Warren A, Helm RF and Potts M (2005) UV irradiation and desiccation modulate the three dimensional extracellular matrix of *Nostoc commune* (cyanobacteria). J Biol Chem. 280: 40271–40281.

Wu H, Gao K, Villafan˜e VE, Watanabe T and Helbling EW (2005) Effects of solar UV radiation on morphology and photosynthesis of filamentous cyanobacterium *Arthrospira platensis*. Appl Environ Microbiol 71: 5004–5013.

Wulff A, Mohlin M and Sundback K (2007) Intraspecific variation in the response of the cyanobacterium *Nodularia spumigena* to moderate UV-B radiation. Harmf Algae 6: 388–399.

Xue L, Zhang Y, Zhang T, An L and Wang X (2005) Effects of enhanced ultraviolet-B radiation on algae and cyanobacteria. Crit Rev Microbiol 31: 79–89.

11

Screening of Photosynthetic Pigments as an Indicator of Heavy Metal Stress in *Euglena gracilis*

Hoda Ahmed, Donat-Peter Häder and Peter Richter

Abstract

This study was conducted to examine the response of photosynthetic pigments to heavy metal stress in a short-term toxicity test using high performance liquid chromatography (HPLC). HPLC pigment analysis of *Euglena* has resulted in 5 separated pigments: chlorophylls *a*, *b* and three carotenoids, which are 9-cis-neoxanthin, diadinoxanthin and β-carotene. The pigment concentrations significantly decreased upon exposure to heavy metals like cadmium and copper. After exposure to copper at 37 and 56 mg l^{-1} for 24 h, the chlorophyll *a* concentration reduced to about 35–40%. Upon longer exposure time of 4 days, all copper concentrations from 0.4 to 56 mg l^{-1} showed significant effects on chlorophyll *a* concentration. Results have shown significant breakdown of photosynthetic pigments like chlorophyll *b*, β-carotene as well as 9-cis-neoxanthin at > 11 mg l^{-1} copper concentrations at all incubation times. The diadinoxanthin content was significantly inhibited after 48 h treatment with 11 and 22 mg l^{-1}, and the effect lasted for the entire experiment. Both chlorophyll *b* and β-carotene were the most sensitive pigments toward cadmium exposure and were significantly reduced even at low concentrations of 4 mg l^{-1} after 24 h of treatment. Chlorophyll a content was significantly reduced upon cadmium treatment at > 25 mg l^{-1} for 4 days. These results prove that exposure to copper and cadmium are significant stress factors of ecological relevance. It further proves that photosynthetic pigments can be used as valuable markers for heavy metal stress.

Keywords: *Euglena gracilis*; Chlorophyll *a, b*; ß-carotene; Copper; Cadmium.

Introduction

Heavy metals pollution resulting from anthropogenic activities is a major environmental problem threatening the entire biosphere. Copper (Cu) is an essential microelement for organism but it is toxic at higher concentrations. Cadmium (Cd) is a highly toxic metal to aquatic organisms. Heavy metal toxins interfere with the synthesis of photosynthetic pigments by inhibition of chlorophyll biosynthesis enzymes has been reported several times (e.g. Clijsters and Van Assche 1985; Macfarlane and Burchett 2001). HPLC separation provides an accurate method for pigments analysis. It can be successfully used for separation of a mixture of pigments and pigment derivatives, which cannot be obtained either by spectrophotometer or fluorometer methods (Jeffery et al. 1997; Wright et al. 1991). Analysis of photosynthetic pigments was used in the last decade in aquatic ecosystem research to determine water quality by indicating the composition of the phytoplankton and consequently the trophic status in the water system (Gilbert et al. 2000; Wilhelm et al. 1995).

Drexlerand and Fletcher (1981) have reported the use of inhibition of chlorophylls and carotenoids in the presence of fiuridone herbicides as a bioassay tool. Reduction of total chlorophyll concentration was observed in *Chlorella* and *Euglena gracilis* in response to Cd, Zn and Hg (De Filippis et al. 1981; De Filippis and Pallaghy 1976). A significant effect of Cd has been observed at a concentration of 2 μM on the chlorophyll content of cyanobacteria (Zhou et al. 2006). Chlorophyll *a* and *b* as well as á, ß carotene content in *Chlorella pyrenoidosa* are affected upon exposure to the heavy metal Cr (IV) (Hörcsik et al. 2006). In this context, estimation of changes in photosynthetic pigments content seems to be essential for determination of harmful effects caused by toxins on photosynthesis and consequently on organism growth and survival rate (Macfarlane and Burchett 2001). This may further support the idea of testing photosynthetic pigments as ecological end points in bioassays. This study examines the variation of the photosynthetic pigments in response to the heavy metals copper and cadmium.

Materials and Methods

Cell culture and growth conditions

All experiments were performed using an axenic culture of the freshwater green flagellate *Euglena gracilis* Klebs (strain Z), obtained from the culture collection in Göttingen,Germany. The cells were grown in a sterilized mixture of a mineral and complex medium described previously (Checcucci et al. 1976; Starr 1964) at 22°C and 20 Wm^{-2} continuous white light in stationary cultures in 100-ml Erlenmeyer flasks.

Heavy metal exposure

Young cell cultures (9 days old) and different concentrations of heavy metals solutions at a ratio (1:1) were transferred into sterile 100-ml Erlenmeyer flasks and mixed well. Cells were treated with Cu at concentrations ranging from 0.4 to 56 mg l^{-1}. Cd was used at concentrations of 0.75 to 220 mg l^{-1}. Samples were taken daily up to 4 days to determine the pigment concentrations. The cells were harvested by centrifugation and washed three times with Millipore water to remove heavy metal traces, then shortly stored in dark tubes at -80°C till pigment analysis.

HPLC analysis

Setting up the extraction method was the first essential step in pigment analysis. The efficiency of pigment extraction is controlled by a number of factors like selection of extraction solvent and time as well as handling of samples before extraction (Buffan-Dubau and Carman 2000; Jeffery et al. 1997). Methanol and acetone have been recommended for phytoplankton pigment extraction (Jeffery et al. 1997; Wright et al. 1991). Cells were harvested by centrifugation (Eppendorf centrifuge 5702, Hamburg, Germany) at 1000 x g for 10 min and pigments were extracted under dim light on ice for 45 s using ultrasound agitation (Branson Sonifer w-450, Gerhard Heinemann, Schwäbisch Gmünd) in 2 ml of 100 % methanol HPLC grade. After extraction aliquots were centrifuged at 5000 x g for 15 min to remove cell debris, and 1 ml of the supernatant was transferred into 1 ml HPLC glass tubes for analysis.

Pigments were successfully separated by a reverse-phase HPLC system (Waters 996 photodiode array detector, Waters, Milford, MA, USA; pump L-7100, Hitachi, Darmstadt, Germany) equipped with an endcappd Licrospher RP 18 column protected by a guard column (5 μm packing; 250 x 4 mm Merck, Darmstadt, Germany) and Millennium32 software. The samples (30 μl) were injected into the HPLC column through a Waters 717 plus autosampler (Waters) and the protocol of Gilbert et al. (2000) was used. Elution was done using a ternary gradient of the solvents A [0.5 M ammonium acetate/methanol: H_2O, 85:15 (v/v)], B [acetonitrile: H_2O, 90:10 (v/v)] and C (100 % ethyl acetate) running 32 min at a flow rate of 1 ml min^{-1}.

The HPLC instrument was calibrated using highly purified standards, chlorophylls *a*, *b* and β-carotene obtained from Sigma–Aldrich Germany. Pigments were identified by spectral data and retention time on the column. Pigment concentrations of samples were calculated using peak area and calibration curves from pigment standards. Pigments were determined at 660 nm for chlorophyll *a* and at 646 nm for chlorophyll *b*. Specific extinction coefficients e (100 ml g^{-1} cm^{-1}) for 9-cis-neoxanthin and diadinoxanthin were obtained from literature (Jeffery et al. 1997) and used for quantification of pigments concentrations.

Results

After examination of several solvents like methanol, ethanol and acetone etc. instant methanol extraction gave satisfactory resolution for all pigments and could be directly injected into the HPLC after filtration. HPLC analysis of *Euglena* has resulted in 5 separated pigments at different retention times which are chlorophylls (Chl *a* and *b)* and three carotenoids. The main carotenoids in *Euglena* cells are 9-cis neoxanthin, diadinoxanthin and ß-carotene (see Fig. 1 and Table 1). Absorption spectra and retention times of peaks 3, 4, 5 exactly match with the standard samples of Chl *a*, Chl *b* and β-carotene. Absorption spectra and retention times of peaks 1 and 2 are nearly identical to the references (Jeffery et al. 1997; Ston′ and Kosakowska 2002).

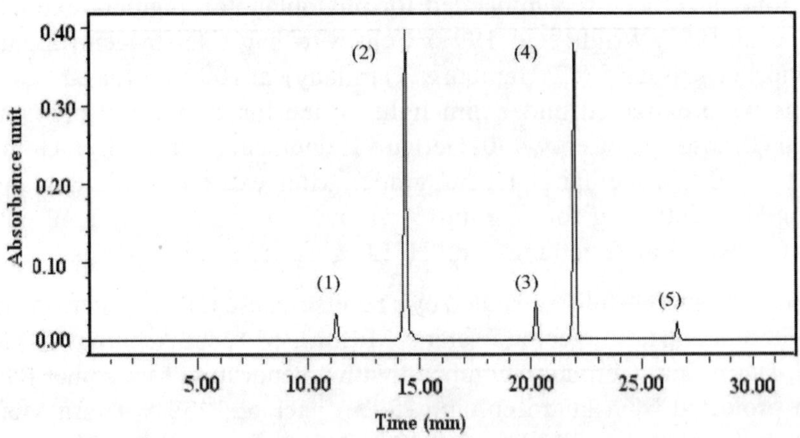

Fig. 1: HPLC chromatogram, detected at 440 nm, of pigments in *Euglena gracilis* extracted with methanol.

Table 1: Retention times (Rt) and spectral characteristics of the HPLC separated pigments.

Peak number	Pigments	Rt (min)	Absorption maxima (nm)		
1.	9-cis Neoxanthin	11.2	415	438.8	466
2.	Diadinoxanthin	14.5	422	448.5	477
3.	Chlorophyll *b*	20.7	458	646	-
4.	Chlorophyll *a*	22.3	431	663	-
5.	ß-carotene	26.6	454	480	-

Effects of copper on the photosynthetic pigments

The effect of copper on the concentration of chlorophylls and carotenoids in dependence of the incubation time are plotted in Figs. 2-6. The pigments compositions are significantly decreased upon incubation with copper concentrations of 11 mg l^{-1} or more.

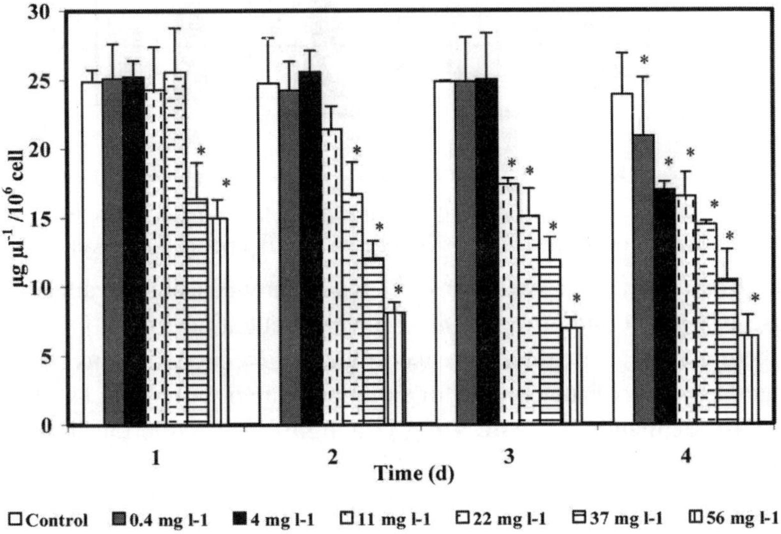

Fig. 2: Changes in chlorophyll *a* concentration in *Euglena gracilis* after exposure to different copper concentrations. Asterisks indicate significance at 0.05 levels (ANOVA).

For chlorophyll *a* (Chl *a*), the concentration declined to about 35 – 40 % of the controls after exposure to copper at 37 and 56 mg l^{-1} for 24 h, respectively. The loss in Chl *a* increased with increasing incubation times. After 2 days of treatment, a concentration of 22 mg l^{-1} began to have a significant effect on Chl *a* concentration. With longer exposure times (4 days), lower copper concentrations (0.4 and 4 mg l^{-1}) started to show a pronounced effect on Chl *a* concentration. In parallel, data from HPLC clearly showed reduction in chlorophyll *b* (Chl *b*) concentration which is time and concentration dependent (Fig. 3).

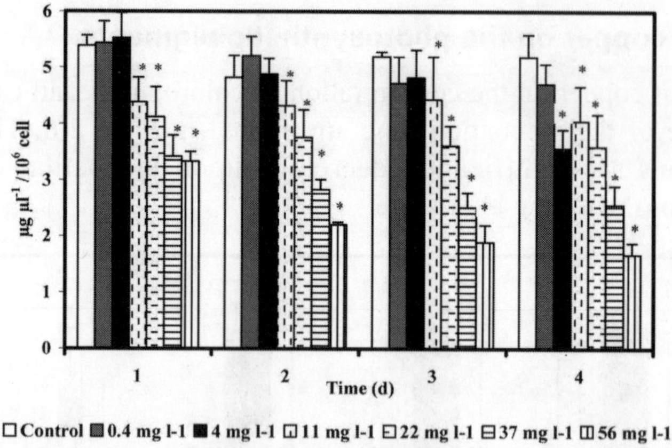

Fig. 3: Changes in chlorophyll *b* concentration in *Euglena gracilis* after exposure to different copper concentrations. Asterisks indicate significance at 0.05 levels (ANOVA).

Samples treated with concentrations >11 mg l^{-1} showed a significant ($p < 0.05$) decrease in Chl *b* content in *Euglena* for all incubation times. On the other hand, there is no significant difference in the Chl *b* concentrations between 1, 2 as well as 3 days of incubation for samples treated with 4 mg l^{-1}. However, after 4 days of incubation with 4 mg l^{-1} a significant reduction in the Chl *b* content was determined.

β-Carotene seems to be more affected by copper. Although this effect was not significant for 4 mg l^{-1} concentrations up to 3 days of incubation it was definitely significant for the higher tested concentrations during the experiment (Fig. 4). Exposure to copper concentrations at >11 mg l^{-1} for one day showed apparent effects on β-carotene content. There are some effects observed for cells treated with 0.4 mg l^{-1} after 3 and 4 days of exposure but these effects were insignificant ($p < 0.05$). In samples treated with 56 mg l^{-1} the concentration of β-carotene decreased by about 90% in comparison to the controls after 3 days of incubation.

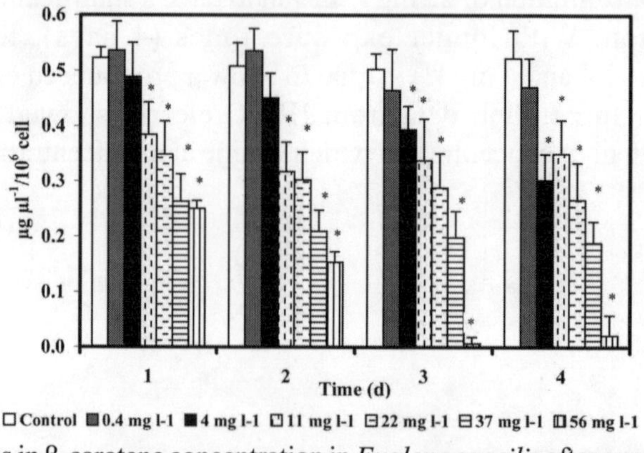

Fig. 4: Changes in β-carotene concentration in *Euglena gracilis* after exposure to different copper concentrations. Asterisks indicate significance at 0.05 levels (ANOVA).

Copper significantly ($p < 0.05$) reduced the level of xanthophylls (diadinoxanthin and 9-cis- neoxanthin) in *E. gracilis*. The toxicity of copper to xanthophylls increased with concentration and treatment duration (Figs. 5 and 6). It was observed that 37 and 56 mg l^{-1} copper significantly ($p < 0.05$) decreased the concentrations of both diadinoxanthin and 9-cis-neoxanthin at all tested exposure times.

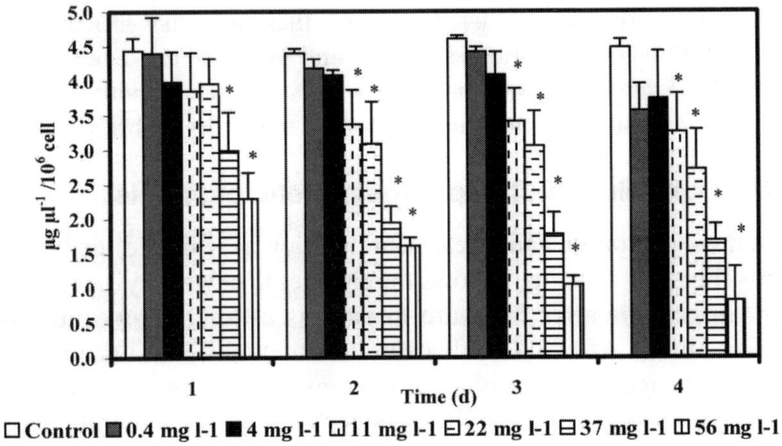

Fig. 5. Changes in diadinoxanthin concentration in *Euglena gracilis* after exposure to different copper concentrations. Asterisks indicate significance at 0.05 levels (ANOVA).

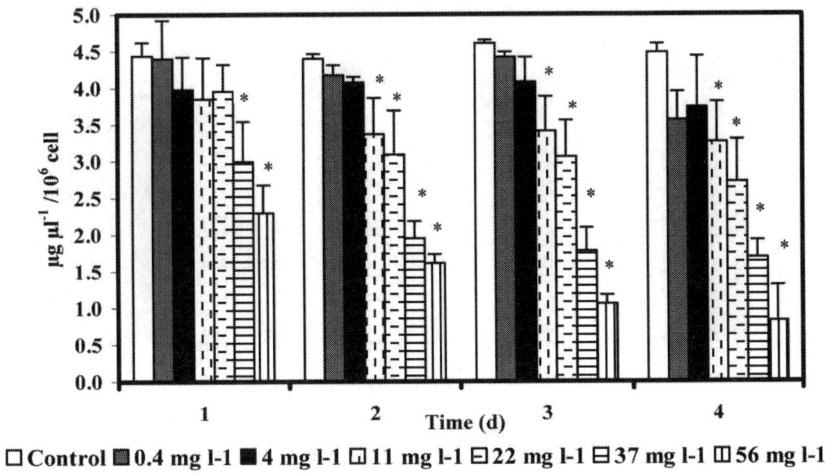

Fig. 6: Changes in 9-cis-neoxanthin concentration in *Euglena gracilis* after exposure to different copper concentrations. Asterisks indicate significant at 0.05 levels (ANOVA).

Both 11 and 22 mg l^{-1} showed drastic inhibition of the diadinoxanthin content after 48 h treatment, and this effect lasted for the entire experiment. After 3 days there were no significant differences between samples exposed to 0.4 and 4 mg l^{-1} and the control. But after 4 days these samples have shown significant effects on the xanthophylls concentrations.

Declines in the 9-cis-neoxanthin content were significant ($p < 0.05$) in samples treated with copper concentrations of >11 mg l^{-1} during the whole treatment time. In samples treated with 0.4 and 4 mg l^{-1} there was insignificant loss in the 9-cis-neoxanthin concentrations after 1, 2 and 3 days of exposure. Subsequent incubation of *Euglena* cells with Cu for 4 days revealed apparent reduction in the 9-cis-neoxanthin concentration compared to the control samples.

Effects of cadmium on the photosynthetic pigments

Cadmium concentrations and treatment duration accelerated the bleaching of pigments (Figs. 7 – 11). Fig. 7 shows the changes in chlorophyll *a* concentration in *Euglena gracilis* after exposure to various cadmium concentrations and incubation times. After 24 h all cadmium-treated cells did not show any significant change in chlorophyll *a* content. When the exposure time was extended to 2 days, chlorophyll *a* concentrations significantly ($p < 0.05$) decreased in both 100 and 200 mg l^{-1} treated cells. Interestingly, a significant decrease in concentration was observed at concentrations >37 mg l^{-1} after 3 and 4 day incubation with cadmium. The addition of 4 or 10 mg l^{-1} cadmium had no considerable effect on chlorophyll *a* content.

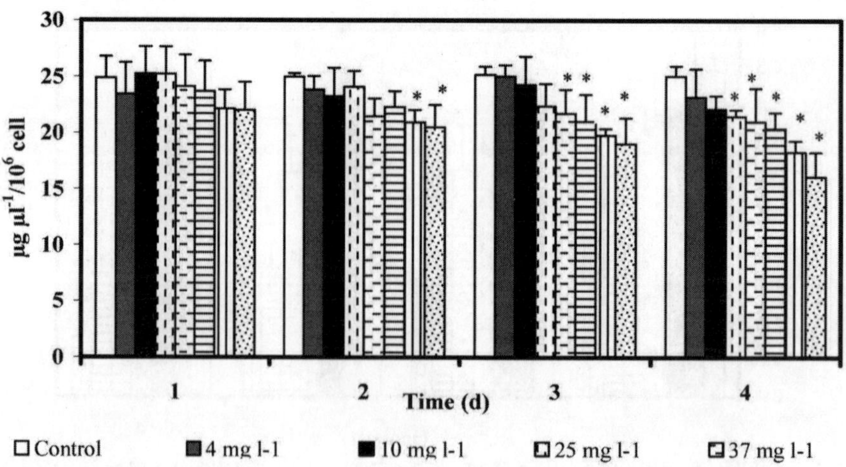

Fig. 7: Changes in chlorophyll *a* concentration in *Euglena gracilis* after exposure to different cadmium concentrations. Asterisks indicate significance at 0.05 levels (ANOVA).

In contrast, a decrease in chlorophyll *b* concentration was observed in cells treated with cadmium at concentrations of 4 mg l^{-1} and more for all incubation times (Fig. 8).

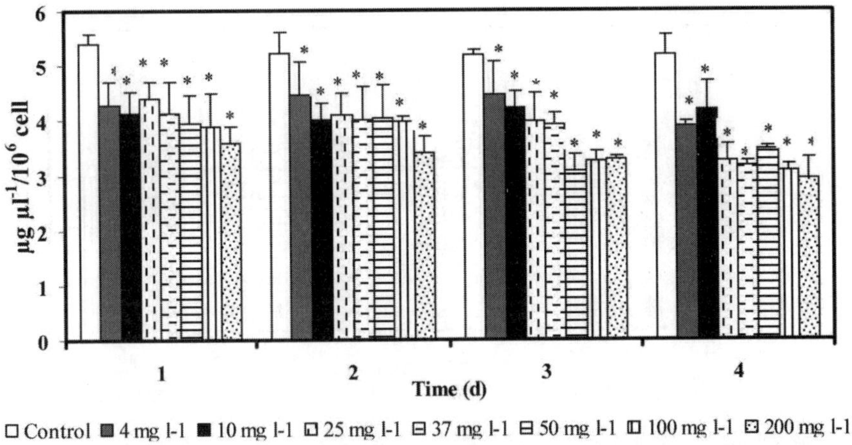

Fig. 8: Changes in chlorophyll *b* concentration in *Euglena gracilis* after exposure to different cadmium concentrations. Asterisks indicate significance at 0.05 levels (ANOVA).

The cadmium concentrations examined had negative effects on the chlorophyll *b* content which was significant ($p < 0.05$) between all treated cells and control cells at all treatment times. Although the effects, produced by different concentrations ranging from 10 to 100 mg l^{-1}; on the chlorophyll *b* concentration on the first and second day were significant, higher cadmium concentration induced similar effects on chlorophyll *b* content. A highly significant ($p < 0.05$) reduction in the chlorophyll *b* concentrations was observed after longer incubation for 3 days with concentrations ranging from 50 to 200 mg l^{-1}.

Changes in ß-carotene concentration in *E. gracilis* exposed to different concentration of cadmium are shown in Fig. 9. The ß-carotene concentration was gradually and linearly decreasing as the cadmium concentration increased and the exposure time was extended. Alterations in ß-carotene concentrations were significant at all tested concentration of cadmium for all exposure times compared to control measurements.

Increasing cadmium concentrations from 4 to 200 mg l^{-1} did not show any effect on the diadinoxanthin concentration after 1 or 2 days of incubation (Fig. 10). However, it was found that toxicity of cadmium to the diadinoxanthin content in *Euglena* became obvious after exposure to concentrations of > 25 mg l^{-1} for 3 or 4 days of incubation. The diadinoxanthin content was found to be reduced to about 25 and 35 % compared to the control upon addition of cadmium to 100 and 200 mg l^{-1}, respectively, after 3 days. The degradation of diadinoxanthin increased with these treatments with further exposure time of 4 days.

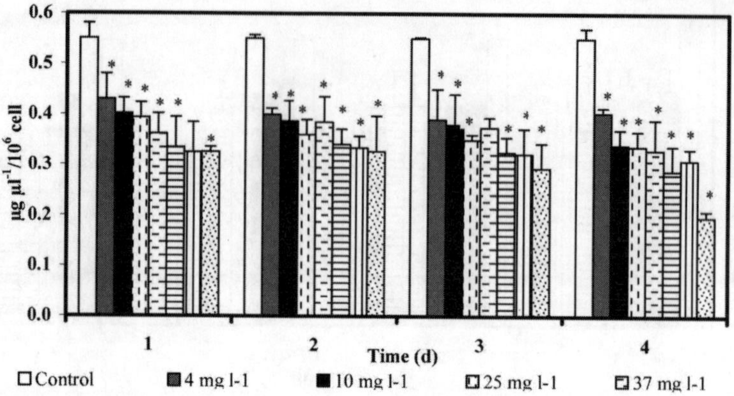

Fig. 9: Changes in β-carotene concentration after exposure to different cadmium concentrations. Asterisks indicate significance at 0.05 levels (ANOVA).

Significant declines in the 9-cis- neoxanthin content were induced by addition of cadmium at 200 mg l^{-1} after 2 days treatment duration (Fig. 11). Moreover, samples treated with 50 or 100 mg l^{-1} had definite effects on the 9-cis-neoxanthin concentration after 3 and 4 days incubation time. Lower cadmium concentrations of < 22 mg l^{-1} did not show any obvious impact on the determined 9-cis-neoxanthin concentration during the whole treatment time.

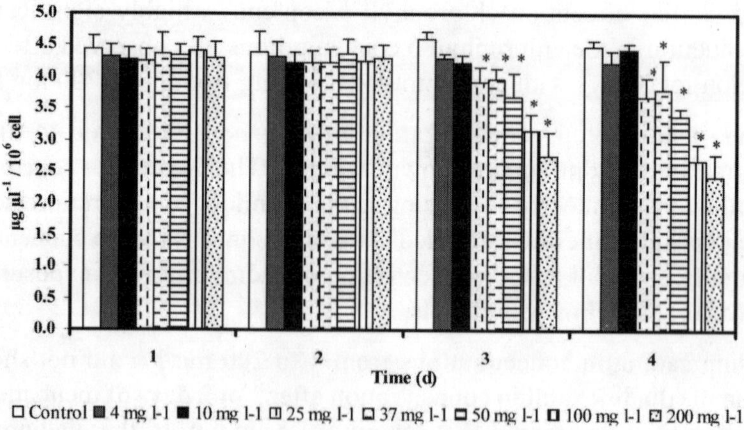

Fig. 10: Changes in diadinoxanthin concentration in *Euglena gracilis* after exposure to different cadmium concentrations. Asterisks indicate significance at 0.05 levels (ANOVA).

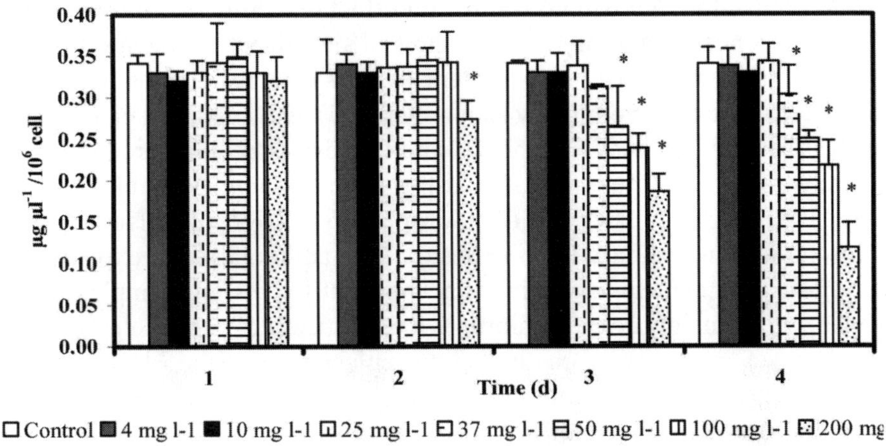

Fig. 11: Changes in 9-cis-neoxanthin concentration in *Euglena gracilis* after exposure to different cadmium concentrations. Asterisks indicate significance at 0.05 levels (ANOVA).

Discussion

HPLC has proven to be applicable in the study of phytoplankton including quantitative and qualitative assessment of photosynthetic pigments providing important information on the biomass (Wright et al. 1991), community composition using pigment fingerprint (chemotaxonomic marker, Ston′ and Kosakowska 2002), physiological state and adaptation to environmental conditions (van Leeuwe and Stefels 1998; van Leeuwe et al. 2006). HPLC provides a short-time analysis of complex samples with high resolution of different pigments without any interference between pigments of similar chemical structure. Heavy metals have been reported to disturb the physiological processes in phototrophic organisms, especially photosynthesis and chlorophyll biosynthesis (Rai et al. 1981). The current experiments especially aimed at evaluating the acute lethal toxicity of cadmium and copper on the photosynthetic pigments in *Euglena gracilis*.

In agreement with (Wright et al. 1997), methanol was used in pigment extraction using ultrasonic disruption which gave highest extraction efficiency. HPLC analysis indicated that heavy metal stress causes significant alterations in the content of photosynthetic pigments in *Euglena*. These changes in the pigment concentrations were not equal for all separated pigments. Higher copper concentrations induced a significant degradation of all measured photosynthetic pigments and at all treatment durations. The lowest concentration of 0.4 mg l^{-1} after 4 days of incubation showed significant effects on Chl *a*, diadinoxanthin and 9-cis neoxanthin but had no significant effect on Chl *b*, β-carotene. After 24 h copper concentration of 11 mg l^{-1} showed significant degradation of the photosynthetic pigments Chl *b*, β-carotene as well as 9-cis-neoxanthin. This is in contrast to the results reported by Hou et al. (2007) who found that chlorophyll *b* was less sensitive to heavy metal (Cu and Cd) exposure compared to Chl *a* and carotenoids in a 4 days bioassay using *Lemna minor*. Manankina et al. (2003) have reported greater loss of chlorophyll *b* in *Euglena* cells compared to chlorophyll *a* in response to Ni exposure. The authors tried to explain this effect of Ni treatment by assuming that the light-harvesting complexes are degraded, which mainly contain chlorophyll *b*. Influence of copper on pigments is based on several mechanisms including replacement of the Mg ion in the chlorophyll molecules and binding to active sites inhibiting electron transfer in the chloroplast (Hou et al. 2007).

Both chlorophyll *b* and β-carotene concentrations were the most sensitive pigments toward cadmium treatment and were significantly reduced even at a low concentration of 4 mg l^{-1} after 24 h of exposure. In contrast, the diadinoxanthin content was insignificantly affected upon treatment with cadmium at all tested concentrations after one or two days of incubation. The effect of cadmium on diadinoxanthin concentration was visible only after 3 days of treatment. The content of chlorophyll *a* and 9-cis-neoxanthin were unchanged by cadmium addition for 24 h. After 4 days of incubation the Chl *a* content was reduced upon treatment with > 25 mg l^{-1}. Inhibition of chlorophyll by Cd may be a result of different mechanisms such as: inhibition of CO_2 fixation through interference with the Calvin cycle as well as inhibition of biosynthesis enzymes (Hou et al. 2007; Mohn and Hosetti 1997; Weigel 1985). Trevors et al. (1986) have reported that algae and cyanobacteria are the most sensitive organisms toward cadmium exposure. Moreover, cadmium has caused severe inhibition of growth, photosynthesis and nitrogen fixation in organisms at concentrations less than 2 ppm. The authors further suggested a mechanism related to Cd toxicity as a result of effects on protein synthesis and cellular organelles such as mitochondria and chloroplasts. Reduction of total chlorophyll content in response to Cd has been proven by Lamaia et al. (2005), suggesting that Cd causes chloroplasts disorganization leading to a decrease of the photosynthetic pigments. In addition, a decrease of chlorophyll *a* content was reported as a symptom of heavy metals toxicity (De Filippis et al. 1981; El-Sheekh et al. 2003).

The mechanism of cadmium toxicity in *Euglena gracilis* was studied by Watanabe et al. (2003) stating that cadmium induces formation of reactive oxygen species (ROS) which negatively affect the thylakoid membrane.

As reported several times, the mechanisms of heavy metal toxicity on the photosynthetic pigments can be grouped into the following:

1) At lower heavy metal concentrations, replacement of Mg ion in the chlorophyll molecules (Küpper et al. 2002).
2) At higher concentrations, they produce oxidative stress through formation of reactive oxygen species causing damage of chloroplast, thylakoid membrane and decreasing the cellular antioxidant ability (Livingstone 2001; Watanabe et al. 2003).
3) Upon accumulation of ROS to higher concentrations, photosynthesis process in general and chloroplast as well as pigment especially in algae are the major target for the peroxidation (Ali et al. 2006; Rodrìgueza et al. 2007).
4) Heavy metals hinder the transport of other essential metal causing inhibition of enzymes like δ-aminolevulinic acid dehydratase and protochlorophyllide reductase involved in chlorophyll synthesis (Aravind and Prasad 2003; Dai et al. 2006).
5) They also enhance pigment degradation through activation of catabolic enzymes (Hou et al. 2007).

In algae, carotenoids are involved in the protection mechanism against environmental stresses like heavy metal and light. As reported elsewhere (Behera et al. 2002; Dai et al. 2006; Okamoto et al. 2001; Rodrìgueza et al. 2007), carotenoids also protect the light harvesting pigments in the antenna complexes against ROS by acting as quenchers. Pancaldi et al. (1996) have stated a fast degradation of xanthophylls and ß-carotene as a result of *Euglena* chloroplasts ageing oxidative stresses. Moreover, several studies have reported involvement of xanthophylls in photoprotection (Demmig-Adams and Adams III 1992; Horton et al. 1996; Jin et al. 2003). This is also supported by the current results that carotenoids are involved in protection of the chloroplast from heavy metal stress. A significant reduction in *E. gracilis* photosynthetic pigments has been also reported upon growing the cells in a medium containing other toxins like 5 μM norflurazon (Tschiersch et al. 2002). From the current work it is clear that chloroplasts are the main target organelles for heavy metal toxicity. This is in agreement to Wong et al. (1994) who also reported inhibition of chlorophyll synthesis in *Chlorella* species assuming destruction of the chloroplasts as primary target for metal toxicity.

Conclusion

Methanol extraction using ultrasonic disruption gave highest extraction efficiency of photosynthetic pigments in *Euglena gracilis*. A rapid degradation of all photosynthetic pigments, mainly chlorophyll *b* and β-carotene, occur in response to Cd treatment at 4 mg l^{-1} after 24 h. Significant alterations in chlorophyll *b*, β-carotene as well as 9-cis-neoxanthin content were found with copper stress at 11 mg l^{-1} The results suggest that carotenoids and xanthophylls are involved in protection of the chloroplast against heavy metal stress. The present study recommends the use of photosynthetic pigments in *Euglena gracilis* as a bioassay endpoint, since they respond sensitively and rapidly to heavy metals at ecologically relevant concentrations.

Acknowledgement

We would like to thank the Deutsche Akademischer Austauschdienst (DAAD) for granting a scholarship to Hoda Ahmed.

References

Ali NA, Dewez D, Didur O, Popovic R (2006) Inhibition of photosystem II photochemistry by Cr is caused by alteration of both D1 protein and oxygen evolving complex. Photosynth Res 89: 81-87.

Aravind P and Prasad MNV (2003) Zinc alleviates Cadmium-induced oxidative stress in Ceratophyllum demersum L.: A free floating freshwater macrophyte. Plant Physiol Biochem 41: 391-397.

Behera RK, Mishra PC, Choudhury NK (2002) High irradiance and water stress induce alterations in pigment composition and chloroplast activities of primary wheat leaves. J Plant Physiol 159: 967-973.

Buffan-Dubau E and Carman KR (2000) Extraction of benthic microalgal pigments for HPLC analyses. Mar Ecol Prog Ser 204: 293-297.

Checcucci A, Colombetti G, Ferrara R, Lenci F (1976) Action spectra for photoaccumulation of green and colorless *Euglena*: evidence for identification of receptor pigments. Photochem Photobiol 23: 51-54.

Clijsters II and Van Assche F (1985) Inhibition of photosynthesis by heavy metals. Photosynth Res 7: 31-40.

Dai L, Xiong Z, Hung Y, Li M (2006) Cadmium-Induced changes in pigments, total phenolics, and phenylalanine ammonia-lyase activity in fronds of *Azolla imbricata*. Environ Toxicol 21: 505-512.

De Filippis LF, Hampp R, Ziegler H (1981) The effect of sublethal concentrations of zinc, cadmium and mercury on *Euglena*. Adenylates and energy charge. Z Pflanzenphysiol 103: 1-7.

De Filippis LF and Pallaghy CK (1976) The effect of sublethal concentrations of mercury and zinc on *Chlorella*. 11. Photosynthesis and pigment composition. Z Pflanzenphysiol 78: 314-322.

Demmig-Adams B and Adams III WW (1992) Photoprotection and other responses of plants to high light stress. Annu Rev Plant Physiol 43: 599-626.

Drexlerand DM and Fletcher RA (1981) Inhibition of photosynthetic pigments in cucumber cotyledons as a principle for a bioassay with fiuridone. Weed Research 2: 71-76.

El-Sheekh MM, El-Naggar AH, Osman MEH, El-Mazaly E (2003) Effect of cobalt on growth, pigments and the photosynthetic electron transport in *Monoraphidium minutum* and *Nitzchia perminuta*. Braz J Plant Physiol 15: 159-166.

Gilbert M, Domin A, Becker A, Wilhelm C (2000) Estimation of primary productivity by chlorophyll a *in vivo* fluorescence in freshwater phytoplankton. Photosynthetica 38: 111-126.

Hörcsik Z, Oláh V, Balogh Á, Mészáros I, Simon L, Lakatos G (2006) Effect of Chromium(VI) on growth, element and photosynthetic pigment composition of *Chlorella pyrenoidosa*. Acta Biol Szeged 50: 19-23.

Horton P, Ruban AV, Walters RG (1996) Regulation of light harvesting in green plants. Annu Rev Plant Physiol 47: 655-684.

Hou W, Chen X, Song G, Wang Q, Chang CC (2007) Effects of copper and cadmium on heavy metal polluted waterbody restoration by duckweed (*Lemna minor*). Plant Physiol Biochem 45: 62-69.

Jeffery WR, Mantoura RFC, Wrigh SW (1997) Phytoplankton pigments in oceanography : guidelines to modern methods. UNESCO Publishing, Paris

Jin ES, Yokthongwattana K, Polle JEW, Melis A (2003) Role of the reversible xanthophyll cycle in the photosystem II damage and repair cycle in *Dunaliella salina*. Plant Physiol 132: 352-364.

Küpper H, Setlik I, Spiller M, Küpper FC, Prasil O (2002) Heavy metal-induced inhibition of photosynthesis: targets of *in vivo* heavy metal chlorophyll formation. J Phycol 38: 429-441.

Lamaia C, Kruatrachuea M, Pokethitiyooka P, Upathamb SE, Soonthornsarathoola V (2005) Toxicity and Accumulation of Lead and Cadmium in the Filamentous Green Alga *Cladophora fracta* (O.F. Muller ex Vahl) Kutzing: A Laboratory Study. Science Asia 31: 121-127.

Livingstone DR (2001) Contaminant-stimulated reactive oxygen species production and oxidative damage in aquatic organisms. Mar Pollut Bull 42: 656-666.

Macfarlane GR and Burchett MD (2001) Photosynthetic pigments and peroxidase activity as indicators of heavy metal stress in the Grey Mangrove, *Avicennia marina* (Forsk.) Vierh Mar Pollut Bull 42: 233-240.

Manankina EE, Mel'nikov SS, Budakova EA, Shalygo NV (2003) Effect of Ni^{2+} on Early Stages of Chlorophyll Biosynthesis and Pheophytinization in *Euglena gracilis*. Russ J Plant Physiol 50: 390-394.

Mohn BS and Hosetti BB (1997) Potential phytotoxicity of lead and cadmium to Lemna minor grown in sewage stabilization ponds. Environ Pollut 98: 233-238.

Okamoto OK, Pinto E, Latorre LR, Bechara EJH, Colepicolo P (2001) Antioxidant modulation in response to metal-induced oxidative stress in algal chloroplasts. Arch Environ Contam Toxicol 40: 18-24.

Pancaldi S, Bonora A, Dall'Olio G, Bruni A, Fasulo MP (1996) Ageing of *Euglena* chloroplasts *in vitro*. I. Variations in pigment pattern and in morphology. J Exp Bot 47: 49-60.

Rai LC, Gaur JP, Kumar HD (1981) Phycology and heavy-metal pollution. Biol Rev 56: 99-151.

Rodrìgueza MC, Barsantib L, Passarellib V, Evangelistab V, Confortia V, Gualtierib P (2007) Effects of chromium on photosynthetic and photoreceptive apparatus of the alga *Chlamydomonas reinhardtii*. Environ Res 105: 234-239.

Starr RC (1964) The culture collection of algae at Indiana University. Am J Bot 51: 1013-1044.

Ston′ J and Kosakowska A (2002) Phytoplankton pigments designation–an application of RP-HPLC in qualitative and quantitative analysis. J Appl Phycol 14: 205-210.

Trevors JT, Stratton GW, Gadd GM (1986) Cadmium transport, resistance, and toxicity in bacteria, algae, and fungi. Can J Microbiol 32: 447-464.

Tschiersch H, Ohmann E, Doege M (2002) Modification of the thylakoid structure of *Euglena gracilis* by norflurazon-treatment: consequences for fluorescence quenching. Environ Exp Bot 47: 259-270.

van Leeuwe MA and Stefels J (1998) Effects of iron and light stress on the biochemical composition of Antarctic *Phaeocystis* sp. (Prymnesiophyceae). II. Pigment composition. J Phycol 34: 496-503.

van Leeuwe MA, Villerius LA, Roggeveld J, Visser RJW, Stefels J (2006) An optimized method for automated analysis of algal pigments by HPLC. Mar Chem 102: 267-275.

Watanabe M, Henmi K, Ogawa K, Suzuki T (2003) Cadmium-dependent generation of reactive oxygen species and mitochondrial DNA breaks in photosynthetic and non-photosynthetic strains of *Euglena gracilis*. Comp Biochem Physiol C Toxicol Pharmacol 134: 227-234.

Weigel HJ (1985) Inhibition of photosynthetic reaction of isolated spinach chloroplast by cadmium. J Plant Physiol 119: 179-189.

Wilhelm C, Volkmarr P, Lohmann C, Becker A, Meyer M (1995) The HPLC-aided pigment analysis of phytoplankton cells as a powerful tool in water quality control. J Water SRT-Aqua (London)132-141.

Wong SL, Nakamoto L, Wainwright JF (1994) Identification of toxic metals in affected algal cells in assays of wastewaters. J Appl Phycol 6: 405-414.

Wright SW, Jeffrey SW, Mantoura RFC, Llewellyn CA, Bjornland T, Repeta DJ, Welschmeyer NA (1991) Improved HPLC method for the analysis of chlorophylls and carotenoids from marine phytoplankton. Mar Ecol Prog Ser 77: 183-196.

Wright SW, Jeffrey SW, Mantoura RFC (1997) Evaluation of methods and solvents for pigment extraction. In: S.W.Jeffrey, R.F.C.Mantoura, and C.W.Wright (ed) Phytoplankton Pigments in Oceanography. UNESCO, Paris, pp. 261-282.

Zhou W, Juneau P, Qiu B (2006) Growth and photosynthetic responses of the bloom-forming cyanobacterium *Microcystis aeruginosa* to elevated levels of cadmium. Chemosphere 65: 1738-1746.

12

Computational Methods and Strategies for Protein Structure Prediction

Minu Kesheri, Swarna Kanchan, Richa and Rajeshwar P. Sinha

Abstract

In order to fill the gap between the protein sequences and their 3D structures available in the public databases, the computational prediction of protein structure has become the need of the hour. Therefore, the present chapter aims to highlight various methods and strategies for computational protein structure prediction by established *ab initio*, threading and homology modeling methods. *Ab initio*, threading and homology modeling methods are mainly based on the low sequence similarity, similarity between 25-30% and similarity more than 30% with the template structure respectively. Hybrid methods are preferred which club together a number of features such as structural alignments, solvent accessibility and secondary structure informations. Moreover, nowadays meta-servers are also available that integrate protein structure predictions performed by various methods and assemble to interpret the results with a consensus prediction. In order to keep pace with recent developments, CASP (Critically Assessment of techniques for protein Structure Prediction) meetings are commenced every two years which critically evaluate the various protein structure programmes and servers besides assigning ranks for the same.

Keywords: Protein structure prediction, *Ab Initio* Modeling, Threading, Homology Modeling, CASP

Introduction

Proteins have a precise 3D structure that directs their function. Determining the structures of various proteins would aid in our understanding of the mechanisms of various proteins in biological systems. Prediction of protein structure from amino-acid sequences has been one of the most challenging tasks in computational biology/bioinformatics for many years (Baker and Sali 2001, Skolnick et al. 2000). Currently, only biophysical experimental techniques such as X-ray crystallography and nuclear magnetic resonance are able to provide precise protein 3D structures. There are 5,16,16,950 protein sequences in the latest release of UniProtKB/TrEMBL as of January, 2014, while the Protein Data Bank (PDB) contained only 97,362 protein structure till then. This gap has widened too much over the last decade, despite the development of dedicated high-throughput X-ray crystallography pipelines (Berman et al. 2000). Thus, the availability of experimental structures is lacking for the vast majority of protein sequences. Solving the protein structure by NMR is limited to small and soluble proteins only. Moreover, X-ray crystallography and NMR are costly and time consuming methods to solve the protein structure. Therefore, the computational prediction of structure of proteins is highly required to fill the gap between the sequences available and their structures. Historically, protein structure prediction was classified into three categories: (i) *ab initio* folding (Liwo et al. 1999, Zhang et. al. 2003, Bradley et al. 2005, Klepeis et al. 2005, Klepeis 2003) (ii) threading (Bowie et al. 1991, Jones et al. 1992, Xu and Xu 2000, Zhou and Zhou 2005, Skolnick et al. 2004) and (iii) comparative modeling (Sali and Blundell 1993, Fiser et al. 2000). The two approaches threading and comparative modeling build protein models by aligning query sequences onto solved template structures by X-ray crystallography or NMR. When close templates are identified, high-resolution models could be built by the template-based methods. If templates are absent from the Protein Data Bank (PDB), the models need to be built from scratch, i.e. *ab initio* folding. Nowadays, these prediction categories are clubbed into two major groups: free modeling (FM) i.e. *Ab initio* folding and template-based modeling (TBM), which includes comparative/homology modeling and threading. The predicted models must be checked for protein structure quality validation by various programmes available.

Ab Initio Modeling

This method is simply based on elementary fundamentals of energy and geometry (Moult and Melamud, 2000). *Ab initio* structure prediction seeks to predict the native conformation of a protein from the amino acid sequence alone. *Ab initio* prediction of protein structures makes no use of information available in databases mainly PDB (Nanias et al. 2005) and the goal of this method is to predict the structure of a protein based entirely on the laws of physics and chemistry. This is the most difficult category of protein structure prediction among all the three different methods of structure prediction which completely predicts a new fold (Skolnick and Kolinski 2002, Floudas et al. 2006). With increasing protein sizes, the conformational space to be searched increases sharply, which makes the *ab initio* modeling of larger proteins extremely difficult (Zhang and Skolnick 2004). Currently, the accuracy of *ab initio* modeling is limited to small proteins having length less than 100 amino acid residues. *Ab initio* structure prediction requires an efficient free energy function to find the conformation of the modeled protein near to native state protein structure with lowest free energy. *Ab initio* structure prediction is challenging because current potential functions have limited accuracy and the conformational space to be searched is vast. Few popular web servers for modeling of the protein structure by *ab initio* folding method are listed in Table 1.

Fold Recognition or Threading

Fold recognition or threading method aims to fit a target sequence to a known structure in a library of folds and the model built is evaluated using residue based contact potentials (Floudas 2007). Although fold recognition will not yield equivalent results as those from X-ray crystallography or Nuclear magnetic resonance yet, it is a comparatively fast and inexpensive way to build a close approximation of a structure from a sequence without the time and costs of experimental procedures. Fold Recognition (FR) was reserved for methods which did not rely on sequence searching and where the sequence identity between target and template was below the so-called "twilight zone" of between 25–30%. The rationale behind the threading method is total number of experimentally solved 3D structure deposited in Protein Data Bank database don't have a new fold. The nature has limited number of basic folds. These basic folds form the framework of most of the protein structures available in PDB.

Table 1: List of web servers for modeling protein structure by *ab initio* folding method along with the webpage URL and the programme description.

Sl. No.	Name of the Web Server/ Group (URL)	Description of the Web Server/Group
1.	ROBETTA (Kim et al. 2004, Bradley et. al. 2005) [http://robetta.bakerlab.org]	Provides both *ab initio* and comparative models of protein domains. Domains having no sequence similarity with PDB sequences are modeled by Rosetta *de novo* protocol.
2.	QUARK (Xu and Zhang 2012)	*De novo* protein structure prediction web server aims to construct the correct protein 3D model from amino acid sequence by replica-exchange Monte Carlo simulation under the guide of an atomic-level knowledge-based force field.
3.	PROTINFO (Hung et al. 2005) [http://protinfo.compbio.washington.edu]	*De novo* protein structure prediction web server utilizes simulated annealing for generation and different scoring functions for selection of final five conformers.
4.	SCRATCH (Cheng et al. 2005) [http://www.igb.uci.edu/servers/psss.html]	Server utilizes recursive neural networks, evolutionary information, fragment libraries and energy to build protein model.
5.	BHAGEERATH (Jayaram et al. 2006)[http://www.scfbio-iitd.res.in/bhageerath]	Energy based methodology for narrowing down the search space and thus helps in building a good protein model.

Generally, similar sequence implies similar structure but the reverse is not true. Similar structures are often found for proteins for which no sequence similarity to any known structure can be detected (Floudas et al. 2006). Using fold recognition or threading, we are able to identify proteins with known structures that share common folds with the target sequences.

Fold recognition methods work by comparing each target sequence against a library of potential fold templates using energy potentials and/or other similarity scoring methods. We first need to define a library of potential folds. Once the library is defined, the target sequence will be fitted to each library entry and an energy function is used to evaluate the fit between the target sequence and the library entries to determine the best possible templates. The template with the lowest energy score (or highest similarity score) is then assumed to best fit the fold of the target protein.

Table 2: List of web servers for modeling the protein structure by Threading or Fold Recognition method along with the webpage URL and the programme description.

Sl. No.	Name of the Web Server/ Group [URL]	Description of the Web Server/Group
1.	I-TASSER (Zhang et al. 2005) [http://zhanglab.ccmb.med.umich.edu/I-TASSER/]	3D models are built based on multiple-threading alignments by LOMETS and iterative templat fragment assembly.
2.	LOOPP (Teodorescu et al. 2004) [http://cbsuapps.tc.cornell.edu/loopp.aspx]	A fold recognition program based on the collection of numerous signals to build the target structure.
3.	PROSPECT (Xu and Xu 2000) [http://compbio.ornl.gov/structure/prospect]	PROSPECT is based on scoring function for a threading alignment, which consists of four additive terms: i) a mutation term, ii) a singleton fitness term, iii) a pairwise-contact potential term, and iv) alignment gap penalties.
4.	MUSTER (Wu and Zhang 2008)[http://zhanglab.ccmb.med.umich.edu/MUSTER/]	Muster generates sequence-template alignments by combining sequence profile-profile alignment with multiple structural information.
5.	PHYRE2 (Kelley and Sternberg 2009)[http://www.sbg.bio.ic.ac.uk/~phyre2/html/page.cgi?id=index]	A server which uses proûle–proûle matching algorithms to build the protein model.

Fold recognition methods also includes various properties of structural environment of the residue. Structural environments are more conserved than the actual type of residue, therefore in the absence of homology, a fold could be predicted by measuring the compatibility of a sequence with template folds in terms of amino acid preferences for certain structural environments. These amino acid preferences for structural environment provide sufficient information to choose among folds. The amino acid preferences for three main types of structural environment comprise of the solvent accessibility, the contact with polar atoms and the secondary structure. The main limitation of this method is high computational cost, since each entry in the whole library of thousands of possible folds needs to be aligned in all possible ways to select the fold(s). Another major bottleneck is the energy function used for the evaluation of alignment. It is not reasonable to expect to find the correct folds in all cases with a single form of energy function. Few popular web servers for modeling the protein structure by threading method is listed in Table 2.

Homology Modeling or Comparative Modeling

Comparative or homology protein structure modeling builds a three-dimensional model for a protein of unknown structure (the target) based on one or more related proteins of known structure. The necessary conditions for getting a useful model are (i) detectable similarity (nearly greater than and equal to 30%) between the target sequence and the template structures and (ii) availability of a correct alignment between them. In practice, homology modeling is a multistep process that can be summarized in five steps:

1. Template search, selection and alignment
2. Alignment correction in core regions
3. Backbone, Loop and Side-chain generation and modeling
4. Model refinement
5. Model evaluation or validation

Step 1: Template Search, Selection and Alignment

Template search is generally done by comparing the target sequence with the sequence of each of the structures in the PDB database. The performance depends on the sensitivity of the comparison of target and template sequences by various programmes e.g. Fasta, Blast or Psi-Blast. The simplest template selection rule is to select the structure with the highest sequence similarity to the target sequence. The quality of a template increases with its overall sequence similarity to the target and decreases with the number and length of gaps in the alignment. Multiple sequence alignment by various freely available programmes e.g. ClustalW, Mufft, Kalign, Probcons etc. and a phylogenetic tree by freely available programmes e.g. Phylip, MEGA etc. can help in selecting the template from the subfamily that is closest to the target sequence. The similarity between the 'environment' of the template and the environment in which the target needs to be modeled should also be considered. The quality of the experimentally determined structure is another important factor in template selection i.e. high resolution X ray crystal structure is more preferred for template selection then that of low resolution crystal structure. Multiple templates rather than selecting a single template, generally increases the model accuracy. A good protein structure model depends on alignment between the target and template.

Step 2: Alignment correction in core regions

An accurate alignment can be calculated automatically using standard sequence-sequence alignment methods e.g. Blast2seq and Needle global sequence dynamic programming. In low sequence identity cases, the alignment accuracy is the most important factor which affects the quality of the predicted model. Alignments can be improved by including structural information from the template protein structure. Gaps should be avoided in core regions mainly in secondary structure elements (which are found to be conserved in most cases), buried regions and between two residues that are far in space. It is important to inspect and edit the alignment manually, especially if the target-template sequence identity is low.

Step 3: Backbone, Loop and Side-chain generation and modeling

Creating the backbone is essential for modeled protein structure. For backbone, we simply copy the coordinates of those template residues that show up in the alignment with the model sequence. If two aligned residues differ, only the backbone coordinates (N, Cα, C and O) can be copied. If they are the same, we can also include the coordinates of side chain.

In comparative modeling, target sequences often have few inserted residues in compared to the template structures. Thus, no structural information about these inserted regions could be obtained from the template structures. These regions are called surface loops. Loops often play an important role in defining the functional specificity of a given protein structure, forming the active and binding sites. The accuracy of loop modeling is a major factor determining the usefulness of comparative models in applications such as protein-ligand docking. There are two main classes of loop modeling methods: (i) Database search approaches, where a small loop of 3-10 amino acid residues are searched in a database of all known protein structures and if such loops fit the criteria of lowest energy, such loops are selected and added to the model structure. All major molecular modeling programs and servers support this approach e.g. Modeller (Sali and Blundell, 1993), Swiss-Model (Guex and Peitsch 1997). (ii) The conformational search approaches mainly depend on efficient energy function to choose the loop with lowest energy. If required, energy of the selected loop is minimized using Monte Carlo or molecular dynamics techniques to arrive at the best loop conformation with lower energy.

Side chain modeling is also one of the essential components in structure prediction of proteins. When we compare the side-chain conformations (rotamers) of residues that are conserved in structurally similar proteins, we can copy coordinates of conserved amino acid residues entirely from the template to the model. But when we have variable residues or loops regions, side chains are added to each amino acid and their all possible rotamers are searched to find the most stable (having least energy) rotamers from rotamer library.

Step 4: Model refinement

One of the major limitations of computational protein structure prediction is the deviation of predicted models from their experimentally derived true, native structures. Refinement of the protein model is required, if there is problem in structural packing of side chains, loops, and secondary structural elements in the complete target model. For any error in backbone or side chain packing, energy minimization is done which requires an enormous precision in the energy function. At every minimization step, a few big errors (like bumps, i.e., too short atomic distances) are removed while many small errors might be introduced. In energy minimization, force ûelds must be fast to handle these large molecules efficiently. Refinement of the low resolution predicted models to high resolution structures are close to the native state, however, it has proven to be extremely challenging. One of the refinement methods called 3Drefine is computationally inexpensive, consuming only few minutes of CPU time to refine a protein of typical length of 300 amino acid residues (Bhattacharya and Cheng 2013).

Step 5: Model evaluation or validation

The predicted model must be checked for (1) Side chain packing errors of the modeled structure. (2) Distortions or shifts in correctly aligned region of target with the template structures. (3) Distortions or shifts of a region that does not align with any of the template structures. (4) Distortions or shifts of a region that is aligned incorrectly with the template structures. Structural model accuracy is mainly based on global distance test (GDT), which is an average percentage of model Cα atoms within a specified distance threshold to actual native conformation (Jauch et al. 2007). TM score is another method to validate the model accuracy to score the topological similarity of target and template structures, where the score near to 1.00 is the best predicted near-native model against the actual experimental structure for a target (Xu and Zhang 2010).

Various programmes are available for checking the quality of the model. One of these is Procheck that generates the Ramachandran Plot, which illustrates the stereo chemical quality of the protein model. Few popular web servers for protein structure quality validation are listed in Table 3.

Table 3: List of web servers for protein structure quality validation along with the webpage URL and the programme description.

Sl. No.	Name of the Web Server/Group [URL]	Description of the Web Server/Group
1.	QMEAN (Benkert et al. 2009)[http://swissmodel.expasy.org/qmean/cgi/index.cgi]	Quality estimate is based on geometrical analysis of single models, and the clustering-based scoring function.
2.	PROSA-WEB (Wiederstein and Sippl 2007)[https://prosa.services.came.sbg.ac.at/prosa.php]	Quality is checked by generation of Z scores and energy plots that highlight potential problems spotted in protein structures.
3.	PROCHECK (Laskowski et al. 1993) [http://services.mbi.ucla.edu/SAVES/]	Stereochemical quality of a protein structure is checked by analyzing residue-by-residue geometry and overall structural geometry.
4.	VERIFY-3D (Bowie et al. 1991; Lüthy et al. 1992)[http://services.mbi.ucla.edu/SAVES/]	Determines the compatibility of an atomic model (3D) with its own amino acid sequence (1D) by assigning a structural class based on its location and environment.
5.	ERRAT (Colovos and Yeates 1993) [http://services.mbi.ucla.edu/SAVES/]	Analyzes the statistics of non-bonded interactions between different atom types and plots the value of the error function.

Thus, *ab initio* folding and threading are computationally expensive methods which also results in protein structural models having low accuracy. *Ab initio* folding and threading methods have limitations to search a vast conformational space and protein structures available in protein databank respectively. Therefore, comparative or homology modeling is the best method for protein structure predication with high accuracy if the sequence identity between target and template sequence is more than 50%. Few popular web servers for modeling the protein structure by homology modeling or comparative modeling method along with the webpage URL and description are listed in Table 4.

Nowadays, a number of fully automated hybrid methods are designed in order to perform rapid, fully automated fold recognition on a proteome wide scale. Few fully automated hybrid methods for fold recognition are GenTHREADER (Jones 1999) and FUGUE (Shi et al. 2001). These hybrid methods club together a number of features such as structural alignments, solvent accessibility and secondary structure information in order to produce a protein model with high accuracy.

Table 4: List of web servers for modeling the protein structure by homology modeling or comparative modeling method along with the webpage URL and the programmes description.

Sl. No.	Name of the Web Server/Group [URL]	Description of the Web Server/Group
1.	GENO3D (Combat et al. 2002) [http://geno3d-pbil.ibcp.fr/]	A web server which builds the model based on distance geometry, simulated annealing and energy minimization algorithms to build the protein 3D model.
2.	CPHMODELS 3.2 (Nielsen et al. 2010) [http://www.cbs.dtu.dk/services/CPHmodels/]	Protein modeling is based on profile-profile alignment guided by secondary structure and exposure predictions.
3.	3DJIGSAW (Bates et al. 2001) [http://www.bmm.icnet.uk/servers/3djigsaw/]	An automated server to build three dimensional models for proteins based on homologues of known structure.
4.	SWISS-MODEL (Guex and Peitsch 1997) [http://swissmodel.expasy.org/SWISSMODEL.html]	A fully automated protein structure homology modeling server.
5.	ESYPRED3D (Lambert et al. 2002) [http://www.fundp.ac.be/sciences/biologie/urbm/bioinfo/esypred/]	This server results in good protein model by using several multiple alignment programs by combining, weighing and screening.

Rather than hybrid methods, several meta-servers are available that integrate protein structure predictions performed by various methods. These servers assemble and interpret the results to come up with a consensus prediction. Pcons.net (Lundstrom et al. 2001) and 3D-Jury (Ginalski et al. 2003) are the meta- servers which focus on the selection of high quality obtained from many different servers. Another meta-server, 3D-SHOTGUN (Fischer 2003) doesn't just select the best model but also refines initial models for building the protein structure model with high accuracy. Moreover, protein structure prediction strategies and methodologies are tested every two years in the CASP (Critically Assessment of techniques for protein Structure Prediction) meeting, which started since 1994. It critically evaluates the various protein structure programmes and servers besides assigning ranks for the same. In CASP10, some 41,740 predicted models submitted by 150 predictor groups were assessed as template-based modeling predictions where Zhang-Server, QUARK, PMS, Leecon and Zhang groups provided the most accurate models for the assessment units targets (Huang et al. 2014). Thus, CASP meeting is the best way to keep updated with the advancement in protein structure prediction strategies and methodologies.

Acknowledgements

Minu Kesheri is thankful to the UGC, Govt. of India, New Delhi, for providing financial assistance in the form of research fellowship. Swarna Kanchan is thankful to Dr. Shibasish Chowdhury for his unremitting motivation and

University Grant Commission, Govt. of India, New Delhi, for providing the financial support in the form of the Basic Science Research Fellowship under UGC SAP programme at Biological Science Department, BITS, Pilani. Richa is thankful to DST, Govt. of India, New Delhi, for providing the financial support in form of the fellowship under a project (No. SR/WOS-A/LS-140/ 2011).

References

Baker D and Sali A (2001) Protein structure prediction and structural genomics. Science 294: 93-96.

Bates PA, Kelley LA, MacCallum RM, Sternberg MJ (2001) Enhancement of protein modeling by human intervention in applying the automatic programs 3D-JIGSAW and 3D-PSSM. Proteins 45: 39-46.

Benkert P, Künzli M, Schwede T (2009) QMEAN Server for Protein Model Quality Estimation. Nucleic Acids Res 37: W510-14.

Berman HM, Westbrook J, Feng Z, Gilliland G, Bhat TN, Weissig H, Shindyalov IN, Bourne PE (2000) The protein data bank. Nucleic Acid Res 28: 235-242.

Bhattacharya D, Cheng J (2013) 3Drefine: consistent protein structure refinement by optimizing hydrogen bonding network and atomic-level energy minimization. Proteins 81: 119-31.

Bowie JU, Lüthy R, Eisenberg D (1991) A method to identify protein sequences that fold into a known three-dimensional structure. Science 253: 164-170.

Bradley P, Misura KM, Baker D (2005) Toward high-resolution de novo structure prediction for small proteins. Science 309: 1868-71.

Cheng J, Randell AZ, Sweredoski MJ, Baldi P (2005) SCRATCH: a protein structure and structural feature. Nucleic Acids Res 33: W72-76.

Colovos C and Yeates TO (1993) Verification of protein structures: patterns of nonbonded atomic interactions. Protein Sci 2: 1511-19.

Combat C, Jambon M, Deleage G, Geourjon C (2002) Geno3D: automatic comparative molecular modelling of protein. Bioinformatics 18: 213-214.

Fischer D (2003) 3D-SHOTGUN: a novel, cooperative, fold-recognition meta predictor. Proteins 51: 434-441.

Fiser A, Do RK, Sali A (2000) Modeling of loops in protein structures. Protein Sci 9: 1753-73.

Floudas CA, Fung HK, McAllister SR, Mönnigmann M, Rajgaria R (2006) Advances in protein structure prediction and de novo protein design: a review. Chem Eng Sci 61: 966-988.

Floudas CA (2007) Computational methods in protein structure prediction. Biotechnol Bioeng 97: 207-213.

Ginalski K, Elofsson A, Fischer D, Rychlewski L (2003) 3D-Jury: a simple approach to improve protein structure predictions. Bioinformatics 19: 1015-1018.

Guex N and Peitsch MC (1997) SWISS-MODEL and the SwissPdbViewer: an environment for comparative protein modeling. Electrophoresis 18: 2714-2723.

Huang YJ, Mao B, Aramini JM, Montelione GT (2014) Assessment of template-based protein structure predictions in CASP10. Proteins 82: 43-56.

Hung LH, Ngan SC, Liu T, Samudrala R (2005) PROTINFO: new algorithms for enhanced protein structure prediction. Nucleic Acid Res 33: W77-80.

Jauch R, Yeo HC, Kolatkar PR, Clarke ND (2007) Assessment of CASP7 structure predictions for template free targets. Proteins 69: 57-67.

Jayaram B, Bhushan K, Shenoy SR, Narang P, Bose S, Agrawal P, Sahu D, Pandey V (2006) An energy based web enabled computer software suite for limiting the search space of tertiary structures of small globular proteins. Nucleic Acids Res 34: 6195-6204.

Jones DT, Taylor WR, Thornton JM (1992) A new approach to protein folds recognition. Nature 358: 86-89.

Jones DT (1999) GenTHREADER: an efficient and reliable protein folds recognition method for genomic sequences. J Mol Biol 287: 797-815.

Kelley LA and Sternberg MJ (2009) Protein structure prediction on the web: a case study using the Phyre server. Nat Protoc 4: 363-371.

Kim DE, Chivian D, Baker D (2004) Protein structure prediction and analysis using the robetta server. Nucleic Acids Res 32: W526-31.

Klepeis JL and Floudas CA (2003) ASTRO-FOLD: a combinatorial and global optimization framework for Ab initio prediction of three-dimensional structures of proteins from the amino acid sequence. Biophys J 85: 2119-2146.

Klepeis JL, Wei Y, Hecht MH, Floudas CA (2005) *Ab initio* prediction of the three-dimensional structure of a de novo designed protein: a double-blind case study. Proteins 58: 560-570.

Lambert C, Léonard N, De Bolle X, Depiereux E (2002) ESyPred3D: prediction of proteins 3D structures. Bioinformatics 18: 1250-1256.

Laskowski RA, MacArthur MW, Moss DS, Thornton JM (1993) PROCHECK: A program to check the stereochemical quality of protein structure. J Appl Cryst 26: 283-91.

Liwo A, Lee J, Ripoll DR, Pillardy J, Scheraga HA (1999) Protein structure prediction by global optimization of a potential energy function. Proc Natl Acad Sci USA 96: 5482-5485.

Lundstrom J, Rychlewski L, Bujnicki J, Elofsson A (2001) Pcons: a neuralnetwork-based consensus predictor that improves fold recognition. Protein Sci 10: 2354–2362.

Lüthy R, Bowie JU, Eisenberg D (1992) Assessment of protein models with three-dimensional profiles. Nature 356: 83-85.

Moult J and Melamud E(2000) From fold to function. Curr Opin Struct Biol 10: 384-389.

Nanias M, Chinchio M, Oldziej S, Czaplewski C, Scheraga HA (2005) Protein structure prediction with the UNRES force-field using Replica-Exchange Monte Carlo-with-Minimization; comparison with MCM, CSA, and CFMC. J Comput Chem 26: 1472-1486. Nielsen M, Lundegaard C, Lund O, Petersen TN (2010) CPHmodels-3.0 - Remote homology modeling using structure guided sequence profiles. Nucleic Acids Res 38: W576-81.

Sali A and Blundell TL (1993) Comparative protein modelling by satisfaction of spatial restraints. J Mol Biol 234: 779-815.

Shi J, Blundell TL, Mizuguchi K (2001) FUGUE: sequence-structure homology recognition using environment-specific substitution tables and structure dependent gap penalties. J Mol Biol 310: 243–257.

Skolnick J and Kolinski A (2002) A Unified Approach to the Prediction of Protein Structure and Function, in Computational Methods for Protein Folding, Vol 120 (ed R. A. Friesner), John Wiley & Sons, Inc., New York, USA. Adv Chem Phys 120: 131-192.

Skolnick J, Fetrow JS, Kolinski A (2000) Structural genomics and its importance for gene function analysis. Nat Biotechnol 18: 283-287.

Skolnick J, Kihara D, Zhang Y (2004) Development and large scale benchmark testing of the PROSPECTOR 3.0 threading algorithm. Proteins 56: 502-518.

Teodorescu O, Galor T, Pillardy J, Elber R (2004) Enriching the sequence substitution matrix by structural information. Proteins 54: 41-48.

Wiederstein M and Sippl MJ (2007)ProSA-web: interactive web service for the recognition of errors in three-dimensional structures of proteins. Nucleic Acids Res 35: W407-10.

Wu S and Zhang Y (2008) MUSTER: Improving protein sequence profile-profile alignments by using multiple sources of structure information. Proteins 72: 547-556.

Xu D, Zhang Y (2012) Ab initio protein structure assembly using continuous structure fragments and optimized knowledge-based force field. Proteins 80: 1715-1735.

Xu J and Zhang Y (2010) How significant is a protein structure similarity with TM score=0.5?. Bioinformatics 26: 889–895.

Xu Y and Xu D (2000) Protein threading using PROSPECT: design and evaluation. Proteins 40: 343-354.

Zhang Y and Skolnick J (2004) Tertiary structure predictions on a comprehensive benchmark of medium to large size proteins. Biophy J 87: 2647-2655.

Zhang Y, Arakaki AK, Skolnick J (2005) TASSER: an automated method for the prediction of protein tertiary structures in CASP6. Proteins 61: 91-98.

Zhang Y, Kolinski A, Skolnick J (2003) TOUCHSTONE II: A new approach to ab initio protein structure prediction. Biophy J 85: 1145-1164.

Zhou H and Zhou Y (2005) Fold recognition by combining sequence profiles derived from evolution and from depth-dependent structural alignment of fragments. Proteins 58: 321-328.

13

Water Parameters of the Chubut River and Andes Lakes (Patagonia, Argentina)

Donat P. Häder, Virginia E. Villafañe, Peter Richter and E.Walter Helbling

Abstract

A transect was performed along the Chubut River (Patagonia) from the Atlantic towards the Andes measuring major water parameters of this drainage area. Samples were taken at different locations of the Chubut River, and also brooks several Andes lakes in the Chubut and Río Negro Provinces as well as from several Andes rivers and mountain creeks. For comparison, the Laguna Cacique Chiquichano in Trelew (Chubut) was investigated. Several physico-chemical parameters were determined (pH, oxygen, nitrogen compounds, conductivity, redox potential, water hardness and spectral transmission), as well as photosynthesis by means of PAM fluorimetry. Water toxicity was assessed with a bioassay based on toxicity-induced changes of the movement behavior of a flagellate. Water quality was very good in the upper part of the river, deteriorating towards its mouth. Transmission spectra of the water indicate the presence of Gelbstoff. The photosynthetic efficiency and chlorophyll fluorescence were considerably high in the FlorentinoAmeghino reservoir and remained high downstream. Toxicity measurements did not indicate significant pollution. The Laguna Cacique Chiquichano differed considerably from all other water samples. The water was high in nitrogen compounds and the conductivity was much higher than in the river samples. Phytoplankton composition varied along the transect as well as in the lakes: diatoms dominated in the Chubut River, whereas Chlorophytes and/or Cyanophytes also contributed an important share in the lakes and brooks samples. Since phytoplankton populations were most abundant in the lower part of the Chubut River, the species composition was intensively evaluated in this part. Here, a total of 38 species was identified belonging to the Bacillariophyceae, Chlorophyceae, Cyanobacteria, Dinophyceae and Euglenophyceae in the order of species diversity.

Keywords: Water pollution; Water quality; Patagonia; Ecotoxicology; Phytoplankton diversity.

Introduction

The Chubut River is the most important river in the Chubut province (Patagonia, Argentina). It originates from the Andes, which are the watershed of this river. It has a yearly average discharge of the 53 m^3 s^{-1}(Villafañe et al. 1991).The main source of water is the area between the Andes cities of Esquel and Bariloche. In addition to the Chubut River, other important rivers of the Province are Río Chico and Senguer. Some large lakes in the western part are Futalaufquen and Puelo. Near the city of Gualjaina the Chubut River enters the plains of the Patagonian Steppe and flows through the sparsely populated Chubut province. Despite the low population, the basin of the Chubut River partially suffers from land degradation due to anthropogenic activities such as herding and wood collection(Miserendino and Archangelsky 2006). Agriculture is developing, using water from the Chubut River or the FlorentinoAmeghino reservoir (see below). Only few settlements are located in the Patagonian plain, such as Las Plumas, Paso de Indios and Los Altares. Two larger cities are located on the river close to the Atlantic, Trelew and Rawson, in which the majority of Chubut's population live (a total of about 250,000 inhabitants). About 120 km upstream from Trelew a dam was constructed for electrical power generation and irrigation (FlorentinoAmeghino reservoir), which forms an artificial lake with an extension of about 70 km^2. This reservoir dams the Chubut River and the Río Chico. Currently the valley of the Chubut River is mainly used for sheep farming, but some companies have interest to investigate and, if applicable, exploit the natural resources such as oil and gold, which are assumed to exist in this area.

In the Río Negro Province, the two most important rivers are Limay and Río Negro, and in the western part of the Province there are many lakes that are generally being used for water consumption and recreation (e.g., NahuelHuapi, Gutierrez, Mascardi, among others). These lakes of glacial origin are generally deep and cover a range of environmental conditions ranging from ultraoligotrophic to oligotrophic; from these originate the largest Patagonian rivers that feed large reservoirs located in the Patagonian Plateau(Quirós and Drago 1999).

The aim of this study was to monitor the physico-chemical water parameters, water pollution, photosynthetic activity as well as phytoplankton composition in order to establish a database for the Chubut River and other water sources in Chubut and Río Negro provinces in a relatively unaffected state.

Materials and Methods

Sample locations

Water samples were collected along the Chubut River from the Atlantic to the Andes, in several tributaries as well as in various Andes lakes of the Chubut and Río Negro Provinces. The locations, altitudes and sampling times are summarized in Table 1.

Determination of physical and chemical water parameters

The following parameters were determined (partially redundant with different methods) directly at the sampling locations: oxygen, pH, temperature, nitrate, nitrite, ammonium, total hardness (GH), carbonate hardness (KH), conductivity and redox value. Additionally, oxygen was determined with a hand-held oxymeter (model DO-5509, Lutron, Taiwan).

A multimeter was used to measure pH, temperature, conductivity and reductivity of the water (HI 98204, Hanna Instruments). The following parameters were measured by means of water test strips from JBL (JBL GmbH & Co. KG, Neuhofen, Germany, water tests nitrate, nitrite, total hardness, carbonate hardness, pH).In addition, nitrate and ammonium were measured with colorimetric tests (Merck, Darmstadt, Germany). Furthermore, pH was measured with a hand-held pH meter and temperature with a sensor from the PAM device (see below).

Table 1: Sample locations. Chubut River samples indicated by *, mountain lake samples by †, river or mountain creek samples by §. DiqueFlorentinoAmeghino is an artificial reservoir of the Chubut River (about 120 km upstream from Trelew). The Laguna Cacique Chiquichano in Trelew is a small lake, which is independent from the water system of the Chubut River.

Location	Sampling Date	Local Time	GPS South	GPS West	Elevation
Chubut River (El Elsa)*	27.09.2006	14:00	43° 19.548'	63° 03.725'	3 m
Chubut River (Rawson)*	06.10.2006	16:00	43° 18.395'	63° 06.248'	7 m
Chubut River (Water intake Rw)*	06.10.2006	15:30	43° 17.523'	65° 09.706'	7 m
Trelew*	06.10.2006	15:00	43° 16.375'	65° 15.061'	11 m
Gaiman*	06.10.2006	14:00	43° 17.494'	65° 29.700'	23 m
DiqueFlorentino Ameghino *	29.09.2006	10:00	43° 41.869'	66° 28.602'	74 m
Las Plumas*	29.09.2006	12:30	43° 43.219'	67° 17.312'	158 m
Los Altares*	29.09.2006	14:40	43° 51.557'	68° 28.775'	240 m
LagoPuelo †	02.10.2006	11:40	42° 5.846'	71° 36.848'	245 m
Río Futaleufú Est. Aforo §	30.09.2006	16:00	43° 08.129'	71° 34.915'	341 m
Río Percey §	30.09.2006	18:00	43° 05.271'	71° 28.176'	375 m
LagoFutalaufquen (Pasarela) †	01.10.2006	16:00	42° 43.569'	71° 44.455'	550 m
LagoFutalaufquen †	01.10.2006	12:00	42° 50.728'	71° 36.179'	568 m
Trout farm §	30.09.2006	19:30	43° 11.377'	71°36.133'	367 m
NahuelHuapi (PlayaBonita) †	03.10.2006	16:00	41° 7.274'	71° 23.917'	774 m
Lago Moreno †	03.10.2006	12:00	41° 4.448'	71° 30.497'	789 m
Laguna ElTrébol†	03.10.2006	14:00	41° 4.221'	71° 29.371'	791 m
Lago Gutierrez †	03.10.2006	15:00	41° 9.463'	71° 24.397'	815 m
Lago Mascardi†	02.10.2006	14:00	41° 21.999'	71° 30.615'	840 m
Laguna C. Chiquichano	07.10.2006	10:00	43 14.544'	65° 18,037'	12 m

Spectral analysis of the water samples

At each sampling location about 31 of water were taken and transported in ice chests until further analysis. The spectral transmission of the water samples was measured, before and after filtration by means of a USB photometer (Ocean Optics USB2000) attached to a laptop computer. Distilled water was used for the blanks.

ECOTOX - Bioassay for the assessment of water toxicity

Principle of the bioassay

To determine water toxicity a manual ECOTOX device (Real Time Computers, Möhrendorf, Germany) was used. This instrument has the same functions as the automatic ECOTOX device described earlier (Ekelund and Aronsson 2004; Millán de Kuhn et al. 2006; Streb et al. 2006; Streb et al. 2002; Tahedl and Häder 1999, 2001) and is based on the computerized analysis of movement parameters of unicellular organisms before and after addition of a test solution. The cells are transferred into a thin observation chamber of a miniaturized microscope and monitored by a CCD camera. To monitor gravitactic behavior the chamber is inserted vertically inside the device. The signal is transferred to a computer, which analyses the movement by means of image analysis software. The latest versions of ECOTOX use a firewire camera (IEEE 1394), which can be directly connected to the firewire port of the host computer. The standard organism, which was also used for the experiments during this project, was the unicellular flagellate *Euglena gracilis*. In the manual ECOTOX all cell-test solution mixtures are made manually by the experimenter. After mixing the suspensions some drops were transferred into a cuvette made from two slides glued together by means of double-sidedscotchtape (depth about 100 μm). First a control measurement is performed (cells and corresponding amount of distilled water); thereafter cells mixed with test solution (in this case water from the sample locations) are monitored. The software analyzes the differences between the control measurement and the sample measurement. The energy consumption of the manual ECOTOX is considerably low, so that it can run on an internal battery and is connected to a laptop.

Organisms and growth conditions of the test organism

The flagellate *Euglena gracilis* Z was obtained from the algal culture collection at the University of Göttingen(Schlösser 1994). The cells were grown in a modified mineral medium: the macro elements were taken from a mineral medium described previously (Checcucci et al. 1976), the micro elements from another mineral medium (f/2) (Guillard and Ryther 1962) in stationary cultures in 100-ml Erlenmeyer flasks at about 20°C under continuous light of about 18 W m^{-2} from mixed cool-white and warm-tone fluorescent lamps. During transport to the sample site the cells were kept inside an ice chest at about 15°C.

Measurement procedure

To save time during on-site analysis, the water samples (about 31 each) were kept in an ice box till the evening, when toxicity measurements, photometric and organism filtration (see below) were performed. In order to determine if photoinduced toxic substances could play a role as pollutants, water samples were irradiated with a solar simulator (Hönle, Martinsried, Germany, Sol 1200) and the samples tested with the ECOTOX instrument.

PAM fluorescence

Photosynthetic efficiency of the cells was determined by means of a PAM (pulse amplitude modulated) fluorimeter (Water-PAM, Walz, Effeltrich Germany). Cell samples were measured immediately on site.The photosynthetic yield was determined according to the following equation(Genty et al. 1989; Weis and Berry 1987):

yield = (Fm'-Ft)/Fm',

with Fm' representing the maximal fluorescence achieved after saturation of the photosynthetic electron transport chain (light pulse) and Ft as temporary fluorescence. Ten measurements were performed for each sample. Data are presented as mean values with standard deviation.

Additionally, light curves (P vs. E) were recorded at following illuminations 0, 65, 96, 144, 226, 327, 482, 737 and 1101 µmol photons/m^2. The electron transport rate (ETR) was calculated according to the following formula (Genty et al. 1989):

ETR = yield * PAR * factor,

where the factor was 0.41.

Phytoplankton species composition

Phytoplankton was sampled along the transect (late September – early October 2006) as well as at four sites along the lower part of the river in the coastal meso-tidal plain, between the estuary and 10 km upstream (Perillo et al. 1999). Samples for phytoplankton species composition in the lower part of the river were taken from March 2006 to October 2007. Samples were collected just below the surface with an acid-clean (1 N HCl) polycarbonate bottle and taken to the Estación de Fotobiología Playa Unión, were analysis and measurements were carried out as described below.

Taxonomic analysis and phytoplankton abundance of cells was performed using an inverted microscope (Leica MD IL, Germany) after overnight settling of 10 mL of phytoplankton samples fixed with buffered formalin (final concentration in the sample, 0.4%). Cell counts were made following the methodology described by Villafañe and Reid (Villafañe and Reid 1995). The biodiversity was calculated with Shannon's index (Zar 1999).

Statistical Analysis

One-way ANOVA was applied to test for statistical significance. Dunnet, Fisher, Tukey and Scheffe were applied as post hoc tests.

Results and Discussion

Physico-chemical parameters and spectral transmission

A transect has been carried out in order to determine physico-chemical and biological parameters along the Chubut River, in some Andes lakes and mountain creeks of the Chubut and Río Negro Provinces. The water quality of the Chubut River and the Andes lakes is exceptionally good. The water at all locations was found to be very low in water hardness (total and carbonate hardness) as well as in conductivity (Table 2).

The conductivity data are in good coincidence with data from an earlier study(Miserendino and Archangelsky 2006). The relatively high conductivity of sample 1 (Chubut river mouth) can most likely be traced back to the fact that at this point the river water is mixed with sea water. This was also seen in the study by Miserendino (Miserendino and Archangelsky 2006), where 1528 $\mu S\ cm^{-1}$ were measured. Oxygen was always high, measured nitrogen compounds low (nitrate in most cases less than 10 mg/l, nitrite mostly below 0.05 mg/l and ammonia not more than 0.2 mg/l). As the ion concentration of the water was close to distilled water the precision of the measured pH and the redox potential is limited (higher ion concentration necessary). Higher values of nitrogen compounds were recorded in LagoPuelo (25 mg/l NO_3^-, 1 mg/l NO_2^-), where the sampling point was near a hotel. The sample was possibly influenced by domesticsewage (compare also PAM data). LagoNahuelHuapi, a large lake near Bariloche, which is an important city and frequented vacation spot (105,000 inhabitants and about 1 million visitors per year), is not visibly affected by domestic waste waters. But one has to consider that during sampling it was early tourist season so that the sewage influx into the lake was limited. The International Lake Environment Committee classifies this lake as seriously eutrophic, but not seriously toxically contaminated(World Lake Data Base - International Lake Environment Committee Foundation (ILEC) and http://www.ilec.or.jp/database/sam/dsam02.html).All Andes lakes are very clear, as shown by spectral transmission measurements; Figure 1 shows the spectral

Table 2: Physical and chemical parameters at the different sampling sites. GH total hardness, KH carbonate hardness.

Location	Nitrate (mg/l)	Nitrite (mg/l)	Ammonium (mg/l)	GH	KH	pH	Temp	Conductivity (µS)	Redox (mV)	Oxygen (mg/l)
Chubut River (El Elsa)*	5	<0.05	0.1	4-7	3	7.2	13.3	331	133	9.4
Chubut River (Rawson)*	10	<0.05	0.1	4-7	4	7.2	11	150	170	6.7
Chubut River (Water intake Rawson)*	<10			4	3-4	6.6	11.4	193	160	6.4
Trelew*	10	<0.05	0.1	3-4	3	7.4	10.4	69	244	5.8
Gaiman*	10	<0.05	0.1	4	3	7.2	9.6	140	265	5.5
Dique F. Ameghino *	0			<3	3	8.3	7	144	198	
Las Plumas*	10	0.05	0.11	4-7	6		n.d.	140	185	8.6
Los Altares*	0			<3	0	7.2	13.8	124	134	6.7
LagoPuelo †	25	0.05	0.2	4	3	7	n.d.	234	194	8.3
Río Futaleufú Est. Aforo §	10		0	4	3	6.4	9.2	28	960	7.5
Río Percey §	10		0.2	<3	0		8.2	56	640	10.8
LagoFutalaufquen (Pasarela) †	0	0	<0.2	0	3	7	n.d.	162	207	
LagoFutalaufquen †	10	0	0.1	0	<3	6.9		420	167	10.4
Trout farm §	5		<0.2	0	0	7.7	7.3	28	220	12.6
NahuelHuapi (Playa Bonita) †	10	<0.05	0	0	0		12.3	31	247	10
Lago Moreno †	<10	n.d	0.2	<3	0	6.4	8.6	43	220	9.2
Laguna El Trébol †	10	<0.05	0	<3	3		9.4	23	264	10.3
Lago Gutierrez †	10	<0.05	0.2	<3	0		8.6	40	258	10.1
LagoMascardi †	0	0	0	0	3	6.6	n.d.	194	204	9
Laguna C. Chiquichano	50	>1	2	>21	5	8.2	12.6	1818	123	3.7

transmission curves for several samples. The Andes River and mountain creek samples were turbid in some cases, but only with filterable suspended particles. After filtration the spectral transmission of the samples was similar compared to the Andes lake samples.

The Chubut River samples, in contrast, contained non-particulate substances, probably Gelbstoff (Fig. 1). The data indicate that the load of particulate substance was increasing towards the Atlantic, because the unfiltered samples closer to the Andes (Los Altares and Las Plumas) show a lower absorption in the UV/blue region of the spectrum compared to other samples closer to the shore, with exception of the last location (El Elsa), where the water was already influenced by the ocean (Fig. 1). Several cadavers of drowned animals (sheep, cattle etc.) were observed in the river, which represent a possible problem for drinking water collection.

As mentioned above, the Laguna Cacique Chiquichanoin Trelew is much more polluted with organic matter as indicated by the nitrogen parameters and by the high conductivity (Table 2). It is a small lake without natural influx or efflux. It is high in microalgae and zooplankton (e.g. water fleas). Many waterfowls inhabit this lagoon making it an important natural reserve. The spectral transmission spectrum indicates a relatively high Gelbstoff concentration (Fig. 1).

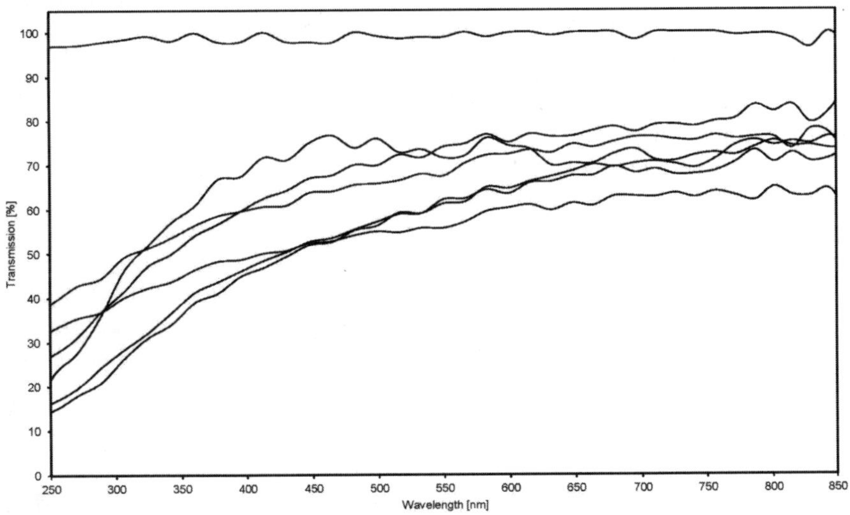

Fig. 1: Spectral transmission of water samples from the Andean lake Moreno, and various sites along the Chubut River. The distance to the Atlantic decreases in the following order: Los Altares, Las Plumas, Trelew, Water intake Rawson and El Elsa, and Laguna Cacique Chiquichano. Lines from top to bottom (left side): Lago Moreno, Los Altares, Las Plumas, El Elsa, Laguna C. Chiquichano, Water intake Rawson, Trelew.

PAM measurements

The PAM device detected photosynthetic activity in all samples, although the signal in some Andes lake was extremely low and only detectable with high amplification. The photosynthetic yield was very low in all Andes samples (Fig. 2a). The difference between the mountain lakes was not significant. The measured light curves (data not shown) indicated that the photosynthetic organisms in the mountain lakes do not show pronounced photosynthetic efficiency and seem not to be well adapted to high light conditions.

The Chubut River samples showed similar values for photosynthetic efficiency at the locations far inside the Patagonian Steppe (Los Altares yield: 0.15 and Las Plumas yield about 0.2) as the mountain river samples. The photosynthetic efficiency increased considerably in the reservoir (highest values measured: > 0.5). Towards the Atlantic the yield decreased from site to site, but not significantly between Gaiman and Trelew (Fig. 2b). The fluorescence intensity, which is a marker for the chlorophyll concentration, was lower in the mid Steppe samples (Los Altares and Las Plumas) and high in the reservoir and all following locations (data not shown). The photosynthetic efficiency was higher in the later samples. It seems that the reservoir increases the quantity of the photosynthetic organisms in the river. These algae are delivered to and increase the cell density of the downstream part of the river. At the last sampling point (El Elsa) close to the shore the photosynthetic efficiency was significantly lower, which is most likely caused by a mixture of the river water with sea water. The difference between all samples is significant ($p<0.01$, one-way ANOVA), with the exception of Los Altares compared with El Elsa (post hoc test Dunnet and Fisher) or Los Altares compared with El Elsa and Gaiman to Trelew (post hoc Tukey and Scheffe).

ECOTOX measurements

The ECOTOX measurements with the test organism *Euglena gracilis* was used to measure the percentage of motile cells (motility), the r-value of gravitactic orientation, the alignment of swimming directions, the percentage of upward swimming cells in the horizontal cuvette, the form factor (cell form) and cell size as well as swimming velocity. All parameters did not indicate any contamination in the water samples. Figure 3 shows the results from Chubut River samples in Trelew, Gaiman, Water intake point and El Elsa. Also after 4 h incubation of the cells in the water samples no relevant changes in the movement parameters were observed.

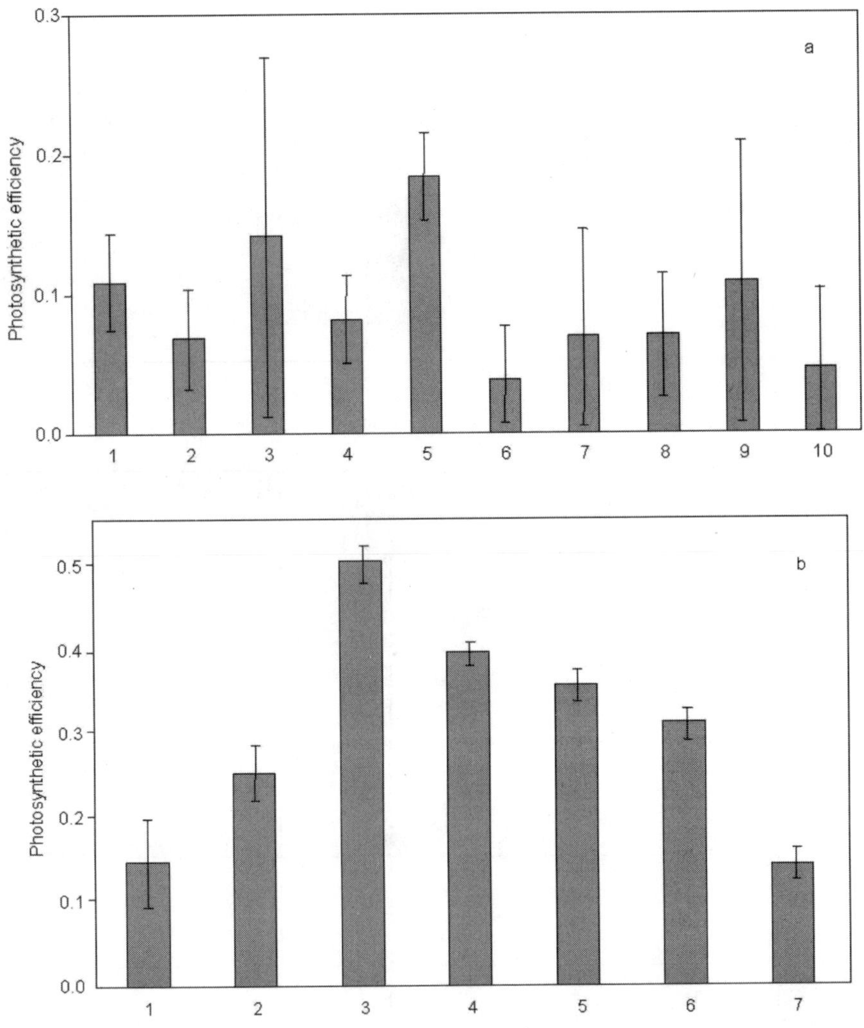

Fig. 2: Photosynthetic efficiency of the organisms in the Andes water samples (a: 1 LagoPuelo, 2 Rio Futaleufu, 3 Rio Percey, 4 LagoFutalaufquen, 5 Trout farm, 6 Lago N. Huapi, 7 Lago Moreno, 8 Lago El Trébol, 9 LagoGuiterrez, 10 LagoMascardi) and Chubut River samples (b: 1 Los Altares, 2 Las Plumas, 3 Dique F. Ameghino, 4 Gaiman, 5, Trelew, 6 Water intake Rawson, 7 El Elsa).

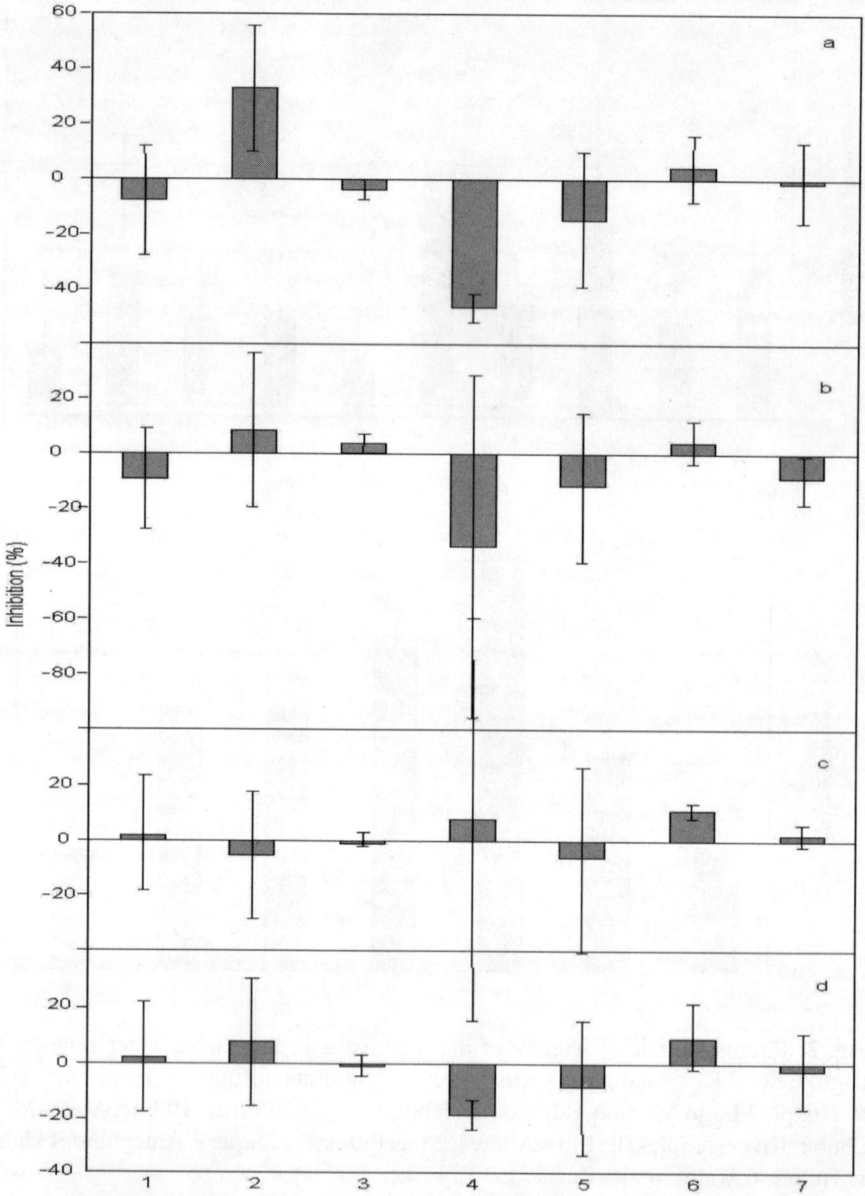

Fig. 3: Changes of movement parameters (in %) compared to controls in samples of the Chubut River in Trelew (a), Gaiman (b), Water intake Rawson (c) and El Elsa (d). Motility (1), r-value (2), alignment (3), upward swimming (4), form factor (5), cell size (6), swimming velocity (7). Positive values indicate inhibition, negative enhancement of movement parameter.

After irradiation of water samples with a solar simulator in order to induce some phototoxic products no clear changes in the movement behavior was found. All deviations from the controls are in the range of statistical noise as shown for samples from the Chubut River and the Laguna Cacique Chiquichano (Fig. 4a), indicating that the water is not impaired by high concentrations of toxic substances, because the ECOTOX system responds very sensitively to a variety of pollutants such as heavy metals, organic solvents or pesticides(Azizullah et al. 2011; Ekelund and Aronsson 2004; Streb et al. 2006; Streb et al. 2002; Tahedl and Häder 1999, 2001).

To detect possible light-induced toxicity of the water (photoactivation), samples were irradiated with solar-like radiation for 12 h. The samples were mixed with *Euglena gracilis* cell suspensions and measured with the ECOTOX. No significant effects on the cells were detected (Fig. 4b). Regarding the results of taxonomic composition in the water bodies sampled, diatoms dominated in the Chubut river samples (Table 3). In the lakes and brooks however, and although diatoms contributed an important share in most cases (Table 3), other groups were also important, e.g., Chlorophytes in lakes Gutiérrez and Puelo, Cyanophytes in PasarelaArrayanes, and Dinophytes in Lake Mascardi.

The species composition of the potamoplankton in the lower part of the Chubut River, identified a total of 38 algal taxa (Table 4). The majority of taxa identified belonged to the divisionHeterokontophyta(Bacillariophyceae, 80%), Chlorophyta (3%) and Cyanophyta (5%), Dinophyceae (3%). Ten taxa were present in more than half of the campaigns. In the order of their abundance, these taxa included *Aulacoseiragranulata*, *Cocconeisplacentula*, *Cyclotellameneghiniana*, *Diatomavulgare*, *Gomphonema* sp., *Melosira* sp., *Navicula* sp., *Synedra* sp., *Closterium* sp. and *Anabaena* sp.

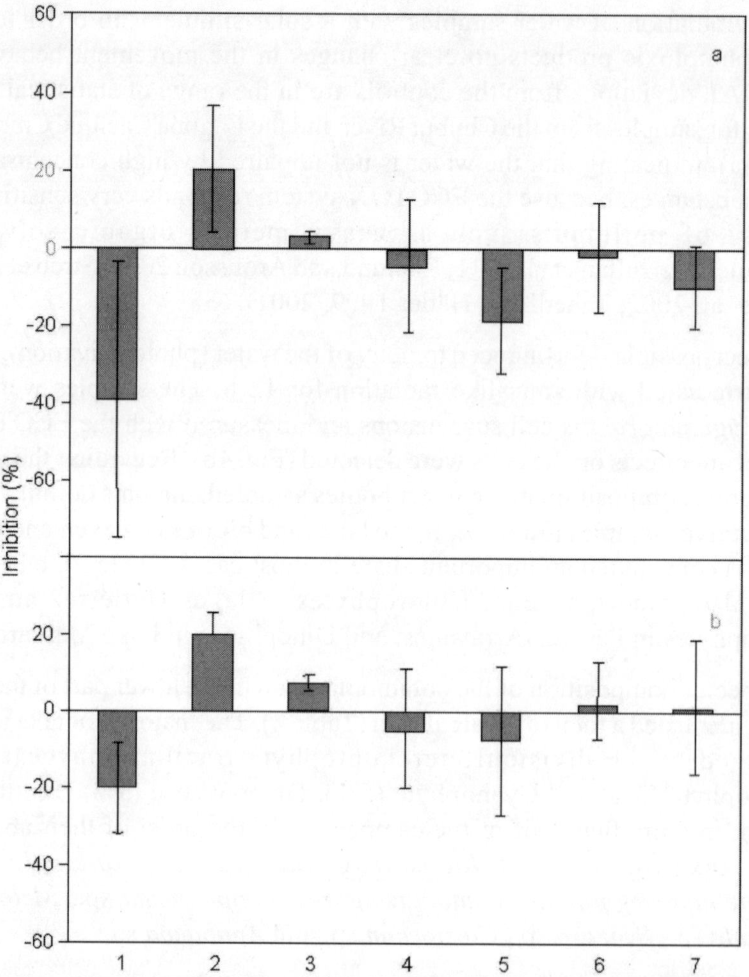

Fig. 4: Changes of movement parameters compared to controls in samples of the Water intake Rawson (a) and Laguna Cacique Chiquichano (b) after photoactivation of the water samples using a solar simulator. Motility (1), r-value (2), alignment (3), upward swimming (4), form factor (5), cell size (6), swimming velocity (7). Positive values indicate inhibition, negative enhancement of movement parameter.

Table 3: Total cell concentration (in cells/ml) of the different phytoplankton groups determined in the transect along the Chubut river, as well as in the Andean lakes sampled during late September – early October 2006. Bac. Bacillariophyceae, Chl. Chlorophyceae, Cya. Cyanophyceae, Din. Dinophyceae.

Location	Bac.	Chl.	Cya.	Din.	Other
Transect along Chubut River					
El Elsa	1954		4		
Water intake					
Rawson	1000		29		
Trelew	312		8		4
Gaiman	417				
D. Florentino					
Ameghino	302	2			6
Las Plumas	129				
Los Altares	221	4	4		
Lakes and brooks of the Andes					
L. Futalaufquen	587	268	360		
R. Futaleufú	646	30	80		
L. Gutierrez	271	281	72		282
Trout farm	471	4	175	26	35
L. Puelo	285	281	24		
L. Mascardi	78	100	110	196	118
L. Moreno	482	40	20		80
L. Futalaufquen (Pasarela)	182	308	444	80	
R. Percey	183				
L. El Trébol	1952	610	406	33	280

The classes in the order of number of species were Bacillariophyceae (23), Chlorophyceae (7), Cyanophyceae (5), Dinophyceae (2) and Euglenophyceae (1), respectively (Table 4). The variation of total phytoplankton abundance was mostly explained by variations in the Bacillariophyceae that formed the predominant group during the whole year. The abundance of this group was always greater than 80% and also correlated with biomass algal. Among the 23 planktonic species of Bacillariophyceae identified, 18 belonged to the order Bacillariales and five to the Biddulphiales. The diversity of the assemblages (Shannon's Index, Table 5) was low in the fall of 2006 (1.6 bits ind.-1) and highest in summer 2007 (2.7 bits ind.-1). The chlorophyll concentration of the potamoplankton assemblages (μg Chl*a* per L) and total algal abundance (cells per mL) were positively correlated with nitrite and nitrate concentrations (data not show). The algal biomass was highest in winter 2007 and smallest in summer 2007. The total algal abundance was highest (2667 cells per mL) in winter 2007 and smallest (300 cells per ml) in summer 2007.

Table 4: Taxonomy of the 32 phytoplankton species encountered during the investigation period (2006-2007) in the Chubut River in 42 samples.

Bacillariophyceae
Biddulphiales
Aulacoseiragranulata
Aulacoseira sp.
Cyclotella meneghiniana
Melosira sp.
Odontella aurita
Bacillariales
Asterionella sp.
Caloneis sp.
Cocconeis placentula
Cymatopleura sp.
Cymbella spp.
Diatoma vulgare
Ephitemia sp.
Gomphonema spp.
Gyrosigma sp.
Navicula sp.
Nitzschia sigmoidea
Nitzschia spp.
Unidentified pennate diatoms
Pinnularia sp.
*Rhoicosphaenia*sp.
Rhopalodia sp.
Synedra sp.
Unidentified species
Chlorophyceae
Closteriumaciculare
Closterium spp.
Unidentifiedcolonialspecies
Pandorina sp.
*Scenedesmus*sp.
Cyanophyta
Anabaena sp.
Lyngbya sp.
Oscillatoria sp.
Dinophyta
Ceratium hirundinella
Prorocentrummicans
Euglenophyta
Not identified

Table 5: Means and SD for transparency, Chl*a* concentration, phytoplankton biomass and diversity measured in the Chubut River. Means and SD are based on four sites per campaign in autumn 2006 (n=3), in winter 2006 (n=2), spring (n=1), summer 2007 (n=2) and winter 2007 (n=2).

Variable	Autumn 2006		Winter 2006		Spring 2006		Summer 2007		Winter 2007	
	Mean	SD	Mean	SD	Mean	SD	Mean	SD	Mean	SD
a_{cdom} (m^{-1})	3.51	1.08	9.07	2.90	6.90	0.45	5.77	1.11	5.22	0.95
Chl*a* (µg L^{-1})	5.82	2.02	7.11	2.81	7.06	1.12	3.23	0.71	30.7	52.5
Abundance (cells mL^{-1})	866	743	706	396	1337	229	411	237	1410	801
Diversity (H')	1.66	0.80	2.40	0.21	2.40	0.13	2.77	0.40	2.30	0.44

Conclusion

The data presented here summarize the water parameters of the catchment area of the Andes region in Chubut and Río Negro provinces as well as from the Chubut River valley. The water has excellent quality from the spring to the shore. Several circumstances prevent the water from being polluted (Quirós 2000).

- Large parts of the Andes region are under protection (Parque Nacional Nahuel Huapi, LagoPuelo, Los Alerces and others),
- The Patagonian Steppe is only sparsely populated,
- At the moment not too much fertilizer is used in agriculture along the Chubut River valley,
- No heavy industry is located in this area.

Planned mining projects can impair the quantity (oil production needs large amounts of water) and quality (cyanide extraction of gold) of the Chubut River water and by this the ecosystems along this important Patagonian lifeline.

Acknowledgements

We thank D. Pérez, C. Malbárez and R. Villanueva for their help in sampling collection. C. Medina performed taxonomic analyses. This work was supported by the United Nations - Global Environmental Fund (PNUD B-C-39), Consejo Nacional de Investigaciones Científicas y Técnicas (CONICET) – Deutsche Forschungsgemeinschaft(DFG), Agencia Nacional de Promoción Científica y Tecnológica – ANPCyT (PICT 2005- 32034 and PICT 2007-1651), and Fundación Playa Unión. This is contribution No. xx of Estación de Fotobiología Playa Unión.

References

Azizullah A, Richter P, Häder D-P (2011) Ecotoxicological evaluation of wastewater samples from Gadoon Amazai Industrial Estate (GAIE), Swabi, Pakistan. Intern J Environ Sci 1: 959-976.

Checcucci A, Colombetti G, Ferrara R, Lenci F (1976) Action spectra for photoaccumulation of green and colorless *Euglena*: evidence for identification of receptor pigments. Photochem Photobiol 23:51-54.

Ekelund NGA, Aronsson KA (2004) Assessing *Euglena gracilis* motility using the automatic biotest ECOTOX application to evaluate water toxicity (cadmium). Vatten 60: 77-83.

Genty B, Briantais JM, Baker NR (1989) The relationship between the quantum yield of photosynthetic electron transport and quenching of chlorophyll fluorescence. Biochim Biophys Acta 990: 87-92.

Guillard RRL, Ryther JH (1962) Studies of marine planktonic diatoms. I. Cyclotella nana Hustedt and *Detonula convervacea* (Cleve) Gran. Can J Microbiol 8: 229-239.

Millán de Kuhn R, Streb C, Breiter R, Richter P, Neeße T, Häder D-P (2006) Screening for unicellular algae as possible bioassay organisms for monitoring marine water samples. Water Res 40: 2695-2703.

Miserendino ML, Archangelsky M (2006) Aquatic coleoptera distribution and environmental relationships in a large Patagonian river. Intern Rev Hydrobiol 91: 423-437.

Perillo GME, Piccolo MC, Pino-Quivira M (1999) What do we know about the geomorphology and physical oceanography of South American estuaries? In: Perillo GME, Piccolo MC and Pino-Quivira M (Eds.), Estuaries of South America. Springer, Heidelberg.

Quirós R (2000) La eutrofización de las aguas continentales de Argentina. In: Fernández A (Ed), El agua en Iberoamérica: Acuíferos, lagos y embalses. Universidad Nacional de Buenos Aires, Buenos Aires, pp. 43-47.

Quirós R, Drago E (1999) The environmental state of Argentinean lakes: An overview. Lakes Reserv Res Manage 4: 55-64.

Schlösser UG (1994) SAG - Sammlung von Algenkulturen at the University of Göttingen, Catalogue of strains 1994. Bot Acta 107: 113-186.

Streb C, Richter P, Häder D-P (2006) ECOTOX - a biomonitoring system for UV-effects and toxic substances. In: Ghetti F, Checcucci G and Bornman JF (Eds.), Environmental UV Radiation: Impact on Ecosystems and Human Health and Predictive Models. Springer, The Netherlands, p. 288.

Streb C, Richter P, Ntefidou M, Lebert M, Häder D-P (2002) ECOTOX-biomonitoring based on real time movement analysis of unicellular organisms. J Gravit Physiol 9: 345-346.

Tahedl H, Häder D-P (1999) Fast examination of water quality using the automatic biotest ECOTOX based on the movement behavior of a freshwater flagellate. Water Res 33: 426-432.

Tahedl H, Häder D-P (2001) The use of image analysis in ecotoxicology. In: Häder D-P (Ed), Image Analysis: Methods and Applications. CRC Press, Boca Raton, pp. 447-458.

Villafañe VE, Helbling EW, Santamarina J (1991) Phytoplankton blooms in the Chubut river estuary (Argentina): Influence of stratification and salinity. Rev Biol Mar 26: 1-20.

Villafañe VE, Reid FMH (1995) Métodos de microscopía para la cuantificación del fitoplancton. In: Alveal K, Ferrario ME, Oliveira EC and Sar E (Eds.), Manual de Métodos Ficológicos. Universidad de Concepción, Concepción.

Weis E, Berry A (1987) Quantum efficiency of photosystem II in relation to the energy dependent quenching of chlorophyll fluorescence. Biochim Biophys Acta 894: 198-208.

World Lake Data Base - International Lake Environment Committee Foundation (ILEC), http://www.ilec.or.jp/database/sam/dsam02.html.

Zar JH (1999) Biostatistical analysis. 4th ed. Prentice Hall, Englewood Cliffs, NJ.

14

Oxidative Stress: Challenges and its Mitigation Mechanisms in Cyanobacteria

Minu Kesheri, Swarna Kanchan, Richa and Rajeshwar P. Sinha

Abstract

Cyanobacteria evolved during the Precambrian era and are believed to be the first one to introduce oxygen into the initially reducing atmosphere by oxygenic photosynthesis. Thus being the main primary producers of organic matter at that time and thereby contributed to a major global transformation leading to the development of aerobic metabolism and the subsequent rise of higher plants and animal forms. Reactive oxygen species (ROS) in cyanobacteria are commonly generated during normal growth and development. However abiotic and biotic stresses enhance the level of ROS which in turn pose the danger of oxidative stress. Ability to perceive ROS and to rapidly initiate antioxidant defenses is crucial for their survival. Therefore cyanobacteria evolved four lines of defense mechanisms to combat the damages caused by oxidative stress. Inretrospect to their ecological adaptivity, ubiquitous presence and efficient source of compounds of biotechnological significance, this chapter aims to elucidate various mechanisms adopted by cyanobacteria to combat oxidative stress.

Keywords: Oxidative stress, Enzymatic antioxidants, Non-enzymatic antioxidants Photoprotective compounds.

Introduction

The imbalance between oxidants and antioxidants in favor of oxidants is termed as oxidative stress. The key hallmarks with reference to the term 'oxidative stress' were classified under four categories (1) increased oxidative load (enhanced ROS production); (2) the potential for uncontrolled oxidation due to rates of production exceeding rates of metabolism; (3) oxidative damage to cellular components, which assumes that the rate of oxidation exceeds repair or replacement; causing (4) the accumulation of damaged cellular components that somehow lead to loss of function and eventual death (Foyer and Noctor 2003).

Reactive oxygen species in cyanobacteria are commonly generated during normal growth and development. However abiotic and biotic stresses enhance the level of ROS which in turn pose the danger of oxidative stress. ROS are highly reactive chemical entities that attack molecules by capturing electrons, thereby modifying their chemical structures and altering their functions. Thus ROS damage cell structures such as carbohydrates, nucleic acids, lipids, and proteins.

Cyanobacteria can occupy various ecotopes by virtue of having high ecological adaptivity (Whitton 1987). Cyanobacterial arsenal to cope with various endogenous as well as exogenous stresses comprises of four lines of defense mechanisms. First line of defense is avoidance strategy which includes mat formation, migration, alterations in morphology, enclosure of filaments in amorphous silica matrices and synthesis of extracellular polysaccharides. Second line of defense encompasses scavengers which includes enzymatic and non enzymatic antioxidants. The presence of enzymatic and non-enzymatic antioxidants as natural arsenal against multiple stresses has been reported in various cyanobacterial species at their natural habitat (Kesheri et al. 2011). Third line of defense includes screening by compounds such as biopterin glucoside and photoprotective compounds. Fourth line of defense comprises of repair and resynthesis mechanisms. Thus with these four lines of defenses, cyanobacteria evolved as organisms with high ecological adaptivity and ubiquitous presence.

Causes of oxidative stress

Cyanobacteria are believed to be the first one to introduce oxygen into the initially reducing atmosphere by oxygenic photosynthesis and thereby being the main primary producers of organic matter at that time. Although oxygen accumulation allowed the development of aerobic organisms that used oxygen as a powerful electron acceptor yet it posed a challenge to cope with the damaging effects of oxygen on the metabolic networks that had originally evolved in an anoxic environment. Atomic oxygen has two unpaired electrons in separate orbits in its outer electron shell. This electron structure makes oxygen susceptible to radical formation. The sequential reduction of oxygen through the addition of electrons leads to the formation of a number of reactive oxygen species (ROS). On the basis of presence and absence of unpaired electrons, ROS can be classified into two groups. First group comprises of free radicals which contain one or more unpaired electrons and thus giving reactivity to the molecule. Second group comprises of nonradical forms created when two free radicals share their unpaired electrons. Thus, ROS comprise of molecules like hydrogen peroxide; radicals such as the hydroxyl radical, the most reactive of all ROS having the capacity to oxidize and react with organic molecules and the superoxide anion which is both ion and radical.

ROS are commonly generated during growth and development as well as during normal cell metabolism related with oxidation and reduction processes, especially in chloroplasts during photosynthesis and mitochondria during respiration. Of all biological pathways, photosynthetic electron transport is particularly prone to produce reactive oxygen species (ROS) due to the very high (positive) redox potential of the primary donor of photosystem II, needed to oxidize water, and the low redox potential of the primary electron acceptor of photosystem I, needed to reduce ferredoxin; ROS encompasses the production of singlet oxygen, superoxide anions, hydrogen peroxide and hydroxyl radicals, all of which are derived through the local energization of O_2. ROS are formed as a result of thylakoids, mitochondria and plasma membrane linked transport and subsequent leaking of electron to the molecular oxygen in the cell. When oxygen comes in contact with metabolic systems it can be transformed into more reactive and toxic form of superoxide ion, hydrogen peroxide, hydroxyl radical and singlet oxygen. Formation of singlet oxygen ($^1O_2$) subsequently stimulates the production of other reactive oxygen species such as hydrogen peroxide (H_2O_2), superoxide anion (O_2^-), hydroxyl (OH^-), and perhydroxyl radicals (O_2H). ROS are also produced by root nodules of N_2-fixers. The enhancement of metabolism and growth by nitrogen may require the presence of additional defense mechanism against ROS (Mullineaux and Criessen 1997). Development of symbiosis and nodule are also triggered by ROS (Frendo et al. 2005). All nitrogen-fixing organisms face a critical dilemma. Nitrogenase, the enzyme that catalyses the conversion of nitrogen into ammonium, is quickly and irreversibly inactivated by oxygen, yet nitrogen fixation is an extremely energy demanding process with at least 16 mol. of ATP being consumed per mol of nitrogen fixed (Ljones et al. 1972). Therefore, relying on aerobic respiration to supply this energy puts the organism at risk of inactivating nitrogenase.

However abiotic and biotic stresses enhance the level of ROS which in turn pose the danger of oxidative stress. Cyanobacteria in their natural habitats are compelled to be exposed to changing external environmental conditions (Fig. 1), such as temperature (Liu et al. 2005, Srivastava et al. 2006), irradiation (Latifi et al. 2009; Zeeshan and Prasad 2009; Singh et al. 2010), heat (Hossain and Nakamoto 2003), desiccation/drought (Rajendran et al. 2007; Higo et al. 2008; Kapoor et al. 2013), salinity (Srivastava et al. 2005; Bhadauriya et al. 2009), and metals (Choudhary et al. 2007). Some environments are characterized by more than one extreme factor. The cyanobacteria dwelling in those habitats are referred to as poly-extremophiles. Higher temperature may go together with acidic media (which is the habitat of thermoacidophilic microorganisms); hot or cold environments are often exposed to UV irradiation (cyanobacteria observed in the deserts or near the Polar Regions), likewise, high temperatures and high pH levels occur in salinity areas.

Exposure to continuous high irradiance during the summer in addition to the low temperature increases the chances of photoinhibition in algae inhabiting the polar environments (Bascuñán-Godoy et al. 2012). The existence of thermophilic cyanobacteria has been extensively documented in the course of the microbiological characterization of hot springs, first in Yellowstone National Park, Wyoming, USA (Brock 1969, 1978). Rastogi et al. (2012) phylogenetically analyzed cyanobacteria inhabiting hot spring namely *Cyanothece* sp., *Nostoc* sp., *Scytonema* sp. and *Rivularia* sp. High temperature enduring unicellular cyanobacteria (*Synechococcus* [*Thermosynechococcus*] sp.) photosynthesize up to 73–74°C, the highest temperatures enabling photosynthesis. Thus different types of unicellular cyanobacteria, classified in the genus *Synechococcus* (*Thermosynechococcus*), are the most thermophilic. Many geothermal springs emit water near the boiling point.

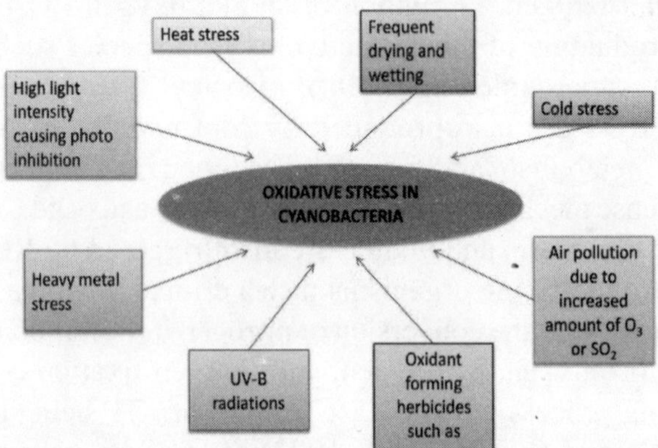

Fig. 1: Exogenous factors leading to oxidative stress in cyanobacteria

The gradual cooling of the water in the outflow channels provides a stable temperature gradient in which populations of microorganisms position themselves according to their temperature requirements. There are distinct differences in the communities found in hot springs in different parts of the world: thermophilic *Synechococcus* (*Thermosynechococcus*) species, abundant in Yellowstone, are absent from similar hot springs in Iceland (Ward and Castenholz 2000). Filamentous cyanobacteria are less thermotolerant: *Mastigocladus laminosus*, *Phormidium* spp., and different thermophilic *Oscillatoria* sp. have their upper temperature limit for growth between 55 and 62°C. Some thermophilic cyanobacteria such as *M. laminosus* (upper temperature limit 58°C) and a strain designated "*Chlorogloeopsis* HTF" (upper temperature limit 64°C) are nitrogen fixers.

Water stress conditions in particular may trigger an increased formation of superoxide radical ($O_2^{\cdot-}$) and hydrogen peroxide (H_2O_2), which can directly attack membrane lipids and inactivate SH-containing enzymes (Navari-Izzo and Izzo 1994). During water depletion, superoxide radical may also react non-enzymatically with H_2O_2, giving rise to products such as hydroxyl radicals and singlet oxygen, which are even more reactive than superoxide itself (Elstner and Osswald 1994). This formation is a consequence of the Mehler reaction, which provides a pathway for the excess removal of electrochemical energy determined by drought stress (Gamble and Burke 1984). Decrease in polyunsaturated fatty acid content coinciding with the increased levels of MDA in response to high osmotic stress is temporarily associated with the increase in electrolyte leakage, suggesting that water stress induces damage at the cellular and subcellular membrane levels *via* lipid peroxidation (Aziz and Larher 1998).

Anthropogenic activities leading to the release of atmospheric pollutants such as chlorofluorocarbons (CFCs), chlorocarbons (CCs) and organobromides (OBs) owe to the continued depletion of stratospheric ozone and has resulted in an increase in ultraviolet-B (UV-B; 280-315 nm) radiation on Earth's surface (Weatherhead and Andersen 2006). Cyanobacteria in their natural habitats are exposed to harmful ultraviolet radiation (UVR) owing to the absorption of solar energy to drive photosynthesis and nitrogen fixation. Lethal doses of UVR reach deep into water column (Häder et al. 2007); down to a depth of 20 m in the clearest oceanic water and to a few centimeters in brown humic lakes and rivers (Kirk 1994). Because of its high energy, UVR radiation leads to the production of ROS in photosynthetic organisms (Apel and Hirt 2004). In cyanobacteria, shorter wavelengths such as ultraviolet radiation (UVR, <400 nm), are known to induce ROS, causing DNA damage, lipid-peroxidation and overall decreased photosynthetic efficiency (He et al. 2002, He and Häder 2002). Behrendt et al. (2013) introduced the first study of ROS generation suggesting that *Acaryochloris marina* can avoid a considerable amount of light-induced stress by using NIR instead of VIS for its photosynthesis, adding further evolutionary arguments to their widespread appearance. Behrendt et al. (2013) further reported the effect of light intensity and spectral composition on ROS generation, as measured in real-time using a laser-photoacoustic gas detector in three different strains of NIR utilizing cyanobacteria belonging to the genus *Acaryochloris*, including a new strain (named *Acaryochloris* CRS), isolated from phototrophic biofilms growing on dead coral branches collected on Heron Island, Australia.

Despite their destructive activity, ROS are well-described second messengers in a variety of cellular processes including tolerance to environmental stresses (Yan et al. 2007). Whether ROS will act as damaging or signaling molecule depends on the delicate equilibrium between ROS production and scavenging. Due to the multifunctional roles of ROS, it is necessary for the cells to control the level of ROS tightly to avoid any oxidative injury and not to eliminate them completely. It was demonstrated that ROS have a role in cell signaling, including; apoptosis; gene expression; and the activation of cell signaling cascades (Hancock 2001). ROS can serve as both intra- and inter-cellular messengers. Reactive oxygen species as second messengers have been described in several plant hormone responses, including stomatal closure, root gravitropism, seed germination, lignin biosynthesis, programmed cell death, hypersensitive responses, and osmotic stress (Sharma et al. 2012).

Damages caused by oxidative stress

Oxidative attack on proteins result in site-specific amino acid modifications, fragmentation of the peptide chain, aggregation of cross-linked reaction products, altered electrical charge and increased susceptibility to proteolysis. Primary, secondary and tertiary protein structures alter the relative susceptibility of certain amino acids. Sulphur containing amino acids, and thiol groups specifically, are very susceptible sites. Activated oxygen can abstract an H atom from cysteine residues to form a thiyl radical that will cross-link to a second thiyl radical to form disulphide bridges. Alternatively, oxygen can add to a methionine residue to form methionine sulphoxide derivatives. Reduction of both of these may be accomplished in microbial systems by thioredoxin and thioredoxin reductase (Farr and Kogoma 1991). Other forms of free radical attack on proteins are irreversible. For example, the oxidation of iron-sulphur centres by superoxide destroys enzymatic function (Gardner and Fridovich 1991). Many amino acids undergo specific irreversible modifications when a protein is oxidized. For example, tryptophan is readily cross-linked to form bityrosine products (Davies 1987).

When photosynthetic organisms are exposed to salt stress, the fatty acids of membrane lipids are desaturated (Allakhverdiev et al. 2001). The reactions of oxygen free radicals with polyunsaturated lipids are the most frequent consequence of oxygen radical production in plant cells. The lipid bilayer membrane is composed of a mixture of phospholipids and glycolipids that have fatty acid chains attached to carbon 1 and 2 of the glycerol backbone by an ester linkage. The peroxidation reactions differ among these fatty acids depending on the number and position of the double bonds on the acyl chain (Frankel 1985). Oxidation of unsaturated fatty acids by singlet oxygen produces distinctly different products than the hydroxyl radical (Bradley and Min 1992). The effect of free radical reactions on the presence of specific lipid-soluble, membrane antioxidants (phenols, flavonoids and quinones), are poorly understood due to differences in their composition. UV irradiation presumed to be an elicitor of multilevel oxidative stress, may cause morphological changes, alter pigment composition, influence adaptive mechanisms, impair photosynthesis and decrease bioproductivity (Vass et al. 1996). UV-B irradiation disturbs the balance of monodehydroascorbate radical (MDA) production and reduction, resulting in increased light-induced MDA signal. Reactive oxygen species produced due to UV-B radiation easily destroys proteins, DNA and other biological molecules (Sinha et al. 1998). A number of physiological and biochemical processes such as growth, survival, pigmentation, photosynthetic oxygen production and phycobiliproteins composition have been reported to be susceptible to UV-B radiation (Wang and Zhang 2001). DNA breaks accumulate during exposure of bacterial cells to superoxide radical and H_2O_2.

DNA seems to be a weak link in a cell's ability to tolerate oxygen free radical attack. Since DNA is effective in binding metals that are involved in Fenton reactions, and can tolerate less damage than other macromolecules, therefore the cell has a number of DNA repair enzymes (Beyer et al. 1991). One reason why eukaryotic organisms have compartmentalized DNA in the nucleus, away from sites of redox cycling that are high in NAD(P)H and other reductants, may be to avoid oxidative damage. Activated oxygen and agents that generate oxygen free radicals, such as ionizing radiation, are responsible for inducing numerous lesions in DNA due to which deletions, mutations and other lethal genetic effects are caused. Characterization of this damage to DNA indicates that both the sugar and the base moieties are susceptible to oxidation thereby causing base degradation, single strand breakage, and cross-linking to protein (Imlay and Linn 1986). Degradation of the base produces numerous products, including 8-hydroxyguanine, hydroxymethyl urea, urea, thymine glycol, thymine and adenine ring-opened and -saturated products. The principle cause of single strand breaks is the oxidation of the sugar moiety by the hydroxyl radical. Cross-linking of DNA to protein serves to be one of the consequences of hydroxyl radical attack on either DNA or its associated proteins (Oleinick et al. 1986).

Cyanobacterial responses to oxidative stress

Cyanobacteria evolved four lines of defense mechanisms (Fig. 2). First line of defense is avoidance strategy which includes mat formation, migration, alterations in morphology, enclosure of filaments in amorphous silica matrices and synthesis of extracellular polysaccharides. Enzymatic and non enzymatic antioxidants form second line of defense. Cyanobacterial antioxidants can be divided into three general classes; (i) non polar solvent soluble carotenoids and membrane associated tocopherols, (ii) polar solvent soluble reductants such as ascorbic acid, glutathione, and phycobiliproteins (iii) enzymes such as superoxide dismutase (SOD), catalase, peroxidase, ascorbate peroxidase and glutathione reductase (Foyer 1993). Screening by photoprotective compounds such as scytonemin and mycosporine-like amino acids (MAAs) forms third line of defense. These secondary metabolites are thought to play multiple roles against several environmental stresses such as UV radiation and desiccation owing to their antioxidative properties (Wada et al. 2013). Some ancestors of living cyanobacteria, the oldest oxygenic organisms, may have evolved in a time when UV radiation fluxes reaching the surface, particularly UVB and UVC, were much higher than at present. The latter wavelength region (UVC) does not reach the Earth's surface at present. Absorption of solar energy to drive photosynthesis and nitrogen fixation exposes cyanobacteria to harmful ultraviolet radiation (UVR) in their natural habitats. Thus, cyanobacteria and other microorganisms in that distant age were compelled to evolve measures to cope with these greater levels of UV radiation, and at present this strategy may demonstrably involve multiple devices, even within one organism.

The presence of a UV-absorbing compound like scytonemin most probably helped cyanobacteria to survive from lethal effects of UV radiation when there was no stratospheric ozone layer. This assumption is supported by the fact that scytonemin has an in vivo absorption maximum at 370 nm whereas purified scytonemin shows maximum absorption at 386 nm, but it also absorbs significantly at 252, 278 and 300 nm (Singh et al. 2010). Recently, Rastogi and Incharoensakdi (2014) elucidated that MAAs may perform a vital role in survival and sustainability of *Gloeocapsa* sp. CU-2556 in harsh environmental conditions by its ability to absorb/screen short wavelength UV radiation and antioxidant function. Fourth and final line of defense comprises of repair and resynthesis mechanisms such as *de novo* protein synthesis, DNA repair and programmed cell death.

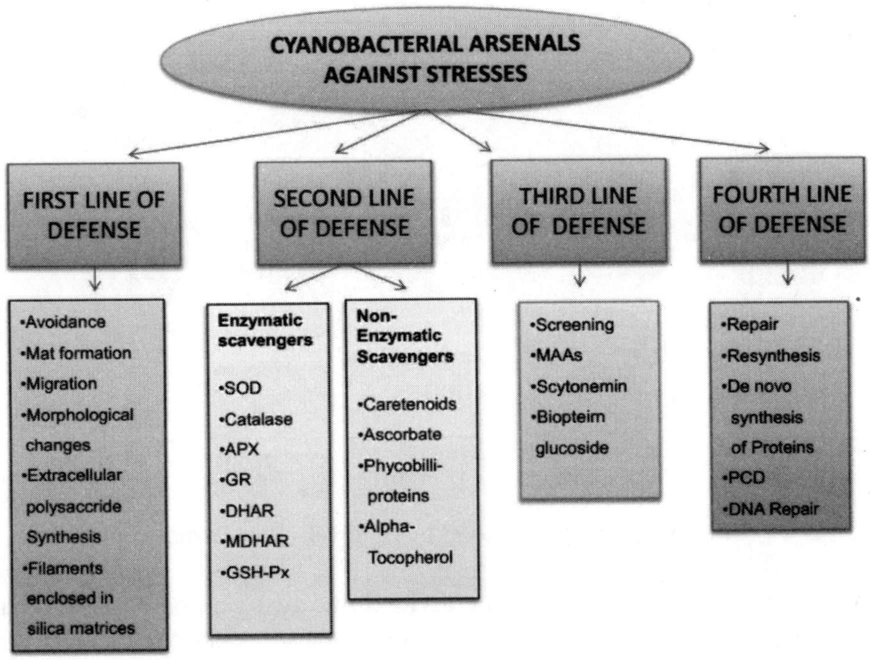

Fig. 2: Four lines of defense in cyanobacteria

Scavenging of ROS has been reported to follow Halliwell-Asada cycle (Halliwell and Gutteridge 1989) as shown in fig 3. Among all the antioxidative enzymes, superoxide dismutase serves to be the first line of defense against nascent oxygen species which mitigates the toxic effect of superoxide ion. There are three distinct types of SOD classified on the basis of the metal cofactor: the copper/zinc (Cu/Zn-SOD), the manganese (Mn-SOD) and the iron (Fe-SOD) isozymes. Prokaryotic cells and many eukaryotic algae contain only Mn-SOD and Fe-SOD isozymes and are believed to be the ancient forms. The presence of SOD is a prominent biomarker of defense against oxidative stress (Priya et al. 2010) and its activity increases with stress as a direct consequence. On the basis of presence and absence of ascorbate peroxidase, the cyanobacteria may be classified into two groups: (i) the one that scavenges hydrogen peroxide with the peroxidase using a photoreductant as the electron donor, and (ii) the group which only scavenges hydrogen peroxide with catalase. Ascorbate is regenerated either via monodehydroascorbate reductase, or glutathione and dehydroascorbate reductase. Glutathione reductase (GR) has been shown to play a role in the protection mechanism which removes oxygen radical in N_2-fixing cyanobacterium *Nostoc muscorum* (Karni et al. 1984).

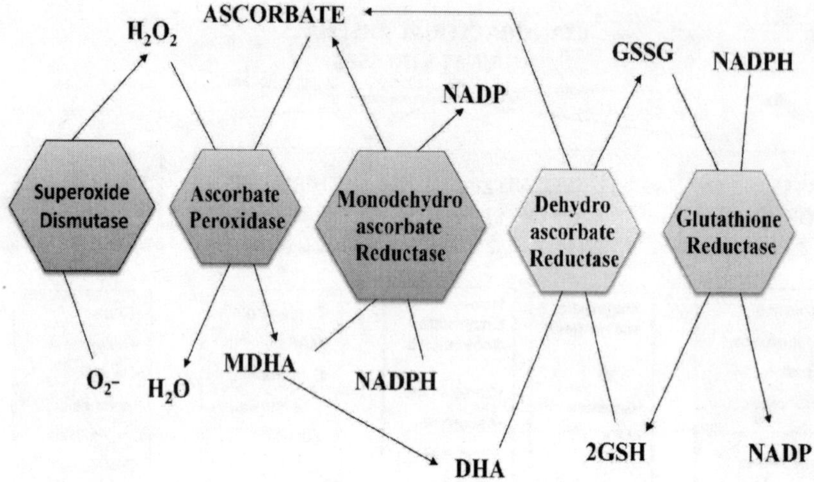

Fig. 3: Scavenging of ROS by Halliwell-Asada pathway

Non-enzymatic antioxidants can be divided into two general classes; (i) non polar solvent soluble carotenoids and membrane associated tocopherols; (ii) polar solvent soluble reductants such as ascorbic acid, glutathione, and phycobiliproteins. Phycobiliproteins are reported to be used in the measurement of peroxy radical damage (DeLange and Glazer 1989). Phycobiliproteins have antimicrobial, antioxidant, anti-inflammatory, neuroprotective and hepatoprotective properties (Richa et al. 2011). In addition, they are being extensively used as natural colorants in food and cosmetics, fluorescent neoglycoproteins, probes for single particle fluorescence imaging fluorescent applications in clinical and immunological analyses. In terms of its antioxidant properties carotenoids can protect the photosystems in one of four ways; (i) by reacting with lipid peroxidation products to terminate chain reactions, (ii) by scavenging singlet oxygen and dissipating the energy as heat, (iii) by reacting with triplet or excited chlorophyll molecules to prevent formation of singlet oxygen and (iv) by the dissipation of excess excitation energy through the xanthophyll cycle. The tripeptide glutathione is a major antioxidant and a low-molecular-weight non-proteinous thiol compound which mainly in its reduced form acts as an intracellular reductant (Galhano et al. 2011).

Thus the ability of cyanobacteria to perceive ROS and to rapidly initiate antioxidant defenses is crucial for their survival (Latiû et al. 2009). With the four lines of defense mechanisms, cyanobacteria evolved as an organism with high ecological adaptivity and ubiquitous presence. Although aging is a multifactorial process, there is significant evidence that shows that oxidative stress is one of the main factors that influences cellular longevity. Moreover, oxidative damage is one of the major contributing factors in the onset of age related degenerative diseases. Interest in the factors that determine longevity has grown recently because in many countries, the main causes of death are currently comorbidities connected to age and oxidative stress. Superoxide dismutase 2 (Mn-SOD or SOD2) has been shown to protect against oxidative stress, a main factor influencing cellular longevity (Mariana et al. 2013). Therefore in the present scenario, it seems relevant to explore cyanobacteria as the source for mass production of antioxidants as well as photoprotective compounds for the benefits of mankind.

Acknowledgements

Minu Kesheri is thankful to UGC, Govt. of India, New Delhi, for providing financial assistance in the form of research fellowship. Swarna Kanchan is thankful to Dr. Shibasish Chowdhury for his unremitting motivation and UGC, Govt. of India, New Delhi, for providing the financial support in the form of the Basic Science Research Fellowship under UGC SAP programme at Biological Science Department, BITS, Pilani. Richa is thankful to DST, Govt. of India, New Delhi, for providing financial support in form of a fellowship under the project (No. SR/WOS-A/LS-140/2011).

References

Allakhverdiev SI, Kinoshita M, Inaba M, Suzuki I and Murata N (2001) Unsaturated fatty acids in membrane lipids protect the photosynthetic machinery against salt-induced damage in *Synechococcus*. Plant Physiol 125: 1842-1853.

Apel K and Hirt H (2004) Reactive oxygen species: metabolism, oxidative stress, and signal transduction. Annu Rev Plant Biol 55: 373-399.

Aziz A and Larher F (1998) Osmotic stress induced changes in lipid composition and peroxidation in leaf discs of *Brassica napus* L. J Plant Physiol 153: 754-762.

Bascuñán-Godoy L, Sanhueza C, Cuba M, Zuñiga GE, Corcuera LJ and Bravo LA (2012) Cold-acclimation limits low temperature induced photoinhibition by promoting a higher photochemical quantum yield and a more effective PSII restoration in darkness in the Antarctic rather than the Andean ecotype of Colobanthus quitensis Kunt Bartl (Cariophyllaceae). BMC Plant Biol 12: 114.

Behrendt L., Staal M., Cristescu SM , Harren FJM , Schliep M., Larkum AWD and Kühl M (2013) Reactive oxygen production induced by near-infrared radiation in three strains of the Chl *d* -containing cyanobacterium *Acaryochloris marina*. F1000 Research 2: 44.

Beyer W, Imlay J and Fridovich I (1991) Superoxide dismutases. Prog Nucl Acid Res 40: 221-253.

Bhadauriya P, Gupta R, Singh S and Bisen PS (2009) NaCl induced metabolic changes in the diazotrophic cyanobacterium *Anabaena cylindrica*. World J Microbiol Biotechnol 25: 341-345.

Bradley DE and Min DB (1992) Singlet oxygen oxidation of foods. Cat Rev Food Sci Nutri 31: 211-236.

Brock TD (1969) Microbial growth under extreme conditions. Symp Soc Gen Microbiol 19: 15-41.

Brock TD (1978) Thermophilic Microorganisms and Life at High Temperatures. Springer Verlag, New York.

Choudhary M, Jetley UK, Khan MA, Zutshi S and Fatma T (2007) Effect of heavy metal stress on proline, malondialdehyde, and superoxide dismutase activity in the cyanobacterium *Spirulina platensis*-S5. Ecotoxicol Environ Saf 66: 204-209.

Davies KJA (1987) Protein damage and degradation by oxygen radicals. I General aspects J Biol Chem 162: 9895-9901.

DeLange RL and Glazer AN (1989) Phycoerythrin fluorescence based assay for peroxy radicals: A screen for biologically relevant protective agents. Anal Biochem 177: 300-306.

Elstner EF and Osswald W (1994) Mechanisms of oxygen activation during plant stress. Proc Roy Soc Edinburgh 102: 131-154.

Farr SB and Kogoma T (1991) Oxidative stress responses in *Escherichia coli* and *Salmonella typhimurium*. Microbiol Rev 55: 561-585.

Foyer CH and Noctor G (2003) Redox sensing and signaling associated with reactive oxygen in chloroplasts, peroxisomes and mitochondria. Physiol Plant 119: 355-364.

Foyer CH (1993) Ascorbic acid. In: Alscher RG, and Hess JL (eds.), Antioxidants in Higher Plants, CRC Press, Boca Raton, 31-58.

Frankel EN (1985) Chemistry of free radical and singlet oxidation of lipids. Prog Lipid Res 23: 197-221.

Frendo P, Harrison J, Norman C, Herandaez-Jimenez MJ, Van de Sype G, Gilabert A and Puppo A (2005) Glutathione and homoglutathione play a critical role in the nodulation process of *Medicago truncatula*. Mol. Plant-Microb Interac 18: 254-259.

Galhano V, Peixoto F and Gomes-Laranjo J (2011) Exposure of the cyanobacterium *Nostoc muscorum* from Portuguese rice ûelds to Molinate (Ordram®): Effects on the antioxidant system and fatty acid proûle. Aquat Toxicol 101: 367-376.

Gamble P and Burke J (1984) Effect of water stress on the chloroplast antioxidant system. I: Alterations in glutathione reductase activity. Plant Physiol 76: 615-621.

Gardner PR and Fridovich I (1991) Superoxide sensitivity of *Escherichia coli* 6-phosphogluconate dehydratose. J Biol Chem 266: 1478-1483.

Häder DP, Kumar HD, Smith RC and Worrest RC (2007) Effects of solar UV radiation on aquatic ecosystems and interactions with climate change. Photochem Photobiol Sci 6: 267-285.

Halliwell B and Gutteridge JMC (1989) Free Radicals in Biology and Medicine. Oxford Clarendon Press.

Hancock JT, Desikan R and Neill SJ (2001) Role of Reactive Oxygen Species in Cell Signaling Pathways. Biochem Biomed Aspects Oxi Modif 29: 345-350.

He YY and Häder DP (2002) Reactive oxygen species and UV-B: effect on cyanobacteria. Photochem Photobiol Sci 1: 729–736.

He YY, Klisch M and Häder DP (2002) Adaptation of cyanobacteria to UV-B stress correlated with oxidative stress and oxidative damage. Photochem Photobiol 76: 188–196.

Higo A, Ikeuchi M and Ohmori M (2008) cAMP regulates respiration and oxidative stress during rehydration in *Anabaena* sp. PCC 7120. FEBS Lett 582: 1883-1888.

Hossain MM and Nakamoto H (2003) Role for the cyanobacterial HtpG in protection from oxidative stress. Curr Microbiol 46: 70-76.

Imlay JA and Linn S (1986) DNA damage and oxygen radical toxicity. Science 240: 1302-1309.

Kapoor S, Tripathi SN and Shrivastava A (2013) Isolation and Purification of Heterotetrameric Catalase from a desiccation Tolerant Cyanobacterium *Lyngbya arboricola*. J Stress Physio Biochem 9: 184-208.

Karni L, Moss SJ and Tel-Or E (1984) Glutathione reductase activity in heterocysts and vegetative cells of *Nostoc muscorum*. Arch Microbiol 140: 215-217.

Kesheri M, Richa and Sinha RP (2011) Antioxidants as natural arsenal against multiple stresses in cyanobacteria. Int J Pharma Biosci 2: B-168-B-187.

Kirk JTO (1994) Optics of UV-B radiation in natural waters. Arch Hydrobiol 43: 1-16.

Latiû A, Ruiz M and Zhang CC (2009) Oxidative stress in cyanobacteria. FEMS Microbiol Rev 33: 258-278.

Liu XJ, Jiang Y and Chen F (2005) Fatty acid proûle of the edible ûlamentous cyanobacterium *Nostoc flagelliforme* at different temperatures and developmental stages in liquid suspension culture. Process Biochem 40: 371-377.

Ljones T and Burris RH (1972) ATP hydrolysis and electron transfer in the nitrogenase reaction with different combinations of the iron protein and the molybdenum-iron protein. Biochim Biophys Acta 275: 93-101.

Mariana Dias Castela de Carvalho and Joelma Freire De Mesquita (2013) Structural Modeling and In Silico Analysis of Human Superoxide Dismutase 2. PLOS ONE 8: e65558.

Mullineaux PM and Criessen GP (1997) Glutathione reductase: regulation and role in oxidative stress. In: Scandalios JG (ed.), Oxidative Stress and the Molecular Biology of Antioxidant Defenses, (Monograph Series vol.34), Cold Spring Harbor Laboratory press, Cold Spring Harbor, 667-713.

Navari-Izzo F and Izzo R (1994) Induction of enzyme activities and antioxidant production in barley plants as a result of SO_2 fumigation. Plant Sci 96: 31-40.

Oleinick NL, Chiu S, Ramakrishnan N and Xue L (1986) The formation, identification, and significance of DNA-protein cross-links in mammalian cells. Brit J Cancer 55: 135-140.

Priya B, Sivaprasanth RK, Jensil VD, Uma L, Subramanian G and Prabaharan D (2010) Characterization of manganese superoxide dismutase from a marine cyanobacterium *Leptolyngbya valderiana* BDU20041. Saline Systems 6: 1-10.

Rajendran UM, Kathirvel E and Anand N (2007) Desiccation-induced changes in antioxidant enzymes, fatty acids, and amino acids in the cyanobacterium *Tolypothrix scytonemoides*. World J Microbiol Biotechnol 23: 251-257.

Rastogi RP and Incharoensakdi A (2014) UV radiation-induced biosynthesis, stability and antioxidant activity of mycosporine-like amino acids (MAAs) in a unicellular cyanobacterium *Gloeocapsa* sp. CU2556. J Photochem Photobiol B: Biol 130: 287-292.

Rastogi RP, Kumari S, Richa, Han T and Sinha RP (2012) Molecular characterization of hot spring cyanobacteria and evaluation of their photoprotective compounds. Can J Microbiol 58: 719-727.

Richa, Rastogi RP, Kumari S, Singh KL, Kannaujiya VK, Singh G, Kesheri M and Sinha RP (2011) Biotechnological potential of mycosporine-like amino acids and phycobiliproteins of cyanobacterial origin. Biotechnol Bioinf Bioeng 1: 159-171.

Sharma P., Jha AB, Dubey RS and Pessarakli M (2012) Defense Mechanism in Plants under Stressful Conditions. Journal of Botany, Article ID 217037: 1-26 .

Singh SP, Häder DP and Sinha RP (2010) Cyanobacteria and ultraviolet radiation (UVR) stress: Mitigation strategies. Ageing Res Rev 9: 79-90.

Singh SP, Kumari S, Rastogi RP, Singh KL, Richa and Sinha RP (2010) Photoprotective and biotechnological potentials of cyanobacterial sheath pigment, scytonemin. Afr J Biotechnol 9: 580-588.

Sinha RP, Klisch M, Gröniger A and Häder D-P (1998) Ultraviolet absorbing/screening substances in cyanobacteria, phytoplankton and macroalgae. J Photochem Photobiol B: Biol 47: 83-94.

Srivastava AK, Bhargava P and Rai LC (2005) Salinity and copper-induced oxidative damage and changes in antioxidative defense system of *Anabaena doliolum*. World J Microbiol Biotechnol 22: 1291-1298.

Srivastava AK, Bhargava P, Mishra Y, Shukla B, and Rai LC (2006) Effect of pretreatment of salt, copper and temperature on ultraviolet-B-induced antioxidants in diazotrophic cyanobacterium *Anabaena doliolum*. J Basic Microbiol 46: 135-144.

Vass I, Sass L, Spetea C, Bakou A, Ghanatakis DF and Petrouleas V (1996) UV-B inhibition of photosystem II electron transport system studied by EPR and chlorophyll florescence. Impairment of donor and acceptor side components, Biochemistry 35: 8964-8973.

Wada N, Sakamoto T and Matsugo S (2013) Multiple Roles of Photosynthetic and Sunscreen Pigments in Cyanobacteria Focusing on the Oxidative Stress. Metabolites 3: 463-483.

Wang JG and Zhang CL (2001) DNA damage and repair of two ecotypes of *Phragmites communis* subjected to water stress. Acta Bot Sin 43: 490-494.

Ward DM and Castenholz RW (2000) Cyanobacteria in geothermal habitats. In: B.A. Whitton, and M. Potts (eds.),Ecology of Cyanobacteria: Their Diversity in Time and Space. Kluwer Academic Publishers, Dordrecht, pp. 37–59.

Weatherhead EC and Andersen SB (2006) The search for signs of recovery of the ozone layer. Nature 441: 39.

Whitton BA (1987) Survival and dormancy of blue-green algae. In: Hennis Y (ed.), Survival and Dormancy of Microorganisms, Wiley, New York. 109-167.

Yan J, Tsuichihara N, Etoh T and Iwai S (2007) Reactive oxygen species and nitric oxide are involved in ABA inhibition of stomatal opening. Plant Cell Environ 30: 1320–1325.

Zeeshan M and Prasad SM (2009) Differential response of growth, photosynthesis, antioxidant enzymes and lipid peroxidation to UV-B radiation in three cyanobacteria. S Afr J Bot 75: 466-474.